# Lecture Notes in Networks and Systems     1027

## Series Editor

Janusz Kacprzyk ⓘ, *Systems Research Institute, Polish Academy of Sciences, Warsaw, Poland*

## Advisory Editors

Fernando Gomide, *Department of Computer Engineering and Automation—DCA, School of Electrical and Computer Engineering—FEEC, University of Campinas—UNICAMP, São Paulo, Brazil*

Okyay Kaynak, *Department of Electrical and Electronic Engineering, Bogazici University, Istanbul, Türkiye*

Derong Liu, *Department of Electrical and Computer Engineering, University of Illinois at Chicago, Chicago, USA*

   *Institute of Automation, Chinese Academy of Sciences, Beijing, USA*

Witold Pedrycz, *Department of Electrical and Computer Engineering, University of Alberta, Alberta, Canada*

   *Systems Research Institute, Polish Academy of Sciences, Warsaw, Canada*

Marios M. Polycarpou, *Department of Electrical and Computer Engineering, KIOS Research Center for Intelligent Systems and Networks, University of Cyprus, Nicosia, Cyprus*

Imre J. Rudas, *Óbuda University, Budapest, Hungary*

Jun Wang, *Department of Computer Science, City University of Hong Kong, Kowloon, Hong Kong*

The series "Lecture Notes in Networks and Systems" publishes the latest developments in Networks and Systems—quickly, informally and with high quality. Original research reported in proceedings and post-proceedings represents the core of LNNS.

Volumes published in LNNS embrace all aspects and subfields of, as well as new challenges in, Networks and Systems.

The series contains proceedings and edited volumes in systems and networks, spanning the areas of Cyber-Physical Systems, Autonomous Systems, Sensor Networks, Control Systems, Energy Systems, Automotive Systems, Biological Systems, Vehicular Networking and Connected Vehicles, Aerospace Systems, Automation, Manufacturing, Smart Grids, Nonlinear Systems, Power Systems, Robotics, Social Systems, Economic Systems and other. Of particular value to both the contributors and the readership are the short publication timeframe and the worldwide distribution and exposure which enable both a wide and rapid dissemination of research output.

The series covers the theory, applications, and perspectives on the state of the art and future developments relevant to systems and networks, decision making, control, complex processes and related areas, as embedded in the fields of interdisciplinary and applied sciences, engineering, computer science, physics, economics, social, and life sciences, as well as the paradigms and methodologies behind them.

Indexed by SCOPUS, INSPEC, WTI Frankfurt eG, zbMATH, SCImago.

All books published in the series are submitted for consideration in Web of Science.

For proposals from Asia please contact Aninda Bose (aninda.bose@springer.com).

Michael E. Auer · Reinhard Langmann ·
Dominik May · Kim Roos

Editors

# Smart Technologies for a Sustainable Future

Proceedings of the 21st International
Conference on Smart Technologies &
Education. Volume 1

 Springer

*Editors*
Michael E. Auer
CTI Global
Frankfurt/Main, Germany

Reinhard Langmann
Edunet World Association e.V.
Blomberg, Germany

Dominik May
University of Wuppertal
Wuppertal, Germany

Kim Roos
Programme Director MScEng
Arcada University of Applied Sciences
Helsinki, Finland

ISSN 2367-3370          ISSN 2367-3389 (electronic)
Lecture Notes in Networks and Systems
ISBN 978-3-031-61890-1          ISBN 978-3-031-61891-8 (eBook)
https://doi.org/10.1007/978-3-031-61891-8

This Springer imprint is published by the registered company Springer Nature Switzerland AG
The registered company address is: Gewerbestrasse 11, 6330 Cham, Switzerland

If disposing of this product, please recycle the paper.

# Preface

It is a great privilege for us to present the proceedings of the 21st "International Conference on Smart Technologies & Education" (STE2024) to the authors and delegates of this event and to the wider, interested audience. The 2024 edition of STE was held under the general theme "Smart Technologies for a Sustainable Future", which was visible throughout the conference program.

The STE conference is the successor of the long-standing annual REV Conferences and the annual meeting of the International Association of Online Engineering (IAOE) together with the Edunet World Association (EWA) and the International Education Network (EduNet). Initiated in 2004, REV has been held in Villach (Austria), Brasov (Romania), Maribor (Slovenia), Porto (Portugal), Dusseldorf (Germany), Bridgeport (USA), Stockholm (Sweden), Brasov (Romania), Bilbao (Spain), Sydney (Australia), Porto (Portugal), Bangkok (Thailand), Madrid (Spain), New York (USA), Dusseldorf (Germany), Bengaluru (India), Georgia (USA), Hong Kong, Cairo (Egypt), and Thessaloniki (Greece).

This year, STE2024 has been organized in Helsinki, Finland as an onsite event supporting remote presentations, from March 6 until March 8, 2024. The co-organizers of STE2024 were the Arcada University of Applied Sciences, the International Association of Online Engineering (IAOE) together with the Global Online Laboratory Consortium (GOLC), the International Education Network (EduNet), and the EDUNET WORLD Association (EWA). STE2024 has been attracted 140 scientists and industrial leaders from more than 40 countries.

STE2024 is an annual event dedicated to the fundamentals, applications, and experiences in the field of Smart Technologies, Online, Remote, and Virtual Engineering, Virtual Instrumentation, and other related new technologies, including:

- Applications & Experiences
- Artificial Intelligence
- Augmented Reality
- Open Science Big Data
- Biomedical Engineering
- Cyber Physical System
- Cyber Security
- Collaborative Work in Virtual Environments
- Cross-Reality Applications
- Data Science
- Evaluation of Online Labs
- Human–Machine Interaction & Usability
- Internet of Things
- Industry 4.0
- M2M Concepts
- Mixed Reality

- Networking, Edge & Cloud Technology
- Online Engineering
- Process Visualization
- Remote Control & Measurements
- Remote & Crowd Sensing
- Smart Objects
- Smart World (City, Buildings, Home, etc.)
- Standards & Standardization Proposals
- Teleservice & Telediagnosis
- Telerobotic & Telepresence
- Teleworking Environment
- Virtual Instrumentation
- Virtual Reality
- Virtual & Remote Laboratories

The conference was opened by the Founding President of IAOE, Michael E. Auer, who underlined the importance to discuss guidelines and new concepts for engineering education in higher and vocational education institutions including emerging technologies in learning. In her greeting, the Rector of Arcada, Mona Forsskåhl pointed out the importance of the digitalization of education and more specifically the engineering education.

STE2024 offered an exciting technical program as well as networking opportunities concerning the fundamentals, applications, and experiences in the field of online engineering and related new technologies.

As part of the conference program, three pre-conference workshops have been organized:

1. Overcoming Traditional Boundaries of STEM Education and Enabling the Engineer of the Future
2. Logiccloud: The Next Generation Of Industrial Control
3. High-Performance Extreme Learning Machines

Furthermore, special sessions have been organized at REV2024, namely

1. Online Laboratories in Modern Engineering Education (OLMEE)
2. Human–Robot Interaction for Sustainable Development (HRI4SD)
3. Advances and Challenges in Applied Artificial Intelligence (ACAAI)

Four outstanding scientists and industry leaders accepted the invitation for keynote speeches:

1. Doris Sáez Hueichapan, University of Chile, Santiago, Chile, talked about "Energy & Water Management Systems for Agro-Development of Indigenous Rural Communities"
2. Dieter Uckelmann, HFT Stuttgart, Stuttgart, Germany, shared his valuable insights to "Why providing a comprehensive IoT education is impossible – but we should nevertheless strive to do so"
3. Roland Bent, Retired CTO Phoenix Contact GmbH & Co.KG, Germany, painted a vision for the future in his talk "The All Electric Society"

4. Hans-Jürgen Koch, Dipl.-Ing. for Communications Engineering, Executive Vice President of the Business Area Industry Management & Automation, Phoenix Contact GmbH, Germany, gave a fascinating introduction to "Innovative and collaborative automation platforms – The key for a sustainable world"

The conference was organized by the Faculty of Arcada University of Applied Sciences and Program Director Kim Roos served as the STE2024 chair. The President of IAOE, Prof. Dominik May has served as STE2024 general chair and Prof. Reinhard Langmann and Prof. Michael E. Auer served as Steering Committee Co-Chairs.

Submissions of Full Papers, Short Papers, Work in Progress, Poster, Special Sessions, Workshops, Tutorials, Doctoral Consortium papers have been accepted.

All contributions were subject to a double-blind review. The review process was extremely competitive. We had to review about to 233 submissions. A team of over 100 program committee members and reviewers did this terrific job. Our special thanks goes to all of them.

Due to the time and conference schedule restrictions, we could finally accept only the best 76 submissions for presentation or demonstration.

The conference was supported by

- Phoenix Contact as Platinum Sponsor
- Air France and KLM as Diamond Sponsor
- As always Sebastian Schreiter did an excellent job to edit this book.

Kim Roos
Dominik May
Michael E. Auer
Reinhard Langmann

# Committees

## STE General Chair

Dominik May — President IAOE, University of Wuppertal, Germany

## STE Steering Committee Co-chairs

Michael E. Auer (Co-chair) — IAOE, Austria
Reinhard Langmann (Co-chair) — EWA, Germany

## STE2024 Chair

Kim Roos — Arcada University of Applied Sciences, Finland

## Program Co-chairs

Erwin Smet — University of Antwerp, Belgium
Valery Varney — RWTH Aachen, Germany
Rizwan Ullah — Arcada University of Applied Sciences, Finland

## Technical Program Chair

Sebastian Schreiter — IAOE, France

## Workshop and Tutorial Chair

Alexander Kist — University of Southern Queensland, Australia

## Special Session Chair

María Isabel Pozzo        National Technological University, Argentina

## Award Chair

Andreas Pester        The British University in Egypt, Cairo, Egypt

## Publication Chair and Web Master

Sebastian Schreiter        IAOE, France

## Local Arrangement Chair

Maria von Bonsdorff-Hermunen        Arcada University of Applied Sciences, Helsinki, Finland

## International Advisory Board

| | |
|---|---|
| Abul Azad | President Global Online Laboratory Consortium, USA |
| Alberto Cardoso | University Coimbra, Portugal |
| Bert Hesselink | Stanford University, USA |
| Claudius Terkowsky | TU Dortmund University, Germany |
| Doru Ursutiu | University of Brasov, Romania |
| Hamadou Saliah-Hassane | Université TÉLUQ, Montréal, Canada |
| Krishna Vedula | IUCEE, India |
| Elio San Cristobal Ruiz | UNED Madrid, Spain |
| Teresa Restivo | University of Porto, Portugal |
| Uriel Cukierman | National Technological University Buenos Aires, Argentina |

# EduNet Forum Committee

## Chair

Klaus Hengsbach                    Phoenix Contact GmbH & Co KG, Germany

## Members

| | |
|---|---|
| Albert Alacorn | Phoenix Contact, Chile |
| Anja Schulz | Phoenix Contact, Germany |
| Christian Madritsch | Carinthia University of Applied Sciences, Austria |
| Christiane Kownatzki | Phoenix Contact, Germany |
| Edmond Wempe | Phoenix Contact, Germany |
| Felipe Mateos Martín | University of Oviedo, Spain |
| Glenn Williams | Harrisburg, USA |
| Hans Lindstrom | Phoenix Contact, Finland |
| Hernan Lopez | Phoenix Contact, Argentina |
| Jana Koenig | Phoenix Contact, Germany |
| Maren Gast | Phoenix Contact, Germany |
| Pascal Vrignat | University of Orleans, France |
| Reinhard Langmann | Edunet World Association, Germany |

## International Program Committee

| | |
|---|---|
| Akram Abu-Aisheh | Hartford University, USA |
| Alexander Kist | University of Southern Queensland, Australia |
| Anastasios Economides | University of Macedonia, Greece |
| Andreas Pester | The British University in Egypt, Egypt |
| Anshul Jaswal | Pennsylvania State University, USA |
| Carlos A. Reyes Garcia | National Institute of Astrophysics, Optics and Electronics, Mexico |
| Chee Sai Stephen Bok | Institute of Technical Education, Singapore |
| Catherine Soh Geok Hong | Institute of Technical Education, Singapore |
| Christian Guetl | Graz University of Technology, Austria |
| Christian Madritsch | Carinthia University of Applied Sciences, Austria |
| Christos Katsanos | Aristotle University of Thessaloniki, Greece |
| Cornel Samoila | University of Brasov, Romania |
| Dario Assante | Universita Telematica Internazionale, Italy |
| David Boehringer | University of Stuttgart, Germany |
| Dieter Wuttke | TU Ilmenau, Germany |
| Erwin Rauch | Free University of Bolzano, Italy |

| | |
|---|---|
| Felipe Mateos Martín | University of Oviedo |
| Gabriel Gomez | National University of Córdoba, Argentina |
| Gabriel XG Yue | International Engineering & Technology Institute, Hong Kong |
| Galip Cansever | Altinbas University, Turkey |
| Glenn P. Williams | Harrisburg University of Science and Technology, USA |
| Gustavo Alves (Vice-President) | Polytechnic of Porto, Portugal & IAOE |
| Horacio Podesta | University of Buenos Aires, Argentina |
| Ian Grout | University of Limerick, Ireland |
| Igor M. Verner | Technion - Israel Institute of Technology, Israel |
| Ingmar Riedel-Kruse | Stanford University, USA |
| James Wolfer | Indiana University South Bend, IN, USA |
| Javier Garcia-Zubia | University of Deusto, Spain |
| Jörg Reiff-Stephan | Wildau Technical University of Applied Sciences, Germany |
| Kalyan Ram B. | Electrono Solutions Pvt Ltd, India |
| Karsten Henke | TU Ilmenau, Germany |
| Katarina Zakova | Slovak University of Technology |
| Katrin Temmen | University of Paderborn, Germany |
| Manuel Castro | UNED Madrid, Spain |
| Maria de los Reyes Poo Arguelles | Universidad de Oviedo, Spain |
| Mathias Döbler | Berlin College of Further Education for Information Technology and Medical Equipment Technology, Germany |
| Matthias Christoph Utesch | Technical University of Munich, Germany |
| Michael Callaghan | University of Ulster, Northern Ireland |
| Nael Bakarad | Grand Valley State University, USA |
| Nikolaus Steffen | University of Flensburg, Germany |
| Olaf Graven | Buskerud University College, Norway |
| Petros Lameras | Coventry University, UK |
| Prajaks Jitngernmadan | Burapha University, Thailand |
| Prasad Ponnapalli | Manchester Metropolitan University |
| Rita Y. M. Li | Hong Kong Shue Yan University, Hong Kong |
| Robert Zhang | Qingdao University of Science and Technology, China |
| Rodrigo Verschae | O'Higgins University, Chile |
| Saman Kumpakeaw | King Mongkut's University of Technology North Bangkok, Thailand |
| Samir El-Seoud | The British University in Egypt (BUE), Egypt |
| Solomon Feleke Aklilu | Debre Berhan University, Ethiopia |
| Stamatis Papadakis | University of Crete, Greece |
| Stavros Nikou | University of Strathclyde, Glasgow, UK |

| | |
|---|---|
| Stefan Marks | Auckland University of Technology, New Zealand |
| Thomas Fischer | University of Applied Sciences Vienna, Austria |
| Thomas Klinger | Carinthia University of Applied Science, Austria |
| Viktoriya Voropayeva | Donetsk National Technical University, Ukraine |
| Yacob Astatke | Morgan State University, USA |
| Younes El Fellah | Hassan II University, Morocco |
| Leonardo Espinosa Leal | Arcada University of Applied Sciences, Finland |
| Matias Waller | Åland University of Applied Sciences, Finland |

# Contents

**Engineering Education of the Future**

Fostering Experiential Learning and Situational Awareness Via Reporting
and Broadcasting About Pilot Projects .................................... 3
  *Prajwal Prabhu and Vishal Singh*

Real Estate Insights: A Boardgame-Based Experiential Learning Approach .... 15
  *Vighneshkumar Rana and Vishal Singh*

Heating System Case Study for Simulation-Based Control System Analysis .... 32
  *Matias Waller, Leonardo Espinosa-Leal, and Kim Roos*

Perceived Importance of Laboratory Learning Objectives by Female
and Male Engineering Students ........................................ 44
  *Konrad Boettcher, Nils Kaufhold, Marcel Schade, Claudius Terkowsky,
  and Tobias R. Ortelt*

Automated Code Readability Feedback on Student Awareness ............... 56
  *Oscar Karnalim, Sendy Ferdian Sujadi, and Rossevine Artha Nathasya*

An Easy-to-Use Experimental Platform for Advanced Control Teaching
in Wuhan University ...................................... 67
  *Xingwei Zhou, Jiuzheng Su, Wenshan Hu, Guo-Ping Liu,
  and Zhongcheng Lei*

LLM Integration in Workbook Design for Teaching Coding Subjects .......... 77
  *Magnus Westerlund and Andrey Shcherbakov*

A Research-Led Contribution of Engineering Education for a Sustainable
Future ............................................................. 86
  *Brit-Maren Block and Marie Gillian Guerne*

IoT Laboratories Based on Reconfigurable Platform with Analog
Coprocessor ......................................... 98
  *Elena-Cătălina Gherman-Dolhăscu, Horia Modran, Cornel Samoilă,
  and Doru Ursuțiu*

Open-Source Software and Digital Sovereignty: A Technical Case Study
on Alternatives to Mainstream Tools .................................. 106
  *Dennis Biström, Kristoffer Kuvaja Adolfsson, and Matteo Stocchetti*

Demand for Future Skills: Education on AI in Comprehensive Digital
Business Development, Big Data Analytics, and Ubiquitous Approach
to Data in Business ............................................... 114
    Martin Zagar, Jasminka Samardzija, Ana Havelka Mestrovic,
    Muhieddin Amer, and Jinane Mounsef

University 4.0: The Future of Higher Education in the Age of Technology ..... 122
    Abdallah Al-Zoubi, María Isabel Pozzo, Monica I. Ciolacu,
    Gustavo R. Alves, and Alexander A. Kist

The Role of Online and Traditional Laboratories in the Context of Modern
Engineering Curricula .............................................. 135
    Alexander A. Kist, Catherine Hills, and Ananda Maiti

Work-in-Progress: Technology-Driven Introductory Phase in Engineering
Sciences for Sustainable Individual Support of Students' Academic Success ... 146
    Brit-Maren Block, Jannis Dethmann, and Benedikt Haus

Will My Job Be Automated? Fathoming Current and Persisting
Impediments for Automation ......................................... 155
    Bastian Prell, Simon Wilbers, Norman Günther, and Jörg Reiff-Stephan

**Industry 4.0 and Education**

Implementation of an Industry 4.0 Module in the International Educational
Project ERASMUS+ WORK4CE ...................................... 169
    Peter Arras, Anzhelika Parkhomenko, and Illia Parkhomenko

Educational Advancement of an Industrial Automation Course: Combining
Simulation with Physical Experience ................................. 178
    Amélie Chevalier

Enhancing Industrial Automation and Control Systems Cybersecurity
Using Endpoint Detection and Response Tools ......................... 186
    Filip Katulić, Stjepan Groš, Damir Sumina, and Igor Erceg

Digital Twins and Model-Based Design for New Vehicle Engineering ........ 198
    Raghuveer Rajesh Dani, Benjamin Geiger, Galyna Tabunshchyk,
    Carsten Wolf, and Friedbert Pautzke

Evaluation Design for Cross-Reality Laboratories ...................... 206
    Konrad Boettcher, Claudius Terkowsky, Marcel Schade, Nils Kaufhold,
    Louis Kobras, Doreen Kaiser, Pierre Helbing, Franziska Herrmann,
    Tobias R. Ortelt, Ines Aubel, Jan Haase, Bernhard Meussen,
    Johannes Nau, and Detlef Streitferdt

Introducing Aviation Technological College Students to Online
Collaborative CAD Design ............................................. 217
   *M. Greenholts, I. Verner, and A. Polishuk*

Education for Industry 4.0: Introducing Engineering Students
to Cloud-Based Collaborative Design .................................. 226
   *D. Cuperman, I. Verner, and U. Rosen*

Predictive Maintenance and Production Analysis in Smart Manufacturing ..... 234
   *B. Kalyan Ram, Nitin Sharma, Abhishek S. Joshi, and Advik Vermani*

Automated Quality Control of 3D Printed Tensile Specimen via Computer
Vision .............................................................. 245
   *Rizwan Ullah, Silas Gebrehiwot, Thumula Madduma Patabendige,*
   *and Leonardo Espinosa-Leal*

## Learning in Virtual Environments

Japanese High School Students' Attitudes and Ways of Learning
with Technology Outside of School After the Pandemic .................... 255
   *Yutaro Ohashi*

Avatars in Immersive Virtual Reality Education: Preliminary Insights
and Recommendations from a Systematic Review ........................ 267
   *Ines Miguel-Alonso, Anjela Mayer, Jean-Rémy Chardonnet,*
   *Andres Bustillo, and Jivka Ovtcharova*

Enhancing Educational Outcomes Through Learning Pathways
and Artificial Intelligence-Driven Learning Analytics ..................... 275
   *Martin Trommer, Susanne Franke, Tobias Teich, Ralph Riedel,*
   *Sven Hellbach, and Daniel Franke*

The Development an IoT-Based Virtual Model of Power Grid System
with Renewable Energy Sources for a Laboratory Practicum of Control
Systems ............................................................. 285
   *Maryna Stupak, Hanna Telychko, Hlib Stupak, and Valerii Potsepaiev*

## Augmented and Virtual Reality

Effectiveness Study of an Augmented Reality App as Preparation Tool
for Electrical Engineering Laboratory Courses ........................... 297
   *Mesut Alptekin and Katrin Temmen*

Enhancing the Immersion of Augmented Reality Through Haptic Feedback .... 305
  Joshua Grodotzki, Benedikt Tobias Müller, and A. Erman Tekkaya

Augmented Reality Based Control of Autonomous Mobile Robots ............ 313
  Benedikt Tobias Müller, Joshua Grodotzki, and A. Erman Tekkaya

Scratch-Based Exergame-Educational Approach in Teaching the Course
"Physical Education" for IT-Specialties .................................. 324
  Oleksandr Blazhko, Viktoriia Podhorna, Anastasiia Kokotieieva,
  and Nataliia Bashavets

A Virtual Lab of Matrix-Assisted Laser Desorption Ionization
MALDI-TOF Protocols for Microbiology Students ........................ 336
  Soma Datta and Ibrahim Ibaad Syed

How Can Pneumatics Be Trained with Augmented Reality in the Context
of Training for Industrial-Technical Professions? ........................ 344
  Jan-Niklas Terschüren, Valérie Varney, and Larissa Müller

Author Index ....................................................... 357

# Engineering Education of the Future

# Fostering Experiential Learning and Situational Awareness Via Reporting and Broadcasting About Pilot Projects

Prajwal Prabhu and Vishal Singh[✉]

Center for Product Design and Manufacturing, Indian Institute of Science
Bengaluru, Bengaluru, India
{prajwalp,singhv}@iisc.ac.in

**Abstract.** This research examines the possibilities of project reporting and broadcasting to improve situational awareness and experiential learning in academic settings. Drawing inspiration from the media industry's use of news reporting to create general situation awareness, this work initially aimed to improve the situation awareness of project teams in Architecture Engineering and Construction (AEC) projects. However, our findings revealed an additional positive impact on experiential learning. Complex design and AEC projects often deviate from planned activities and timelines due to intricate task interdependencies and stakeholder influences. Effective project information management is crucial in such contexts. This work presents the results of two pilot studies conducted in an academic setting to assess the impact of regular project progress updates provided by a Situation Monitoring Analysis and Reporting Team (SMART) through broadcasts. The Qualitative findings of the study indicate multiple benefits of broadcasting and reporting. Firstly, recipients of project progress updates reported enhanced situational awareness, providing them with an improved understanding of project complexities. The video broadcast offered flexibility for the users to listen to the status of the renovation project in their comfort, allowing them to listen to a narrative instead of making efforts to read and find information. Secondly, participants in SMART reported increased attention to project details, leading to improved practical knowledge in design and project management. Lastly, the research highlighted that students often overlook experiential learning opportunities in their surroundings due to a lack of access to relevant details. SMART-led reporting and broadcasting capitalize on these missed opportunities by using familiar and desirable media formats, such as video reporting, to engage students and professionals alike. This research suggests that project reporting and broadcasting promote situation awareness and act as an experiential learning tool, coinciding with current trends in content sharing and learning preferences in an era where 21st-century skills are crucial.

**Keywords:** Experiential Learning · Situational awareness · Reporting · Broadcasting · AEC projects

M. E. Auer et al. (Eds.): STE 2024, LNNS 1027, pp. 3–14, 2024.
https://doi.org/10.1007/978-3-031-61891-8_1

# 1  Introduction

In an academic setting, the management of Architecture Engineering and Construction (AEC) projects, such as the campus development activities, not only presents an engaging educational opportunity but also closely mirrors the challenges encountered in the workplace. These projects, characterized by complex job interdependence and diverse stakeholder interactions, offer an ideal environment for experiential learning, a critical component in preparing students for their future professions [1]. Experiential learning is increasingly understood to play a vital role in preparing students for their future professions by providing them with practical skills, encouraging critical thinking, and fostering adaptability, in addition to imparting traditional project management information. However, the implementation of experiential learning in the academic setting is not without its challenges. These challenges encompass resource allocation, the imperative to provide students with access to genuine real-world scenarios, and the delicate equilibrium required to uphold a structured educational environment while exposing students to the complexities and uncertainties inherent in genuine projects [2]. By immersing students in dynamic, authentic scenarios and allowing them to grapple with the intricacies of actual projects, students are better prepared to face the dynamic challenges awaiting them beyond the classroom.

AEC projects on campuses, big or small, demand a good understanding of architectural planning, budget management, and procurement while ensuring minimal disruption to ongoing academic activities. AEC projects inherently exhibit dynamic characteristics, often deviating significantly from meticulously planned activities, resulting in quality issues, budget overruns, scheduling delays, and the potential for legal disputes and contract conflicts [3]. These challenges can largely be attributed to coordination difficulties exacerbated by information bottlenecks, hindering the timely collection, evaluation, and dissemination of project updates. Effective management of such initiatives necessitates not only technical expertise but also strong interpersonal skills and heightened situational awareness [4]. Endsley's model, a foundational concept in the field of situational awareness, defines situational awareness as "the perception of elements in the environment within a volume of time and space, the comprehension of their meaning, and the projection of their status in the near future" [5].

Attaining situational awareness is an active cognitive process, requiring individuals to rapidly gather relevant data from reliable sources and efficiently exchange information with stakeholders to understand and anticipate evolving circumstances. This concept can be further extended in the academic context to include experiential learning, where students actively engage with real-world situations instead of unconscious knowledge acquisition. They tackle complex tasks, gaining a deeper appreciation for their intricacies and learning to adapt to unexpected challenges, much like professionals in real-world AEC projects. Within this educational framework, Situation Monitoring Analysis and Reporting Teams (SMART) can assume a pivotal role by facilitating the rapid exchange of critical data and information. They can be likened to "media room" scenarios, where dedicated teams share situational awareness. In this academic context, these teams provide real-time updates on project status, issues, and implications to all stakeholders, including students, professionals, and academic leadership.

The need for enhanced situational awareness in AEC projects is well-established. Recent research acknowledges the importance of situational awareness within the context of AEC projects [4, 6]. This research draws inspiration from the media industry's adept use of news reporting to create a shared understanding of complex situations and resolve information flow challenges in AEC projects. In this context, this work suggests that project reporting and broadcasting serve a dual role, extending beyond their primary function of promoting situational awareness. They emerge as powerful tools for nurturing experiential learning, aligning with current trends in content sharing and learning preferences, especially in an era where 21st-century skills are emphasized. By harnessing these innovative approaches, academic settings can better equip students and professionals with the practical insights and knowledge necessary to excel in dynamic AEC projects and work scenarios.

## 2 Literature Review

In this literature review, we will explore two interconnected aspects. Firstly, we briefly review the critical role of information flow in AEC projects and its potential impact on situational awareness. Effective information flow is pivotal in AEC, influencing productivity, project timelines, and quality. Secondly, we briefly review the academic context, focusing on situational awareness and its link to experiential learning. Within academic settings, situational awareness extends beyond professionals to students, encouraging active engagement in real-world scenarios.

### 2.1 Information Flow and its Potential Impact on Situational Awareness in AEC Projects

AEC projects involve interdisciplinary teams coordinating and sharing information to ensure the project's success. Information flow issues often manifest in reduced productivity, increased waste, increased rework, and increased design errors. The information flow between the site and the organization was the focus of Nielsen and Sayar's study from a productivity standpoint, and they discovered that a significant amount of time was lost in this process [7]. Studies have found that inadequate communication and unavailability of information majorly contribute to increased project completion times [8]. As per Hicks [9], 'the notion of waste within the context of information management can be considered to include the additional actions and any inactivity that arise as a consequence of not providing the information consumer immediate access to an adequate amount of appropriate, accurate and up-to-date information.' Waste is often attributed to design errors and project managers' inability to identify critical information among teams as the root cause [10]. From a planning perspective, construction teams coordinate with different stakeholders to decide the physical order of the work that needs to be done in a sequential short-term planning window. However, the planning teams face several obstacles that include construction project status, frequent changes to resource availability, instability in production rates, design changes, and limitations on their scope of action [11]. Achieving good information flow reduces deviation in the actual and the as-planned tasks. Mohsini and Davidson (1992) have highlighted that the sufficiency of information largely accounted for the variance in the project cost and quality [12]. Bubshait et al. have shown that construction stakeholders often operate within their silos, limiting project data sharing among project teams and through the lifecycle [13]. Siloed information often translates into collaboration issues and often blocks the integration of the design and the construction activities, which creates inefficiencies in the project. The issues discussed above show the critical importance of achieving good information flow in AEC projects and the impact of poor information flow on the awareness of the project stakeholders. While recent technological advancements have looked at solving the information flow problems of construction, this has not helped break the information silos. Technology development has meant that the AEC industry can now automate data collection through different hardware like sensors and indoor positioning systems. These developments have increased the data quality and volume, but the different data streams have been used individually, which retains the information silos of the construction industry.

### 2.2 Situational Awareness in Academic context and the role of Experiential Learning

Within academic settings, the convergence of situational awareness and experiential learning represents an approach that enhances students' cognitive capabilities and practical skills. In the educational and learning paradigm, situational awareness is defined as a student's capacity to perceive, comprehend, and anticipate circumstances within their learning environment [5]. This increased awareness enables students to actively participate in intricate, real-world scenarios, which promotes the development of critical thinking, flexibility, and problem-solving abilities that are essential for handling difficult academic tasks. It also increases their awareness of the learning environment at the same time. Experiences are the fundamental source of learning and personal development, according to the theory of experiential learning, which Kolb promoted in his seminal work from 1984 [2]. This concept advocates for active student involvement in real-world settings that closely mirror the professional realms of their disciplines within the academic framework. Experiential learning effectively closes the gap between theory and practice by drawing students into these real-world experiences and deepening their understanding of complex academic issues. Additionally, by fostering a mutually advantageous connection with situational awareness, this immersive pedagogical method creates a symbiotic interaction that has several benefits. Situational awareness and experience learning work together to provide students with the tools they need to solve problems effectively, adapt, and think critically [14]. This method, which has its roots in Endsley's 1995 study on situational awareness, recognizes the significance of dynamic systems awareness, particularly in high-stakes situations. Implementing a novel experiential learning model in an academic setting is challenging. It requires coordination, faculty training, and resources. However, practical solutions such as case studies, simulations, interactive learning techniques, and real-world projects can help overcome these obstacles [15]. By using these techniques, students can apply their knowledge in real-world situations and strengthen their situational awareness under the supervision of mentoring relationships and cooperative team environments. Educational technology offers varied opportunities to improve situational awareness and experiential learning through immersive experiences like virtual and augmented reality[16].

## 3 About the Case

### 3.1 Pre-pilot study – SMART Team Monitoring a yearlong AEC Project

As a precursor to the pilot studies discussed in this paper, a pre-pilot study was conducted as the first trial case. The trial project was a year-long AEC project in which the SMART team was reporting the project updates to the stakeholders. The stakeholders included the department chair, faculty, and students. Using the multimodal broadcasting options, the SMART team broadcasted the updates about the trial case project via presentation slides, videos with narration, and a dedicated web page. A survey was conducted to capture the stakeholders' feedback and insights. Though the number of survey responses was limited, the responses unanimously showed that video-based narratives, combined with web-based information sharing, were preferred modes of seeking situation awareness. The findings from this trial case consolidated the research hypotheses and provided the basis for the pilot studies reported in this paper.

### 3.2 Pilot Study 1 – SMART Team to Monitor the Project Status of a Renovation Project

Pilot Study 1 focused on the practical implementation of the SMART (Situation Monitoring Analysis and Reporting Teams) approach in monitoring ongoing renovation projects within the authors' institution and reporting to the stakeholders. Based on the findings of the trial case study, this pilot's objective was to implement video-based broadcasting complemented by a web-based information portal. The renovation project, spanning over two months, offered a unique setting for experiential learning and research. In contrast to the traditional classroom setting, this project demanded continuous vigilance, effective communication, and the ability to adapt to evolving circumstances from the students, which are the key elements of experiential learning.

The project stakeholders include the students in the department, the faculty in charge, the department chair, and the architect handling the renovation project. The SMART team was comprised of two researchers and a project assistant who were tasked with regular progress monitoring and updating the relevant stakeholders. The SMART team was further tasked with developing a project work breakdown structure, which was central to project tracking and reporting. The SMART team regularly visited the renovation site to monitor the progress and report any issues hindering the progress of the renovation project. To facilitate communication and situational awareness, onsite video recording was implemented, capturing stakeholders' perspectives on the project's status. This approach is illustrated in Figs. 1 and 2, which demonstrate how broadcasting was used to inform and communicate with different stakeholders and to report the site activities and status.

Additionally, unstructured stakeholder interviews were conducted to capture the concerns, expectations, and potential issues relevant to the renovation project. The interview

| Activity | Planned | Actual |
|---|---|---|
| Civil works (finalizing vendors caused delay) | 7th Nov. | 14th Nov. |
| Arrival of new furniture (finished installation in 2 days as planned) | 17th Nov. | 17th Nov. |
| Installation of new Computers | 21st Nov. | 21st Nov. |
| U.P.S arrival (installed on the same day) | NEW | 24th Nov. |
| Installation of Printers with new table, Electrical and data point connections | NEW | 9th Dec. |
| Installation of Plaque | 10th Dec. | 10th Dec. |
| Inauguration of the Lab (delay due to increased scope of project) | 24th Nov. | 12th Dec. |

Broadcast on : 30th Dec, 2022   New Furniture arrived / Installation works began

**Fig. 1.** Broadcast includes narration and review of planned vs actual schedules.

Broadcast on : 15th Nov, 2022   Architect :

**Fig. 2.** Broadcasting the Architect's updates to the different stakeholders

touched on broad aspects of project management and the potential role that broadcasting can play in solving the issues that can be attributed to managing these types of projects. Interview excerpts showed that video broadcasts were useful in understanding the status of the renovation project for the stakeholders. For example, the project architect stated that 'These Video Broadcasts can be used as a reference to catch up on the status of the construction projects'; the faculty-in-charge stated, 'The video broadcasts don't require any technical learning, giving stakeholders the flexibility to catch up on', and the department chair said that 'The idea of broadcasting creates accountability on the person in charge of the responsible tasks'.

The video broadcast offered flexibility for the users to listen to the status of the renovation project in their comfort, allowing them to listen to a narrative instead of making efforts to read and find information. Through the pilot, the SMART team was able to learn the different considerations that they need to adopt while broadcasting and creating videos. This meant that the SMART team was able to learn about the effective broadcasting strategy to improve information dissemination within the construction project. In addition, putting research students and the project assistants in the role of the SMART team also enhanced their knowledge of project management and experiential learning about the renovation project.

### 3.3 Pilot Study 2 – SMART Team to Monitor the Project Status of a Simulated Lab-based Construction Project

A simulated lab-based construction project was carried out for the second pilot research. Six project engineers (students), four project managers (researchers), and a project leader in charge of the lab were among the stakeholders participating in this controlled experiment. The main objectives of the pilot were to evaluate the coordination within the existing project by applying a diagnostic methodology and to assess the impact of frequent information dissemination on stakeholders' situational awareness. The project addressed different construction aspects, including assembly procedures, integration of the Internet of Things (IoT) in the project context, and the design and analysis of structural components. Exposure to a hands-on simulated AEC project provided the project engineers (students) with a means of applying their theoretical knowledge in an applied context grounded in experiential learning, which was the main learning goal. The initial baseline assessment, conducted prior to information broadcasting, laid the groundwork for subsequent evaluations and insights. The data was captured through surveys utilizing the Situational Awareness Rating Technique (S.A.R.T) [17], as shown in Table 1. The questionnaire included qualitative questions that gave students the chance to discuss project-related difficulties. This simulated project offered an experiential learning opportunity, allowing hands-on experience and insights into project coordination challenges and strategies. A customized broadcasting configuration was created to enhance experience learning and situational awareness. A range of channels for sharing information and communicating was used: an interactive dashboard accessible via the lab's platform; video presentations showcasing project progress; real-time WhatsApp communication for prompt stakeholder engagement; periodic email updates to all stakeholders; and the use of MS Project for comprehensive project status tracking (Fig. 3).

The survey results are shown in Fig. 4, which captures the students' responses to the different questions asked in the survey. There were some early project delays because the student team needed time to gain hands-on experience and practice in crucial software tools needed to complete their tasks.

To overcome the delays from lack of hands-on experience with the tools, the SMART team proactively planned town hall meetings, a type of broadcasting that creates a system for the SMART team to address the issues faced by the students at one go, where students who had faced similar problems with the software were given the platform to air their queries which were answered by the SMART team. The outcome was exponential growth in the student's learning, which helped the project recover pace and helped them get comfortable with the software tools. Unstructured Interviews with students confirmed the effectiveness of the experiential learning strategy. The interviews covered a broad range of topics that included the role clarity of the students, the effectiveness of the project communication and the effectiveness of the broadcasting strategies. Students

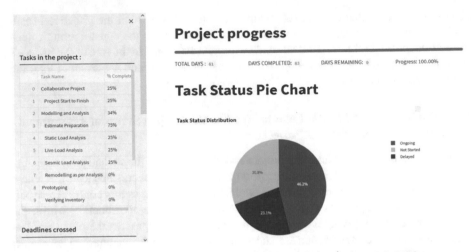

**Fig. 3.** Web-based dashboard to provide status updates to project stakeholders.

**Table 1.** Baseline Assessment Questions Asked During Broadcasting

| Sl. No | Survey Question asked |
|---|---|
| 1 | On a scale of 1 to 5, please rate your understanding of the project's objectives and goals. (Very Poor, Poor, Neutral, Good, Excellent) |
| 2 | How confident are you in your ability to identify critical project information sources? (Not Confident, Somewhat Confident, Moderately Confident, Very Confident, Extremely Confident) |
| 3 | How well do you understand your role within the project team and the roles of other team members? (Poorly, Fairly Well, Well, Very Well, Extremely Well) |
| 4 | How often do you proactively anticipate potential developments or issues affecting the project? (Rarely, Occasionally, Often, Very Often, Always) |
| 5 | How effective is the communication within the project team regarding project-related information? (Ineffective, Somewhat Effective, Moderately Effective, Effective, Highly Effective) |
| 6 | Can you provide examples of situations where you felt that you were not adequately informed or aware of critical project information? |
| 7 | Describe any instances where you felt uncertain about your role within the project team or faced difficulties understanding other team members' roles. |
| 8 | Were there any specific situations or events where you needed to proactively anticipate potential developments or issues within the project? Please describe. |

expressed sentiments such as "Our doubts got effectively cleared (Student 1)" and "It is very productive (Student 3); we are able to apply what we are learning in the process (Student 4)" in the interviews and through the qualitative part of the survey.

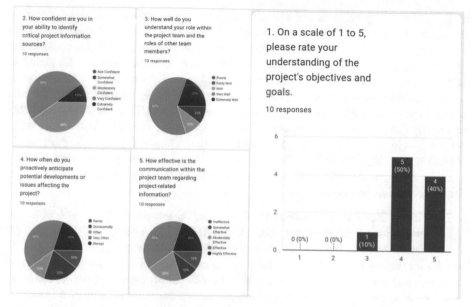

**Fig. 4.** Survey results from Pilot study-2

From the SMART Team's perspective, gaining awareness of the project engineers' struggles with software tools led to proactive decisions, ultimately facilitating the recovery of the project schedule after the initial delays and lack of progress. This pilot study underscored the effectiveness of experiential learning, broadcasting, and coordination assessment within a simulated lab-based construction project, providing valuable insights for academic settings.

## 4 Discussions and Limitations for the Study

The results of the pilot studies demonstrate how broadcasting may support experiential learning and improve situational awareness in AEC projects. The study also revealed a common oversight among students– the tendency to overlook possibilities for experiential learning that might be present around them. For example, there are several AEC-related projects ongoing on the authors' institute campus, but even the AEC-related students hardly ever take notice of such projects or consider these as opportunities for site visits or learning about the AEC process. In their preoccupation with theoretical concepts and classroom activities, students often fail to seek active opportunities for domain-related learning available around them. This means that potential experiential learning opportunities are missed that are available in their environment. Drawing parallels with contemporary content-sharing platforms like YouTube Shorts, which seem to be attractive to the younger audience, broadcasting emerges as a dynamic educational catalyst, seamlessly integrating theoretical concepts into real-world scenarios for swift and accessible learning experiences. Utilizing the fundamental feature of broadcasting, i.e., the multimodal approach, these pilot studies utilize a wide range of broadcasting methods, including interactive dashboards and on-site recordings, to actively engage the stakeholders of the construction project. These broadcasts are readily accessible to stakeholders at their convenience, facilitating self-directed and hands-on learning opportunities. The multimodal feature of broadcasting seamlessly aligns with the latest developments in technology and education, combining interactive, visual, and auditory components to accommodate diverse learning styles. By employing familiar and desirable media formats, especially video reporting, the SMART teams engage students and professionals alike. This intentional use of accessible mediums transforms mundane details into experiential learning moments. Recognizing the ever-changing needs of modern learning, this approach serves as a bridge, closing the gap between abstract concepts and the real world. Implications of the pilot study from an experiential learning perspective are as follows:

- **Enhanced Student Engagement:** The research underscores the potential for enhanced student engagement through SMART-led reporting and broadcasting, making experiential learning an integral part of their educational journey.
- **Closing the Awareness Gap**: SMART-led reporting effectively closes the awareness gap that often leads to missed experiential learning opportunities because of SMART-led reporting's capacity to engage multiple senses and stimulate problem-solving, adaptation and critical thinking abilities. This aligns with the broader goal of creating a learning ecosystem that is both accessible and engaging.

  The limitations of the study are as follows:

- The findings and conclusions of the current study are context-specific to AEC Projects. The possibility of creating experiential learning through pilot studies in other contexts needs to be explored, which is not within the scope of this study.
- Further, there was a limited number of stakeholders giving feedback in these pilot studies. In pilot study 1, even though the stakeholder group was large, the received responses with limited to single digits. Pilot study 2 simulated a typical construction organizational structure that can be seen in a construction project. Hence, the study needs further testing and validation.
- The learning curve connected with new software was the cause of early project delays in Pilot Study 2. However, the study does not go into the details of the nature of these learning curves, which may vary depending on the project.

## 5   Conclusions for the Study

In conclusion, the integration of broadcasting in AEC projects and academic settings proves effective in enhancing information flow, situational awareness, and experiential learning. The real-time updates through broadcasting fill a critical gap in conventional technology-based approaches, ensuring stakeholders have access to the latest project status. This dynamic, multimodal method narration actively involves stakeholders, including students, in AEC projects, aligning with experiential learning principles. Institutions can meet the demand for technology-enhanced experiential learning while bridging the gap between theory and practice. Broadcasting's ability to engage multiple senses, leverage narration and storytelling, stimulate problem-solving, and offer real-time exposure to complex concepts positions it as a powerful educational tool. While this approach aligns with current trends, it's essential to consider context specificity, the nature of learning styles, and the need for further empirical validation. The long-term sustainability and effectiveness of broadcasting methods in diverse contexts warrant continued investigation to advance educational practices and the construction industry.

**Acknowledgement.** The research presented in this paper is part of the SMART research project [SRG/2021/002285] funded by the Science and Engineering Research Board, Department of Science and Technology, India.

# References

1. Jallow, A.K., Demian, P., Baldwin, A.N., Anumba, C.: An empirical study of the complexity of requirements management in construction projects. Eng. Constr. Archit. Manag. **21**, 505–531 (2014). https://doi.org/10.1108/ECAM-09-2013-0084
2. Kolb, A.Y., Kolb, D.A.: Learning styles and learning spaces: enhancing experiential learning in higher education. Acad. Manag. Learn. Educ. **4**, 193–212 (2005). https://doi.org/10.5465/AMLE.2005.17268566
3. Abdul-Rahman, H., Berawi, M.A., Berawi, A.R., Mohamed, O., Othman, M., Yahya, I.A.: Delay mitigation in the malaysian construction industry. J. Constr. Eng. Manag. **132**, 125–133 (2006). https://doi.org/10.1061/(asce)0733-9364(2006)132:2(125)
4. Reinbold, A., Seppänen, O., Peltokorpi, A., Singh, V., Dror, E.: Integrating indoor positioning systems and BIM to improve situational awareness. In: 27th Annual Conference International Group Lean Construction IGLC 2019, pp. 1141–1150 (2019). https://doi.org/10.24928/2019/0153
5. Endsley, M.R.: Toward a theory of situation awareness in dynamic systems. Hum. Factors J. **37**(1), 32–64 (1995)
6. Akinci, B.: Situational awareness in construction and facility management. Front. Eng. Manag. **1**(3), 283 (2014). https://doi.org/10.15302/J-FEM-2014037
7. Nielsen, Y., Sayar, T.: Web-based information flow modelling in construction. Assoc. Res. Constr. Manag. **1**, 5–7 (2001)
8. Hewage, K.N., Ruwanpura, J.Y.: A novel solution for construction on-site communication – The information booth. Can. J. Civ. Eng. **36**, 659–671 (2009). https://doi.org/10.1139/L09-026
9. Hicks, B.J.: Lean information management: Understanding and eliminating waste. Int. J. Inf. Manage. **27**, 233–249 (2007). https://doi.org/10.1016/j.ijinfomgt.2006.12.001
10. Love, P.E.D., Edwards, D.J., Irani, Z.: Forensic project management: an exploratory examination of the causal behavior of design-induced rework. IEEE Trans. Eng. Manag. **55**, 234–247 (2008). https://doi.org/10.1109/TEM.2008.919677
11. Martinez, J.G., Yeung, T., Sacks, R.: Production planners' scope of action in the context of digital twin construction. In: Twelth International Conference Construction 21stCentury (2022)
12. Mohsini, R.A., Davidson, C.H.: Determinants of performance in the traditional building process. Constr. Manag. Econ. **10**, 343–359 (1992). https://doi.org/10.1080/01446199200000030
13. Bubshait, A.A., Siddiqui, M.K., Al-Buali, A.M.A.: Role of communication and coordination in project success: case study. J. Perform. Constr. Facil. **29**, 1–7 (2015). https://doi.org/10.1061/(asce)cf.1943-5509.0000610
14. Duron, R., Limbach, B., Waugh, W.: Critical thinking framework for any discipline. Int. J. Teach. Learn. High. Educ. **17**, 160–166 (2006)

15. Burch, G.F., Giambatista, R., Batchelor, J.H., Burch, J.J., Hoover, J.D., Heller, N.A.: A meta-analysis of the relationship between experiential learning and learning outcomes. Decis. Sci. J. Innov. Educ. **17**, 239–273 (2019). https://doi.org/10.1111/dsji.12188
16. Schott, C., Marshall, S.: Virtual reality and situated experiential education: a conceptualization and exploratory trial. J. Comput. Assist. Learn. **34**, 843–852 (2018). https://doi.org/10.1111/jcal.12293
17. Endsley, M.R., Selcon, S.J., Hardiman, T.D., Croft, D.G.: A comparative evaluation of SAGAT and SART for evaluations of SA. Hum. Factors Ergon. Soc. Annu. Meet. **42**, 82–86 (1998)

# Real Estate Insights: A Boardgame-Based Experiential Learning Approach

Vighneshkumar Rana and Vishal Singh

Indian Institute of Science, CV Raman Rd, Bengaluru, Karnataka 560012, India
vighneshkuma@iisc.ac.ina, singhv@iisc.ac.in

**Abstract.** Real estate asset management demands specialized skills and knowledge for success, even though a large section of society makes real estate investment decisions at one point or the other. However, the traditional learning methods in real estate often lack adequate experiential and immersive learning scenarios that are observable in competitive real-world real estate decision situations. Consequently, many individuals fail to develop investment strategies in real estate, limiting their ability to make informed decisions. This research addresses this gap by introducing a board game to bridge these educational shortcomings in the real estate investment context. The board game offers players a simulated environment that immerses them in the complexities of real estate asset management, enabling them to make strategic decisions, optimise resources, and navigate challenges such as property acquisition, financing, tenant management, maintenance, and risk assessment. By conducting gameplay sessions and collecting participant feedback, this research evaluates the effectiveness of the board game in enhancing knowledge, skills, and understanding of real estate asset management concepts and investment strategy. Thematic analysis of pregame and postgame questionnaires reveals a substantial enhancement in participants' comprehension of real estate concepts through iterative discussions following the post-game session.

**Keywords:** Real estate industry · Investment strategies · Decision-making

## 1 Introduction

As an asset class, real estate encompasses various components that contribute to its overall value. It includes land, buildings, and natural resources, creating a comprehensive category of property. This definition not only considers the physical structures and land but also recognizes the emotional significance that buildings and housing hold for individuals and families as they become cherished homes [1]. Real estate investment is undergoing continuous transformation, with notable growth and appeal observed in recent years, particularly throughout the 2020s. Investors have come to recognize the reliability and potential for wealth accumulation offered by real estate as an asset class [2]. Real estate investment decision-making is a multifaceted, complex, and diverse process involving various factors. Factors such as demographic trends, cultural preferences, and community dynamics are crucial in shaping real estate decisions [3]. Furthermore,

M. E. Auer et al. (Eds.): STE 2024, LNNS 1027, pp. 15–31, 2024.
https://doi.org/10.1007/978-3-031-61891-8_2

real estate investment entails financial risk. Investors must carefully evaluate and manage these financial risks to make informed investment decisions. Proper risk assessment and mitigation strategies are crucial in maximising returns and minimising potential financial setbacks in the real estate sector. Traditional views on equity real estate portray it as an inefficient market where success hinges on the individual's skills and knowledge in selecting and negotiating investments [4]. However, true success lies in the pragmatic execution of property investment strategies. This entails adopting a strategic approach that goes beyond property selection alone. Real estate investment extends beyond personal homeownership, enabling investors to leverage their expertise across diverse real estate assets. Evaluating market conditions, potential returns, and risk mitigation strategies is crucial in the property selection process. One significant challenge in real estate investment is the lack of understanding and common knowledge about the factors involved in decision-making, risk management, and investment strategies [5]. Many individuals may not fully grasp the intricacies of real estate investment and struggle to apply effective strategies.

Recognizing this knowledge gap, one potential solution to enhance people's understanding of real estate and make it more accessible is to provide them with practical, hands-on experiences. These experiences should be designed to be engaging and easy to understand, allowing participants to make decisions, face consequences, and learn from their actions in a safe and controlled environment. This can be achieved through various means, such as interactive workshops, simulations, or even board games that simulate real estate investment and management scenarios. This research employs game-based learning techniques to make the concepts and principles of real estate property investment strategies more engaging and accessible to a wider audience. Additionally, incorporating real-world data and examples into these experiences can bridge the gap between theory and practice, making the complex world of real estate more relatable and accessible to a broader audience. By gamifying real estate investment principles, individuals can learn about value-based investments in a fun and interactive way. This approach can enhance people's understanding of value-based investments, leading to the early adoption of value-seeking behaviors. Ultimately, this can help individuals make more informed investment decisions and moderate their overall investment spending in the long term.

The research question of the following study is as follows:

"How can game-based learning and practical, hands-on experiences, such as board games, be effectively employed to enhance individuals' understanding of real estate investment principles and strategies?"

Games can be utilized to enhance people's understanding and knowledge of decision-making, risk management, and investment strategies to create a more enjoyable and easily understandable learning experience. By transforming complex concepts into interactive games, we can make them accessible to a broader audience. "Serious games" can be employed, focusing on educational purposes rather than entertainment, to improve people's comprehension of real estate strategies [6]. Various types of games, including board games, video games, simulations, and puzzles, can be utilized to convey serious themes. Among these are specific board games designed to explain the real estate market dynamics and financial management [7]. These games aim to enhance understanding by simulating real estate factors in a controlled environment. These serious board games provide players with practical insights into the complexities of real estate investment, property valuation, risk assessment, and financial decision-making [8]. Through hands-on experiences and strategic gameplay that simulate real-world scenarios, individuals can better comprehend the factors that influence the real estate market and improve their ability to make informed investment decisions. Serious board games effectively expand knowledge and foster a practical understanding of real estate dynamics and financial management among players.

This research will leverage the concepts of serious games and game-based learning techniques to develop a board game. The goal of the boardgame is to create an engaging and enjoyable learning experience while accurately reflecting real estate scenarios within the game's internal environment. By incorporating serious game elements and game-based learning techniques, Boardgame aims to provide players with a fun and exciting way to learn and understand the complexities of real estate management.

## 2 Literature Review and Identification of Factors

### 2.1 Unveiling the Dynamics of Real Estate Investment – the Literature Review

The real estate market has traditionally lacked transparency, with limited access to information regarding property transactions and specific house details. In recent times, there has been a significant increase in access to real estate data through online platforms, providing greater transparency and options for property transactions. Literature extensively delves into the complex behaviours and influences that shape individual property purchase decisions, housing preferences, and purchase intentions. A diverse array of factors, spanning multiple dimensions, contributes to individuals' choices in the real estate market. This surge in information access has reshaped the landscape of real estate decision-making.

The location dimension emerges as a significant factor, with attributes such as proximity to workplaces, city amenities, and neighbourhood features playing a crucial role in decision-making. Additionally, factors related to the environment, including health safety, air pollution, climate change, and access to services, contribute to the overall desirability of a location. Financial considerations also hold substantial importance, with factors like property prices, financing options, loan availability, and monthly household income influencing individuals' ability to purchase a property. The affordability and perceived value of the property are key factors in this dimension. A property's design and structural attributes, including the type of house, number of bedrooms, floor level, ventilation, and availability of facilities, also influence buyers' decisions. These factors contribute to the overall comfort and functionality of the property. Social and lifestyle factors, such as social environments, neighbourhood perception, buying power, social status, and desired lifestyle, are additional dimensions that impact property purchase decisions. The experience of previous purchases, family expansion, job changes, and the desire for a larger home also come into play. The research literature further highlights the significance of market dynamics, including prices, market trends, transaction frequency, and the presence of bubbles. Factors related to migration patterns, economic growth, demographics, construction costs, and energy efficiency considerations also shape property purchase decisions [9–11].

To develop a board game, which simulates asset management, it is crucial to integrate these influential factors into gameplay mechanics. By doing so, players can gain a realistic and immersive understanding of the complexities involved in real estate investment. This approach aims to enhance players' strategic thinking and analytical skills, empowering them to make well-informed decisions across various dimensions of property investment.

## 2.2 Integrate Real Estate Factors in Gameplay

Following the identification of factors in the previous section, participants are provided with a feedback form after playing the game. They are asked to rate each specific factor on a scale of 1 to 3, where 1 signifies the factor's least importance and 3 represents its highest importance. Question asked in feedback form: "To what extent does the following factor play a significant role in the process of making decisions regarding property purchase?" Simultaneously, participants are also prompted to evaluate the effectiveness of the integration of these factors into the game, using the same 1 to 3 scale. Question asked in feedback form: "How well do you think the factor below is included and working in the game's rules and game mechanics?". The details of this feedback are presented in Table 1.

**Table 1.** Factors identification and integration in the board game

| Factor Name | Importance of factor (On a scale of 1 to 3) | Integration into game (On a scale of 1 to 3) | In what manner is factor integrated into the game? |
|---|---|---|---|
| Location of Property | 1.83 | 2.83 | Features 10 different locations in the city |
| Finance for purchase | 2.5 | 2 | Loan support |
| Social Environment | 1.83 | 1 | Community building, social responsibility |
| Neighbourhood | 1.67 | 1.67 | Amenities in some areas |
| Property Infrastructure | 2.33 | 2.33 | Types of housing options |
| Structural attributes | 2.33 | 1.17 | Varying maintenance cost |
| Price and value | 2.83 | 2.67 | Different prices for different housing |
| Property Design | 2.33 | 1.33 | Type of housing |

(*continued*)

**Table 1.** (*continued*)

| Factor Name | Importance of factor (On a scale of 1 to 3) | Integration into game (On a scale of 1 to 3) | In what manner is factor integrated into the game? |
|---|---|---|---|
| View and Ventilation | 2.83 | 1 | Specific features of the house |
| Workplace distance | 2.67 | 2.33 | Daily commute with charges |
| City Amenities | 2.67 | 2.17 | Amenities with charges |
| Millennials Socialising | 2 | 1 | Community building |
| Property Quality | 2.33 | 1.33 | Type of properties based on quality |
| Family Income | 2.67 | 1.83 | Jobs and Salary |
| Property types | 2.33 | 1.67 | Bungalows, Row houses, and apartments |
| Mobility facilities | 2.5 | 1.83 | Commute with a specific charge |
| Community experience | 2 | 1.17 | Community building |
| Parents fund | 2.67 | 1 | Starting amount given |
| Property Rent | 2.67 | 2.17 | Specific rent of each property |
| Search time | 2.67 | 1.83 | Player time bound – 2 min |
| Buyer seller ratio | 2.17 | 1.33 | Property selling between players |
| Price Trends | 2.5 | 2.5 | Inflation (15%) |
| Purchase frequency | 2 | 2.17 | Multiple property purchases |
| Market Volatility | 2.5 | 1.67 | Scenario cards |
| Migration of person | 2.17 | 1.5 | Job change, |
| Area growth | 2.17 | 1.5 | Inflation, Scenario cards |
| Demographic trends | 2.5 | 1 | Scenario cards |

Disclaimer: The data samples for Table 1 are limited; nevertheless, ongoing research is actively addressing this constraint. Moreover, the game is currently in its initial phase, and not all the factors mentioned are incorporated. The primary focus is on creating a simulation that represents certain aspects of real estate. In future iterations, additional factors may be introduced to enhance the game's complexity and realism.

## 3 Information About Game

**Design Objectives:**
The design objectives of the board game encompass a variety of purposes, aiming to:

- Deliver a positive learning experience through enjoyable gameplay.
- Simulate the dynamics of the real estate market.

  The game also should serve as a tool:

- Provide a practical platform for participants to experience the complexities of real estate investment in a controlled environment.
- Determine the impact of the game on enhancing real estate knowledge and decision-making.

**Game Components:**
The game consists of a board featuring ten distinct areas within Bengaluru, each with a varying number of property options, dice, job-cards, and scenario cards. These properties encompass a range of values, including high-value bungalows, medium-value row houses, and low-value apartments. The areas themselves are characterized by differing price ranges, mirroring patterns observed on actual Bengaluru real estate websites. Within select areas, players can find workplaces where they can earn money to invest in additional properties. Players have the flexibility to change their jobs and locations during the game. Additionally, there are associated costs such as city amenities and house maintenance, which players can spend money on to access these services. Property acquisition allows players to generate rental income, contributing to their passive income. The game introduces scenario cards that simulate real-world situations, including economic downturns, inflation, market crashes, demand surges, government policy changes, and natural disasters, thus replicating real-life scenarios.

**Winning Conditions:**
The winner is determined based on their property portfolio's overall valuation, including any remaining money and the value of all properties they have acquired throughout the game (Fig. 1).

Prior to the formal research, the game underwent a trial run with a group of researchers in a controlled laboratory setting with the intention of formally releasing it to a behaviour workshop. This preliminary testing served the purpose of identifying potential issues and determining the appropriate duration for the game. Following these trial runs, we concluded that the game typically spans an average of 2.5–3 h, aligning with the timeframe of one practical lab/lecture session, ensuring that the game can be comfortably completed within this allotted time.

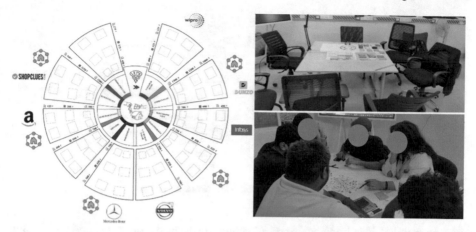

**Fig. 1.** Game Board and experiments setup

## 4  Research Methodology

This study adopted a qualitative research approach to collect and analyse data, focusing on master's grade students and research scholars. In this study, we employ a comprehensive approach to evaluate the participants' understanding of real estate concepts and their learning experiences with pre-game and post-game assessments. Before engaging in the real estate-themed board game, participants are asked to provide their baseline understanding of real estate concepts, knowledge, and confidence levels. Subsequently, after completing the game, they are asked to reflect on their experiences and learning outcomes. Thematic analysis was applied to discern recurring patterns in the collected data, involving an entirely interpretative and subjective analysis of interview conversations. The thematic analysis framework in psychology by Braun et al. [12] serves as the basis for this research. Figure 2 delineates the research framework and the sequential steps undertaken in performing this thematic analysis. The research methodology involves a systematic progression starting with the anonymization of interview data to safeguard individual privacy. Following this, an initial qualitative review is conducted to assess content, leading to the development of a coding scheme for a more structured analysis (Step 2 and 3 in Fig. 2). This process is iterative, informed by related classifications and terms relevant to the targeted coding scheme, allowing modifications based on ongoing review (Step 3). Application of the coding scheme facilitates a qualitative analysis of patterns in the data (Step 4). Further, an in-depth qualitative analysis is pursued through a subjective and interpretive approach, delving into the discussion's various levels of granularity (Steps 5 and 6). The final step (Step 7) involves synthesizing observations from qualitative interpretations to make sense of the findings. Throughout the study, ethical principles were upheld, ensuring informed consent and safeguarding participant confidentiality.

This paper introduces initial findings from the outlined research steps, emphasizing the need for additional research to validate the coded segments and results. To enhance reliability, independent coding by multiple researchers and analysts for cross-validation

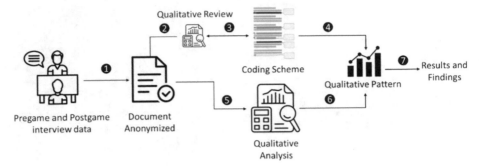

**Fig. 2.** The research framework and steps

is recommended. The findings discussed herein are derived from data coded by a single researcher. Therefore, it is advised to interpret these results with caution until further research involving diverse perspectives in the coding process is undertaken for a more robust validation of the study's outcomes.

## 5 Discussions

The appendix section lists pregame and postgame questions. The pregame questionnaire delves into participants' expectations, motivations, exposure to real estate games, prior investment knowledge, and decision-making for property purchases. It also includes self-assessment. The postgame questionnaire explores post-engagement learning experiences, emphasizing game features contributing to learning. Participants compare experiences, rate changes in understanding and confidence, reflect on challenges, and suggest improvements. Open-ended inductive thematic analysis on these questionnaires involves extracting meaning and identifying themes without predetermined assumptions, ensuring an exploratory, unbiased approach. The coding scheme details in Table 2 outline eight identified code categories, providing brief descriptions and coded segment samples, facilitating a comprehensive and flexible analytical process for nuanced insights.

**Table 2.** Coding scheme applied with illustrative examples

| Coding Scheme | Description | Representative comments (random samples) |
|---|---|---|
| **Financial Literacy in Real Estate** | | |
| Money managing techniques | Focused on budgeting, saving, creating an emergency fund, and expense tracking | How to manage money? How to manage cash flow? |

*(continued)*

**Table 2.** (*continued*)

| Coding Scheme | Description | Representative comments (random samples) |
|---|---|---|
| Profit making | Earning surplus income through successful property investment endeavours | while buying profiting. How much will be in profit? Invest in a profit in again |
| Loans and interest rates | Loans: Borrowed funds, Interest Rates: Cost of borrowing, expressed as a percentage | People buy real estate using loans. And so how they use interest, How do they pay interest on that loan? |
| Portfolio management | Optimizing a collection of investments for maximum return and risk mitigation | Net worth or how much you have. And so all those calculation that you get at the end that actually tells like which one strategy is better if you want to try your hand in real state. So this is very like. Nice simulation of the real state economics |
| **Property purchase/investment** | | |
| Property renting structure | Arrangements governing the terms, conditions, and financial aspects of renting a property | we can rent also but which one at the end is really you know, How rental works or how rental income works in these 2? |
| Property development | Enhancing property value through construction, renovation, or redevelopment of land/buildings | I have played similar kind of game where we can all all the property we can build upon that then |
| Property purchase | Acquire property, transfer ownership for a specified monetary sum | they have to pay your money to buy property like monopoly kind of thing |
| Property investment and returns | Property investment: deploying capital for income and appreciation. Returns: gains, yields | The first thing is the same one that we have to invest in the higher value site or the place where which is the one which has lot of value |
| Value and price of property | Value: Inherent worth. Price: Amount set in a transaction | The value of that site. Yeah. The price of that site matters a lot. That's what I think |

(*continued*)

**Table 2.** (*continued*)

| Coding Scheme | Description | Representative comments (random samples) |
|---|---|---|
| Types of property | Residential, commercial, industrial: diverse real estate classifications for various purposes | what is the difference between commercial property and? Commercial or residential? |
| Finding rented property | Locating leased residence: search, view, negotiate, and secure rental accommodation successfully | I have rented a property. During that time, I faced difficulties in finding rented property. We are living in Bangalore, where rented properties are high. Lot of people are coming |
| **Property Purchase Strategies/Planning** | | |
| Strategies | Strategic acquisition: research, assess, and negotiate for optimal real estate investment | Seeing how people pay loans and like what kind of loan you the strategies which can be used for loan |
| Multiple property management | Managing multiple properties efficiently for optimal investment and tenant satisfaction | Also we can buy multiple properties |
| Observing people's purchase | Analyzing people's buying behaviour for insights into their own preferences and decisions | I saw that. Yeah there, there is involvement of a lot of status. So yeah from next round, I started to Apply those strategy |
| Location selection | Choosing ideal site: assess, demographics, accessibility, amenities, market trends | Uncertainty, I think, is. So first thing is picking up a right location to buy real estate. Because that will give you appreciation on your investment |
| Requirements Analysis | Evaluate needs, constraints, and objectives for informed decision-making or system development | Like what value I'm looking. So what is my requirement for my family side? OK how many people are there and what kind of jobs they do if there are children then where is the school? And if there is I'm working or someone is working in my house home then? |

(*continued*)

**Table 2.** (*continued*)

| Coding Scheme | Description | Representative comments (random samples) |
|---|---|---|
| Risk management | Identify, assess, and mitigate potential risks for optimal decision-making and outcomes | Implement mostly like I won't take a lot of risk. Like let's say I will not rather go into fully debt or things like that |
| Best property selection | Evaluate: location, amenities, and value for optimal real estate investment decision | Theoretically I was like all fine. But like when it when it comes to like practically you'll have to implement. And there were like so many options and I got confused a bit. OK |
| **Property facilities and Amenities nearby** | | |
| Nearby amenities | Proximity: services, facilities, and conveniences in the surrounding vicinity of the property | what and all of the amenities nearby |
| House amenities and payments | Features provided; payment terms in a residential property leasing agreement | Then what kind of facilities are there? If it is an apartment building like whether I need those facilities or I don't want those facilities |
| **Real estate market dynamics** | | |
| Inflation and returns | Rise in property values, reflecting increased market demand and scarcity | Like if that area is growing in 2 years or 3 years, How much? |
| Demand of property | Market desire for real estate, influencing property value and pricing | Even for your residential space also. If you are in a growing. Location which has high demand and only you will get appreciation on your end |
| Legal aspects | Legal considerations, rules, and regulations influencing decision-making or procedural compliance | Like I just know some basic things like OK for example if I have to buy property then I mean I know something about the documents and all which was involved in civil engineering where I used. To apply for permissions and all |

(*continued*)

**Table 2.** (*continued*)

| Coding Scheme | Description | Representative comments (random samples) |
|---|---|---|
| Competition in market | Rivalry among sellers for consumer attention, influencing market dynamics | like in each term there are6 people in market then there is competition also in the market like anyone can buy and rent them then |
| **Real estate issues** | | |
| Issues related to tenants | Tenant issues: maintenance, payments, disputes, adherence to lease agreements, communication | In during the tenure there will be issues like landload issues. Most of the IT depends if you are living in the apartments kind of thing then there are some property guidelines. You also get privacy and freedom and if you are into. What you say private properties like you have a landlord |
| Security issues | Concerns related to safety, protection, and potential threats or breaches | Threats. External threat? Maybe I can refrain this external threat that if something you try to get in the more controlled area like district development authorities |
| Infrastructure related issues | Concerns with the development, maintenance, or functionality of essential public facilities | They made the construction according to whatever the requirements were at that time, so these staircases are narrow Indian there are. A lot of Indian toilets are not updated toilets like a lot of people nowadays prefer Western style dialects. Infrastructure is |
| **Urban development** | | |
| Urban development | City growth, planning, and improvement for enhanced living and functionality | But property is relatively very cheap because it was not a developed area but now fast for wearing that to like 13 years the same area which was technically a forest and nobody was there and all is now very crowded |

(*continued*)

**Table 2.** (*continued*)

| Coding Scheme | Description | Representative comments (random samples) |
|---|---|---|
| Future Perspect of area | Area's anticipated development, value, or trends for future considerations | I used to study that area and try to use contacts to get for whatever more future prospects of their. After you invest. Don't be in a hurry to sell. OK. But if you want to have live |
| **Unexpected scenarios in real estate** | | |
| Unexpected scenarios | Unforeseen events impacting property value, investment, or market dynamics unexpectedly | It might change anytime and anywhere, anyhow. Yeah. So while investing we have to keep like the other factor is risk |

# 6   Preliminary Results and Findings

The current research aims to identify thematic patterns in pregame and postgame interview data, substantiating learning through theme counting and coding schemes of significant topics. The increase in the count of coding schemes and themes indicates a surge in discussions on particular topics after gameplay. This increase is observed especially when participants are prompted to identify the significant areas they learned from the gameplay experience. It reflects the participants' perceived learning experiences, suggesting that the post-game discussions are enriched with more detailed and focused conversations on topics that have made a notable impact during the gameplay. This heightened engagement and discourse on these specific themes underline the depth of understanding and the effectiveness of the learning process facilitated by the gameplay experience.

Table 3 visually represents the shift in interview discussions to specific themes postgameplay, shedding light on the evolving understanding of participants across various topics. The Code Count and Theme Count columns quantitatively capture the increased focus on specific themes. For instance, discussions on financial literacy within real estate themes substantially increased from 12 to 20. Post-game, participants conveyed gaining insights into money management techniques (3 to 7) and finance (6 to 9), reflected in a noticeable surge in the code count. Similarly, discussions on property purchase strategies and planning witnessed a significant surge, rising from 15 to 43, indicating a profound deepening of understanding. The table further highlights noteworthy increments in discussions about strategies (6 to 22), risk management (1 to 11), and real estate dynamics (6 to 19). These shifts indicate that participants, post-game engagement, demonstrated heightened familiarity with intricate aspects of real estate dynamics. Additionally, the data unveils emergent discussions on various topics absent in pregame interviews. These newfound understandings include matching with reality (count 4), unexpected scenarios in real estate (count 4), and the value and price of property (count 5). This emphasizes the game's impact in stimulating conversations and broadening participants' awareness of nuanced real estate concepts not previously addressed in pregame discussions.

This analysis highlights that post-game discussions centred around specific topics significantly contributed to participants' enhanced understanding. Key areas, such as financial literacy in real estate, property purchase strategies and planning, and real estate dynamics, saw notable increases in discussion frequency. Participants also explored novel themes like matching the real estate game with reality, the value and price of property, and unexpected scenarios in real estate. Engaging with the game led participants to realize the importance of these topics more profoundly than in the initial stages. The gameplay experience serves as a foundational knowledge base, providing a starting point for participants to build upon in the future. It initiates their understanding, allowing them to further develop and refine their knowledge based on this experiential learning.

**Table 3.** Comparison of Count of Themes and Coding Schemes in Pregame and Postgame

| Themes | Coding Scheme | PreGame | | PostGame | |
|---|---|---|---|---|---|
| | | Code Count | Theme Count | Code Count | Theme Count |
| Property facilities and Amenities nearby | Nearby amenities | 4 | 6 | 4 | 4 |
| | House amenities and payments | 2 | | 0 | |
| Financial Literacy in Real Estate | Money managing techniques | 3 | 12 | 7 | 20 |
| | Profit making | 3 | | 3 | |
| | Loans and interest rates | 6 | | 9 | |
| | Portfolio management | 0 | | 1 | |
| Property purchase/investment | Property renting structure | 3 | 21 | 1 | 13 |
| | Property development | 4 | | 0 | |
| | Property purchase | 1 | | 0 | |
| | Property investment and returns | 9 | | 7 | |
| | Value and price of property | 0 | | 5 | |
| | Types of property | 3 | | 0 | |
| | Finding rented property | 1 | | 0 | |

(*continued*)

**Table 3.** (*continued*)

| Themes | Coding Scheme | PreGame | | PostGame | |
|---|---|---|---|---|---|
| | | Code Count | Theme Count | Code Count | Theme Count |
| Property Purchase Strategies/Planning | Strategies | 6 | 15 | 22 | 43 |
| | Multiple property management | 1 | | 0 | |
| | Observing people's purchase | 1 | | 0 | |
| | Location selection | 4 | | 3 | |
| | Requirements Analysis | 2 | | 0 | |
| | Risk management | 1 | | 11 | |
| | Best property selection | 0 | | 7 | |
| Real estate issues | Issues related to tenants | 3 | 5 | 0 | 0 |
| | Security issues | 1 | | 0 | |
| | Infrastructure related issues | 1 | | 0 | |
| Real estate market dynamics | Inflation and returns | 4 | 6 | 12 | 19 |
| | Demand of property | 1 | | 0 | |
| | Legal aspects | 1 | | 0 | |
| | Competition in market | 0 | | 7 | |
| Unexpected scenarios in real estate | Unexpected scenarios | 0 | 0 | 4 | 4 |
| Urban development | Urban development | 6 | 10 | 0 | 1 |
| | Future Perspect of area | 4 | | 1 | |
| Other | Distance from workplace | 2 | 2 | 5 | 8 |
| | Matching with reality | 0 | | 3 | |

Currently, the research is focused on calibrating and refining the game, particularly in terms of timings and gameplay dynamics. Following the completion of the game design, the next phase will involve experimentation in a behaviour workshop, where the assessment of participants' learning experiences will be analysed in more detail.

## 7 Future Work

This paper marks the initial phase of evaluating participants' understanding through pregame and postgame inquiries. Subsequent research endeavours aim to delve into the game's impact on participants' comprehension and learning through a thorough approach. This includes evaluating participants' comprehension using game scores and drawing insights from invigilators' observations during discussions. Thematic analysis of recorded gameplay adds depth by exploring nuanced aspects. Moreover, the board game will serve as a testbed for future research, facilitating utilitarian value comparisons and analyses across various categories of real estate assets.

## Appendix

### Pre-Game Questionnaire

- Is there anything specific you hope to learn or gain from this experience?
- Have you ever played a real estate-themed board game or simulation before? If yes, please briefly describe your experience with real estate games.
- Have you had any previous experience or knowledge related to real estate investments?, If yes, please briefly describe your previous experiences or knowledge related to real estate investments.
- What challenges or uncertainties do you currently face in the realm of real estate investments?
- Are you currently a property owner or involved in any real estate-related activities (e.g., buying, selling, renting)?, If yes, please provide a brief description of your current real estate activities.
- What specific real estate topics or concepts are you familiar with?
- List any real estate concepts or strategies that you are familiar with. For each concept, provide a short explanation of what it entails.
- How would you typically approach the decision to purchase a property?
- What factors are most important in your decision-making process?
- On a scale of 1 to 10, with 1 being no knowledge and 10 being expert-level knowledge, rate your current understanding of real estate concepts.

### Post-Game Questionnaire

- What features or mechanics of the game do you believe contributed the most to your learning experience regarding real estate?
- How do you feel the game influenced your understanding of real estate concepts and strategies compared to your pre-game assessment? Please describe any changes or insights you gained.
- How has your understanding of real estate concepts changed after playing the game? Please rate it on a scale of 1 to 5, with 1 being significantly decreased and 5 being significantly increased.

- Do you feel more confident in making real estate-related decisions after playing the game? Please rate your confidence on a scale of 1 to 5, with 1 being not confident and 5 being very confident.
- What were the most challenging or complex aspects of the game in terms of real estate concepts? Did the game help clarify these areas for you?
- Can you identify any specific real estate concepts or strategies that you learned during the game?
- Were there any moments during the game when you felt particularly engaged or challenged in applying real estate concepts? Please describe those moments and how they affected your learning.
- What, if any, real-world applications or decisions do you think the lessons learned from the game could be relevant to? Please provide examples.
- What are the negative points in the game?
- Would you recommend this game to others interested in improving their understanding of real estate concepts and decision-making? Why or why not?

# References

1. Salzman, D., Zwinkels, R.C.J.: Behavioral real estate. J. Real Estate Lit. **25**(1), 77–106 (2017)
2. Hudson-Wilson, S., Gordon, J.N., Fabozzi, F.J., Anson, M.J., Giliberto, S.M.: Why real estate? The J. Portf. Manag. **31**(5), 12–21 (2005)
3. Del Giudice, V., De Paola, P., Francesca, T., Nijkamp, P.J., Shapira, A.: Real estate investment choices and decision support systems. Sustainability **11**(11), 3110 (2019)
4. Firstenberg, P.M., Ross, S.A., Zisler, R.C.: Real estate: the whole story. J. Portf. Manag. **14**(3), 22 (1988)
5. Grum, B., Grum, D.K.: A model of real estate and psychological factors in decision-making to buy real estate. Urbani izziv **26**(1), 82–91 (2015)
6. Michael, D.R., Chen, S.L.: Serious games: Games that educate, train, and inform. Muska & Lipman/Premier-Trade, 2005
7. Trass, K., McMenmy, S.: "Board game." U.S. Patent Appl. No. 10/522,981
8. Bayeck, R.Y.: Examining board gameplay and learning: a multidisciplinary review of recent research. Simul. Gaming **51**(4), 411–431 (2020)
9. Hassan, M.M., Ahmad, N., Hashim, A.H.: Factors influencing housing purchase decision. Int. J. Academic Res. Bus. Soc. Sci. **11**(7), 429–443 (2021)
10. Kurniawan, C., Dewi, L.C., Maulatsih, W., Gunadi, W.: Factors influencing housing purchase decisions of the millennial generation in Indonesia. Int. J. Manag. **11**(4), 350–365 (2020)
11. Sundrani, D.M.: Factors influencing home-purchase decision of buyers of different types of apartments in India. Int. J. Hous. Markets Anal. **11**(4), 609–631 (2018)
12. Braun, V., Clarke, V.: Using thematic analysis in psychology. Qual. Res. Psychol. **3**(2), 77 (2006)

# Heating System Case Study
# for Simulation-Based Control System
# Analysis

Matias Waller[1,2]([⊠]), Leonardo Espinosa-Leal[1], and Kim Roos[3]

[1] Graduate School and Research, Arcada University of Applied Sciences,
Jan-Magnus Janssons plats 1, 00560 Helsinki, Finland
{matias.waller,leonardo.espinosaleal}@arcada.fi
[2] Åland University of Applied Sciences, Mariehamn, Åland, Finland
[3] School of Engineering, Culture and Wellbeing, Arcada University of Applied
Sciences, Jan-Magnus Janssons plats 1, 00560 Helsinki, Finland
kim.roos@arcada.fi

**Abstract.** General nonlinear black-box models, hybrid first-principles
combined with statistical nonlinear models, and physical component
models with block diagrams can often provide detailed and accurate
descriptions useful for simulating dynamical systems. The opaque nature
of such models often makes them less suitable for traditional (linear) approaches to control system design and analysis. We explore a
simulation-based approach to controller design for a heating process as
an alternative. As models, we use a system of nonlinear differential equations presented in a recent paper and a linear (multiple-input multiple-
output) model identified in the present study. Based on defined criteria, PI-controllers for a key quality variable are designed by optimizing
simulations. Simulated and experimental results for the identified linear
model are compared to a nonlinear grey-box model. Based on this comparison, we propose to include simulation-based control system analysis
for developing more reliable criteria for simulation-based design.

**Keywords:** MIMO systems · nonlinear systems · PI-control · control
system analysis · heating process

## 1 Introduction

Possibilities created with nonlinear, hybrid, physical components, and other
types of increasingly accurate but, perhaps, opaque models in combination with
readily available off-line computational power for simulations motivated us to
pose the question of whether new approaches to controller design are warranted.
In the recent paper [5], a simulation-based approach to determine a linear and
a nonlinear PI-controller for controlling a central quality variable in a multiple-
input multiple-output (MIMO) laboratory-scale heating process was presented.
The approach was based on simulations of a system of six coupled (nonlinear)

differential equations with some uncertain parameters fitted to experimental data, i.e., a traditional (nonlinear) grey-box model. Practical evaluations of the control systems thus designed provided very encouraging results. The challenge of sensitivity analysis of nonlinear systems was briefly discussed, along with educational perspectives. In this paper, we explore the simulation-based approach further by identifying and using a linear MIMO model for the system and comparing it to results using the nonlinear grey-box model. Since the present work is a direct continuation of previous research, it is unavoidable to incorporate some discussion from our earlier paper.

## 2    Case Study: Heating Process

Figure 1 illustrates a schematic view of the heating process under study. The laboratory-scale process is manufactured by Armfield but has been rebuilt for the purpose of gaining improved insights into the operation of the heat exchanger. From a control perspective, it is a challenging process and has three control signals: these control signals are $u_c$ (pump on the cold side), $u_h$ (pump on the hot side), and $u_p$ (power to the heater). The temperature of the incoming flow on the cold side entering the heat exchanger, $T_{ci}$, and the surrounding temperatures, $T_{surr}$, can be considered external disturbances. A model describing the physics of the process should, given an initial state and values for control signals and disturbances, simulate at least $P$ (delivered power to the heater), $\dot{V}_c$ (flow on the cold side), $\dot{V}_h$ (flow on the hot side), $T_{hi}$ (the temperature entering the hot side of the heat exchanger), $T_{ho}$ (exiting the hot side), and $T_{co}$ (the temperature leaving the heat exchanger on the cold side). The main variables to control considering possible downstream processes, i.e., the quality variables, are $\dot{V}_c$ and $T_{co}$. Naturally, the hot side is necessary to enable control of $T_{co}$, but from a quality perspective less important. On the other hand, a high temperature on the hot side will correspond to increased heat losses. Therefore, control of, e.g., $T_{hi}$ can be motivated from an energy efficiency point of view in addition to possible benefits for controlling the quality variables. In the following two subsections, two different mathematical models for the process are presented.

### 2.1    Grey-Box Model

In a recent paper [5], we used a simplified model based on a combination of energy balances and empirical step-response models for the heating process. In brief, the model is given by

$$\frac{dT_{hi}}{dt} = \frac{1}{C_p}\left(P - \dot{V}_h\rho c_p(T_{hi} - T_{ho}) - \dot{Q}_{loss}\right) \tag{1}$$

$$\frac{dT_{ho}}{dt} = \frac{1}{C_h}\dot{V}_h\rho c_p(T_{hi} - T_{ho}) - \dot{Q}_{he} \tag{2}$$

$$\frac{dT_{co}}{dt} = \frac{1}{C_c}\dot{Q}_{he} - \dot{V}_c\rho c_p(T_{co} - T_{ci}) \tag{3}$$

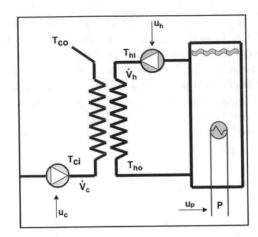

**Fig. 1.** Schematic view of the heating process under study. Red denotes the hot side, while blue denotes the cold side.

For details, explanations and values for $C_p$, $C_h$, $C_c$, $\rho$, $c_p$, $\dot{Q}_{he}$ and $\dot{Q}_{loss}$ we refer to our recent paper. Furthermore, the empirical models

$$\frac{dV_c}{dt} = \frac{1}{T_c}(K_c \max(u_c - 19, 0) - \dot{V}_c) \tag{4}$$

$$\frac{d\dot{V}_h}{dt} = \frac{1}{T_h}(K_h \max(u_h - 19, 0) - \dot{V}_h) \tag{5}$$

$$\frac{dP}{dt} = \frac{1}{T_p}(K_p u_p - P) \tag{6}$$

are included to describe the effect of the control signals on the physical signals. Values for $T_c$, $K_c$, $T_h$, $K_h$, $T_p$ and $K_p$ are also provided in the recent paper. The uncertain parameters of the model were fitted to an identification experiment using random-phase multisines [4]. For estimating the parameters, the criterion defined by Eq. 7 was introduced.

$$V_f = \frac{1}{m} W^T \text{diag}(X^T X) W \tag{7}$$

In the equation, $m$ is the number of observations, $W$ enables relative weight for different signals, and $X$ contain simulation errors. For the grey-box model, the $k$-th row in $X$ has $T_{co,s}(k) - T_{co}(k)$, $T_{hi,s}(k) - T_{hi}(k)$, $T_{ho,s}(k) - T_{ho}(k)$ and $\dot{V}_{c,s}(k) - \dot{V}_c(k)$ as columns. Since control of flow $\dot{V}_c$ is relatively easy, describing $T_{co}$ is our focus and $W^T = (\sqrt{10}\ 1\ 1\ 1)$ is our (heuristic) choice. Equation (7) was minimized using constrained nonlinear optimization with `fmincon` in MATLAB.

## 2.2    Linear Black-Box Model

Originally, our intentions were to use the same 6000 s long multisine identification experiment used to determine the uncertain parameters in the grey-box model to identify various data-based models. Despite significant variations in, especially, $u_c$ and $u_h$, we were not, with the exception of controlling $\dot{V}_c$ with $u_c$, able to determine linear models useful for simulation-based controller design of the PI-controllers under study. Possible explanations could be offered, but for now a discussion of features important to capture in the experiment is postponed to future work.

For the task at hand, however, we designed and performed a new identification experiment using pseudo-random binary signals (PRBS). For $u_c$, $u_h$ and $u_p$, mean values used were 70%, 60% and 25%, amplitudes were 10%, 20% and 20%, and switching times, $4T_s$, $8T_s$ and $20T_s$, respectively. $T_s$ denotes the sampling period. For all experiments and simulations, $T_s = 1$ s. A 200 s long segment of the 1200 s long experiment is illustrated in Fig. 2.

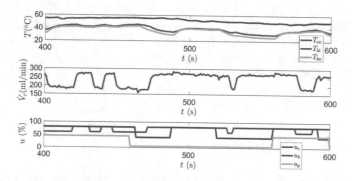

**Fig. 2.** Segment of the PRBS-experiment used to determine the model of Eq. (8). Upper panel: Temperatures $T_{\text{co}}$ (blue), $T_{\text{hi}}$ (red) and $T_{\text{ho}}$ (yellow). Middle panel: $\dot{V}_c$. Lower panel: The three control signals $u_c$ (blue), $u_h$ (red) and $u_p$ (yellow).

Based on the experiment, parameters in discrete-time ARX-type models were fitted. In matrix form, the model is given by $A(q)y(k) = B(q)u(k)$. Specifically,

$$\begin{pmatrix} A_{11}(q) & 0 & 0 \\ A_{21}(q) & A_{22}(q) & A_{23}(q) \\ A_{31}(q) & A_{32}(q) & A_{33}(q) \end{pmatrix} \begin{pmatrix} y_1(k) \\ y_2(k) \\ y_3(k) \end{pmatrix} =$$

$$\begin{pmatrix} B_{11}(q) & 0 & 0 \\ B_{21}(q) & B_{22}(q) & B_{23}(q) \\ B_{31}(q) & B_{32}(q) & B_{33}(q) \end{pmatrix} \begin{pmatrix} u_1(k-2) \\ u_2(k-2) \\ u_3(k-6) \end{pmatrix} \quad (8)$$

In Eq. (8), outputs $y_i(k)$ and inputs $u_i(k)$, $i = 1, 2, 3$, denote normalized values for the outputs $\dot{V}_c$, $T_{\text{co}}$ and $T_{\text{hi}}$ and the inputs $u_c$, $u_h$ and $u_p$. All elements in matrices $A(q)$ and $B(q)$ are polynomials in the shift operator $q^{-1}$, i.e., $q^{-1}y(k) = y(k-1)$. Since only the pump on the cold side, $u_c$, should affect the flow on the cold side, $\dot{V}_c$, the zero elements in the matrices $A(q)$ and $B(q)$ were predetermined. In $A(q)$, the diagonal elements all have the form $A_{ii}(q) = 1 + a_{ii,1}q^{-1} + a_{ii,2}q^{-2} + a_{ii,3}q^{-3}$ while the off-diagonal elements have the form $A_{ij}(q) = a_{ij,1}q^{-1} + a_{ij,2}q^{-2} + a_{ij,3}q^{-3}$, $i \neq j$. All elements in $B(q)$ have the form $B_{ij}(q) = b_{ij,0} + b_{ij,1}q^{-1}$. The delays in $u_i$ are discussed below. In these polynomials, the value $a_{11,3} = 0$ was set, since it was considered reasonable that the dynamics of the flow on the cold side would be easier to model than the temperatures. This means that in total, 34 parameters were fitted to the data yielding a model complexity comparable to the $6^{\text{th}}$ order (nonlinear) state-space model. Naturally, fewer parameters were fitted to the grey-box model, but instead physics-informed nonlinearities are included. Still, the complexities of the models are considered modest and risks of over-fit low. It can be noted that choosing orders and an ARX-type (MIMO) model is quite arbitrary. Our choices are motivated by an objective of our research: Is the model used important for defining suitable criteria for controller design?

In addition to the main experiment, step changes were used to estimate delays. As seen in Eq. (8) these were set to 2, 2, and 6 s with $u_c(t-2)$, $u_h(t-2)$ and $u_p(t-6)$ used not only in Eq. (8) but also replace $u_c$, $u_h$ and $u_p$ in the equations for the grey-box model. However, it can be noted that the flexibility of the linear black-box model easily enables different delays in all $B_{ij}(q)$ polynomials. This is not as easily achieved in the grey-box model as it would require introducing new variables that might be difficult to verify experimentally. For now, this possible advantage of linear black-box models is not used.

## 3   Designing Controllers

Controller design traditionally relies on rules-of-thumb approaches based on simple parametric representations of assumed linear process behavior. More sophisticated linear models enables addressing various factors, such as stability margin, control performance criteria, disturbance rejection, and robustness for stable control under uncertainties. These approaches typically require skills in linearizing (and discretizing) differential equations, linear algebra, block diagrams and frequency analysis, and optimal control, among others. Therefore, gaining expertise in these areas is a significant focus of control engineering courses.

As models become more accurate, techniques based on simplified linear models may seem not to utilize all available knowledge. Additionally, many nonlinear control approaches that rely on local linearity, such as gain scheduling, are mainly developed by adjusting controller design to traditional methods instead of adjusting the design techniques to the models. As an alternative example, model predictive control (mpc) [3] instead determine control signal sequences by optimizing the simulated response over a predictive horizon. Optimal control signal sequences are then updated in real-time. Major advantages of this approach include the use of nonlinear models, implementing constraints and does not require choosing a specific controller. A disadvantage is the undertainty regarding the required computational complexity a priori. Thus, simplifications (linearizing) are often used to ensure practical implementations. As a trade-off, the computational efficiency required can thus render full use of nonlinear models unfeasible, sadly excluding an important and attractive possibility of mpc.

As an alternative, we use a static, off-line simulation based controller design. Accordingly, computational power is not a reason for concern. The controllers are determined by optimization, and developing a suitable criterion is one of the factors we discuss in the presentation. As a first approach, we define a criterion similar to Eq. (7) and is given by

$$V_c = \frac{1}{m}(Q^T \text{diag}(X_c^T X_c)Q + R^T \text{diag}(\Delta u^T \Delta u)R \qquad (9)$$

where $X_c$ includes simulated control errors, $(r - y)$, i.e., the $k$-th row is given by

$$X_c(k) = \big( r_1(k) - y_1(k) \; r_2(k) - y_2(k) \; \cdots \big) \qquad (10)$$

Change in control signal, $\Delta u = u(k+1) - u(k)$ is used to quantify control signal activity.

Naturally, many approaches to designing controllers can be considered as simulation-based. Our ambition, however, is a more general discussion of the implications for this approach than we have encountered in the literature. This change of perspective, is a key contribution of our research. In addition, we present possibilities for including feedback systems analysis within our definition of performance criteria.

The present focus of the case study is to control $T_{co}$ using $u_h$. In Eq. (9), $X_c = T_{co,sp} - T_{co,s}$ and $\Delta u = \Delta u_h$ accordingly. The choice of $Q = 20\sqrt{2}$ and $R = 1$ were used when designing the controllers for the grey-box model[5], and accordingly also evaluated in the current paper. The flow, $\dot{V}_c$, and the temperature, $T_{hi}$, are controlled with $u_c$ and $u_p$, using fixed, non-optimized, PI-controllers with values provided in our recent paper [5]. The well-known equation,

$$u_i(k) = u_i(k-1) + K_i\left(\left(1 + \frac{T_s}{T_{I,i}}\right)e_i(k) - e_i(k-1)\right) \qquad (11)$$

with the constraint

$$u_i(k) = \min(\max(u_i(k), 0), 100)$$

is implemented.

To determine the optimal PI-controller, significant setpoint changes for $\dot{V}_c$ and $T_{co}$ are explored and can be seen in Fig. 3 for $T_{co}$. The disturbances seen clearly in $u_h$ are due to changes in setpoints for $\dot{V}_c$ corresponding to medium flow, 150 ml/min ($t < 1000$), high flow, 220 ml/min ($1000 < t < 2000$) and low flow, 100 ml/min ($t > 2000$). All changes are ramp-shaped with ramps stretching over 5 or 10 s. The setpoint for $T_{hi}$ is 55 °C.

## 3.1   PI Control

Minimizing Eq. (9) for $K_h$ and $T_{I,h}$ gives the results $K_h = 42\%/°C$ and $T_{I,h} = 7.8$ s. The corresponding simulations for the linear ARX-model are illustrated in Fig. 3 (left panel). For reference and comparison, the statistic metric used is the criterion of Eq. (9), with $V_c = 8.1$ in this case.

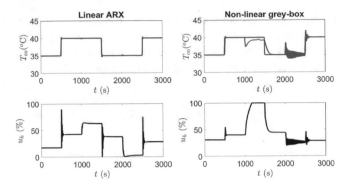

**Fig. 3.** Upper panels: Setpoint changes in of $T_{co}$ (blue) and simulated control (red). Lower panels: Simulated control signal $u_h$. The left panels illustrates simulations using the linear ARX-model and the right panels illustrates simulations using the nonlinear grey-box model.

These can be compared to $K_h = 4.7\%/°C$ and $T_{I,h} = 23$ s obtained using the same criterion with the grey-box model yielding the result $V_c = 510$. Resulting simulations are given in the right panels of Fig. 3. Some very noticeable differences can be discerned. Firstly, in the simulations the PI-controller based on the linear ARX-model clearly outperforms the controller based on the nonlinear grey-box model. Secondly, in the simulations of the grey-box model, $T_{co}$ does not reach the setpoint with high flow and high temperature and the control signal saturates. For practical evaluations, a possible reason for concern is the high gain $K_h$ and short integral time $T_{I,h}$ obtained for simulations of the linear ARX-model compared to the nonlinear grey-box model. We will return to this point in Subsect. 3.2.

More extensive use of (nonlinear) models is an important motivator for our approach. In addition, simulation-based controller design can also be used for nonlinear controller design. A comparison between the panels in Fig. 3 suggests that the nonlinear grey-box model captures some possible challenges. Although a full exploration of general nonlinear controllers is infeasible, a nonlinear PI-controller with linear dependencies in the gain and integration time is proposed as a first approach. Specifically, $K_h + au_h(k-1)$ and $T_{I,h} + bu_h(k-1)$.

Minimizing our criterion for $K_h$, $a$, $T_{I,h}$ and $b$ provides some intriguing results. Firstly, despite the higher degree of freedom but, perhaps, not surprisingly, nothing is gained when a nonlinear PI-controller is considered with the linear ARX-model. Secondly, for the nonlinear grey-box model, the results $K_h = 0.1\%/°C$, $a = 46$ $1/°C$, $T_{I,h} = 0.18$ s, $b = -0.46$ s/% and the value $V_c = 240$ suggests significant improvement over the linear PI-controller. It seems unlikely that this model-dependent difference is coincidental and theoretical analysis could provide further insights whether only a nonlinear model motivates the use of nonlinear control based on simulations. This analysis is postponed for future research.

## 3.2    Experimental Evaluation

Our concerns regarding the high gain $K_h$ and short integral time $T_{I,h}$ obtained for simulations of the linear ARX-model, were indeed substantiated by an experimental evaluation. The resulting variations in $u_h$ were not acceptable, and the clearly audible stress on the pump that could be discerned were painful to listen to while conducting the experiment. For illustrative purposes, a segment is illustrated in Fig. 4.

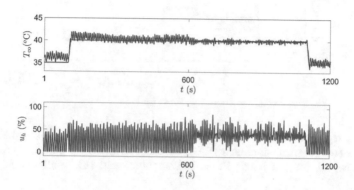

**Fig. 4.** Upper panel: Setpoint for $T_{co}$ (blue) and measurements (red). Lower panels: Corresponding control signal $u_h$. Simulation-based controller design using ARX-model of Eq. (8) and $Q = 20\sqrt{2}$ in Eq. (9) resulting in $K_h = 42\%/°\mathrm{C}$ and $T_{I,h} = 7.8$ s.

In order to still achieve experimental results that could be used for comparing to controllers designed by simulations of the nonlinear grey-box model, the heuristically chosen $Q = 20\sqrt{2}$ in Eq. (9) was replaced by $Q = 1$. This equal weight to control error and control signal activity results in $K_h = 4.3\%/°\mathrm{C}$ and $T_{I,h} = 6.6$ s. This results in $V_c = 100$ with the value $Q = 20\sqrt{2}$ used to calculate $V_c$, but it can be noted that the controller did not specifically minimize Eq. (9) for this value of $Q$. In practice, this controller works well and results are illustrated in Fig. 5 along with the corresponding results using the nonlinear PI-controller designed based on the nonlinear grey-box model. The mean square error for the control error, $T_{co,sp} - T_{co}$, are similar for the two cases, while the nonlinear PI-controller has a mean square error on control signal activity, $\Delta u_h$, 4 times as large as the linear PI-controller.

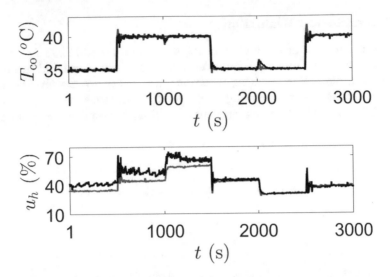

**Fig. 5.** Upper panel: Setpoint for $T_{co}$ (blue), measurements using $K_h = 4.3\%/^\circ C$ and $T_{I,h} = 6.6$ s (red) and the nonlinear PI-controller (black). Lower panels: Corresponding control signal $u_h$ for the linear PI-controller (blue) and the nonlinear PI-controller (black).

## 4    Control System Analysis

A key observation in this paper is that the definition of the cost-function when designing controllers by simulation is central for practical success. Encouraging results were obtained using the nonlinear grey-box model, but for achieving comparable results using the linear ARX-model it was necessary to modify the cost function. It would thus seem that in order to better define a suitable cost-function for simulation based design, control system performance might need to be considered in greater detail.

Approaches for design of controllers "focus on one or two aspects of the [control] problem, and the control-system designer then has to check that the other requirements are also satisfied" [1]. With our focus on simulated control performance for setpoint changes, "other requirements" could include sensitivity to disturbances. For a general feedback system with our structure, the corresponding block diagram is depicted in Fig. 6.

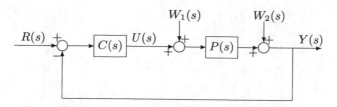

**Fig. 6.** Block diagram for a process $P(s)$ controlled by (negative) feedback. The diagram includes setpoint $R(s)$, control signal $U(s)$, load disturbance $W_1(s)$, measurement noise $W_2(s)$, process value $Y(s)$ and controller $C(s)$.

Correspondingly relevant transfer functions are,

$$Y(s) = \frac{CP}{1+CP}R(s) + \frac{P}{1+CP}W_1(s) + \frac{1}{1+CP}W_2(s) \tag{12}$$

and

$$U(s) = \frac{C}{1+CP}R(s) - \frac{CP}{1+CP}W_1(s) - \frac{C}{1+CP}W_2(s) \tag{13}$$

The general control problem can thus be analyzed through the "Gang Of Four" [2], i.e., the four transfer functions $\frac{CP}{1+CP}$, $\frac{P}{1+CP}$, $\frac{1}{1+CP}$ and $\frac{C}{1+CP}$. For linear systems, a standard approach is to analyze the control system by inspecting the gains of the "Gang Of Four". This can easily be achieved for our closed-loop system using the linear ARX-model and the parameters in the controller $K_h = 42\%/°C$ and $T_{I,h} = 7.8$ s determined by minimizing our original cost function, i.e., Eq. (9) using $R = 1$ and $Q = 20\sqrt{2}$. Although a large number of transfer function for MIMO-systems could be explored, the definition of our simulation-based criteria could motivate further analysis of model uncertainty and disturbance sensitivity. Here, we focus on how simulated sinusoidal measurement noise in $T_{co}$ for different frequencies affect $T_{co}$ and $u_h$ for both our linear and our nonlinear model.

As a word of warning, nonlinear systems can exhibit, e.g., amplitude- and frequency-dependent frequency spreading, i.e., possible harmonics. Also, as our simulations illustrate, the principle of superposition cannot automatically be applied and different regions of operation can have different characteristics. A full exploration of nonlinear MIMO systems is accordingly a formidable task we aim to explore. In our case, we use constant setpoints, 37.5 °C for $T_{co}$, 50 °C for $T_{hi}$ and 160 ml/min for $\dot{V}_c$. The amplitude of the simulated sinusoidal "noise" in $T_{co}$ was set to 1 °C for simulating the nonlinear model, but had to be decreased to 0.1 °C for simulating the linear black-box model. Without this reduction, the high gain for certain frequencies seen in the right panel of Fig. 7 results in the hard limits on $u_h$, i.e., $0 \leq u_h \leq 100\%$ dominating the analysis. Recording only amplitudes as function of frequency in the simulations to estimate the gains of the transfer functions $\frac{1}{1+CP}$ and $\frac{C}{1+CP}$, results corresponding to the three alternatives i) ARX-model with $Q = 20\sqrt{2}$ in Eq. (9), ii) ARX-model with $Q = 1$ and iii) nonlinear grey-box model with nonlinear PI-control and $Q = 20\sqrt{2}$ are illustrated in Fig. 7.

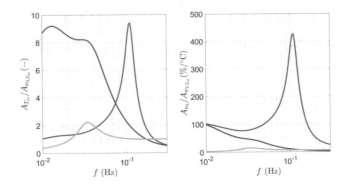

**Fig. 7.** Amplitude-ratios resulting from sinusoidal noise on $T_{co}$ with amplitude $1\,^\circ$C ($0.1\,^\circ$C for the linear model). Left panel, amplitude-ratios for $T_{co}$ for ARX-model with $Q = 20\sqrt{2}$ in Eq. (9) (blue), ARX-model with $Q = 1$ in Eq. (9) (red) and nonlinear grey-box model with nonlinear PI-control and $Q = 20\sqrt{2}$ in Eq. (9) (yellow). Right panel: same as in the left panel but with amplitude-ratios for $u_h$.

Based on the figure, it would be easy to dismiss the possibility of using the controller based on the ARX-model with $Q = 20\sqrt{2}$: The corresponding control signal activity in $u_h$ for certain frequencies is clearly unacceptable. This was also verified by practical evaluation as illustrated in Fig. 4. From these three alternatives, it would also be easy to choose the controller based on the nonlinear grey-box model with nonlinear PI-control and $Q = 20\sqrt{2}$. This choice, however, is naturally affected by the model used in the analysis. Given access to multiple models, it could therefore also be motivated to perform the analysis for several models: this could shed light on the balancing efforts between modelling and developing control performance criteria.

Accordingly, the figure illustrates that it seems to be of central importance to include some form of minimizing or constraining of amplitude-ratios for the "Gang Of Four" in general criteria for simulation-based controller design. This, perhaps, is one of the most important contributions of our research and will be explored in future work.

## 5   Conclusions and Future Work

In this paper, we used two different models of a heating system for simulation-based controller design. The paper illustrated that the choice of model was, perhaps not surprisingly, of central importance for achieving acceptable control performance in practice. In addition, the design of cost-functions for the general approach of controller design by simulation might require incorporating a more detailed understanding of control system performance. Further experiments, studies, analysis, and comparisons to alternative controllers will be pursued in order to explore the design of cost-functions and, in extension, the general applicability of the presented approach.

# References

1. Åström, K.J., Wittenmark, B.: Computer Controlled Systems: Theory and Design, 3rd edn. Prentice-Hall, Hoboken (1997)
2. Åström, K.J., Murray, R.M.: Feedback Systems: An Introduction for Scientists and Engineers, 2 edn. Princeton University Press (2021)
3. Hewing, L., Wabersich, K.P., Menner, M., Zeilinger, M.N.: Learning-based model predictive control: toward safe learning in control. Annual Rev. Control Robot. Auton. Syst. **3**, 269–296 (2020)
4. Pintelon, R., Schoukens, J.: System Identification–A Frequency Domain Approach, 2nd edn. Wiley, Hoboken (2012)
5. Waller, M., Leal, L.E.: Approaching simulation-based controller design: heat exchanger case study. In: Proceedings of the 64th International Conference of Scandinavian Simulation Society, SIMS 2023, Västerås, Sweden, 25–28 September 2023. Linköping University Electronic Press (2023). https://doi.org/10.3384/ecp200010

# Perceived Importance of Laboratory Learning Objectives by Female and Male Engineering Students

Konrad Boettcher[(⊠)] [iD], Nils Kaufhold [iD], Marcel Schade, Claudius Terkowsky [iD], and Tobias R. Ortelt [iD]

TU Dortmund University, 44227 Dortmund, Germany
Konrad.Boettcher@tu-dortmund.de

**Abstract.** This paper is about the perceived importance of learning objective from female and male bachelor students of a process engineering program after participating in an immersive virtual reality laboratory experiment. The focus is set to the importance of the learning objectives regarding their future work. For the evaluation of the laboratory a questionnaire with the 13 fundamental learning objectives by Feisel and Rosa as well as the extension reported by the CrossLab joint project are used. The evaluation was carried out in an online survey using a 7-point Likert scale. The questionnaire contains two qualitative questions as well. The learning objectives are all perceived as more or less important but not neutral or unimportant. In contrast to preliminary findings, the statistical analysis shows that generally only the learning objective itself, and not gender, has a significant effect on the perceived importance of the learning objectives. According the gender there are only small difference between male and female students in a few learning objectives. The learning objectives data analysis, working style and teamwork are perceived to be most important, sensory awareness is the least important.

**Keywords:** Laboratory Education · Engineering Education · Learning Objectives · Students Perception

## 1 Introduction

### 1.1 Constructive Alignment

Constructive Alignment (CA) is an instructional framework for a holistic planning of teaching and learning [1] and it is basing on constructivist learning theory [2] to align learning objectives, teaching-learning activities and learning assessment. According to one of the basic assumptions of modern knowledge psychology [3], knowledge is acquired constructively as a function of prior knowledge, perception, the context of action and affect. The learning objectives address this and must therefore be presented transparently in the CA before the teaching-learning activity. The learning objectives are a sentence that tells learners what they will learn, how and why. It seems to be beneficial if the learning objectives follow the SMART rules: specific, measurable, attainable, relevant, and time-bound [4, 5].

M. E. Auer et al. (Eds.): STE 2024, LNNS 1027, pp. 44–55, 2024.
https://doi.org/10.1007/978-3-031-61891-8_4

## 1.2  Working World 4.0 and Laboratory Education

The world of work is also undergoing profound change as a result of digitization. Additionally, the transformation of the economy and of social life towards sustainability begins, and many new engineers are needed. This seems to be achievable only by reducing the number of students dropping out and by increasing the number of students and, in particular, by increasing the proportion of women. New requirements of Industry 4.0 and Working World 4.0 result in new or modified learning objectives. Teaching in laboratories seems to be a suitable environment for these learning objectives. Feisel and Rosa developed 13 fundamental laboratory learning objectives [6], which were extended by the CrossLab project [7] by seven more for the Working World 4.0 in a preliminary work [8] and both are shown in Table 1 with a short definition. It has already been observed that many fundamental learning objectives are rarely addressed in many laboratory courses [9]. A redesigned laboratory experiment [10] has also revealed that men and women perceive the importance of learning objectives significantly differently in some cases [11], especially in ethical (sustainable) behavior.

## 1.3  Goal of this Work

The motivation is to gather information about the perceived importance of learning objectives. According to constructive alignment [2] the teachers have to set the learning objectives and the best teaching-learning activity which gives several possibilities in laboratory education [6]. Here, it is a useful information whether students think the learning objectives are important or not. Students with a better attitude towards a learning content achieve better results (effect size $d = 0.48$, Hattie study [12]). Deci and Ryan [13] identify in their self-determination theory "Amotivation results from not valuing an activity (Ryan, 1995) or not expecting it to yield a desired outcome (Seligman, 1975)". Therefore students may be more likely to be amotivated if they do not see a value. [14] showed that the perceived importance of a training session is a factor that influences motivation.

This could provide answers to questions that teachers ask themselves: do teachers have to motivate and explain some laboratory learning objectives quite a lot whilst the importance of other objectives is quite clear? Is there a general rule which learning objectives are thought to be more important?

We also want to get information about missing important learning objectives from the point of view of the students and if we have to distinguish between man or women in designing or motivating the learning objectives. We intend to use this information in the attempt to acquire new students. The research questions are:

- RQ1: What laboratory learning objectives do students treat to be important in their future work?
- RQ2: What additional learning objectives do students think are missing?
- RQ3: Are there differences in RQ1 and RQ2 between male and female students?

These results can be used by teaching staff: Do different learning objectives need to be motivated differently for men and women and their significance explained differently? Revealing such differences and aspects of the world of work that are considered important

**Table 1.** Fundamental learning objectives in laboratory learning 1–13 by Feisel and Rosa [6] and 14–20 extensions to Work 4.0 [8]

| LO: Learning objective | | Short definition |
|---|---|---|
| 1 | Instrumentation | Apply sensors, instruments, or software for measuring physical quantities |
| 2 | Models | Evaluate theoretical models, assess strengths and limitations, and validate relationships between data and physical principles |
| 3 | Experiment | Devise, execute, and interpret experiments to characterize engineering materials, components, or systems |
| 4 | Data Analysis | Collect, analyze, and interpret data, make judgments, and support conclusions |
| 5 | Design | Create, build, or assemble parts, products, or systems. Test prototypes using appropriate tools |
| 6 | Learn from failure | Identify and fix issues due to faulty equipment, parts, code, construction, process, or design |
| 7 | Creativity | Demonstrate independent thought, creativity, and problem-solving skills |
| 8 | Psychomotor | Show competence in selecting, modifying, and using engineering tools and resources |
| 9 | Safety | Identify and address health, safety, and environmental issues responsibly |
| 10 | Communication | Effectively convey laboratory work orally and in writing for various audiences |
| 11 | Teamwork | Collaborate in teams, assign roles, monitor progress, meet deadlines, and integrate contributions |
| 12 | Ethics in the lab | Adhere to the highest ethical standards, report information objectively, and maintain integrity |
| 13 | Sensory Awareness | Use human senses to gather information and make informed engineering judgments |

(*continued*)

**Table 1.** (*continued*)

| LO: Learning objective | | Short definition |
|---|---|---|
| 14 | Personality development | Address personal and team issues, from guided support to independent problem-solving |
| 15 | Improved Working style | Enhance learning and working, focusing on balance, target-oriented work, and resilience |
| 16 | Critical thinking and sustainable action | Evaluate consequences and seek ethical, environmental, and socially responsible alternatives |
| 17 | Innovative contextual thinking | Identify gaps, develop solutions, and implement independently |
| 18 | Self-directed learning skills | Reflect, find resources, and adapt learning independently |
| 19 | CPS proficiency | Work with cyberphysical systems (CPS), utilizing decentralized networking and autonomous decisions |
| 20 | Data organization and management | Document, organize, and link research data using new methods |

could make engineering courses more interesting for everyone or specifically for women at events to recruit young students by not focusing on learning objectives that students consider unimportant, e.g., not the handling to produce a uniform chocolate layer around a peanut or the knowledge how a mixer-settler apparatus works, but rather working on a digital twin and using AI to reduce the pollution in a plant. This list can also be used to revise own learning objectives for laboratory experiments.

## 2 Method

### 2.1 Student Sample

Exactly 100 Bachelor third semester students (44 female, 56 male) of Biochemical and Chemical Engineering (both process engineering) of the TU Dortmund University in Winter term 2021/22 were asked after completing an immersive laboratory experiment in virtual reality which learning objectives they think are important in their future work using a mixed method approach. This includes the 13 fundamental learning objectives by Feisel and Rosa [6] as well as the extension reported by the CrossLab joint project [8]. The quantitative online-survey consists of 21 questions about the learning objectives with a 7-point Likert-scale, as LO1 was divided into hard- (LO1a) and software (LO1b). The results are quantified from unimportant (-3) to important (+3) with a neutral element (0). All learning objectives are briefly explained in advance of each question.

Two qualitative questions were also asked after the experiment. The first question aims to get students to reflect on their own lab experiences and possibly reveal learning objectives that should also be directly addressed (in terms of the CA). The second question is intended to encourage students to think about what skills they expect in a vocational environment:

- Q1 What did you learn in this experiment?
- Q2 What other skills do you think are important (i.e. apart from learning objectives 1–13) in a professional engineering environment?

## 3 Results

### 3.1 Descriptive Statistics

The mean value, median and standard deviation of the 20 learning objectives is presented in Table 2. LO1 is divided into hardware (1a) and software (1b) tools. The learning objectives are all perceived as more or less important but not neutral or unimportant. It is shown in Table 2 that each learning objective was rated with a 1 by at least one student and with a 7 by no student. Thus, regardless of gender, the mean value of the perceived importance of the learning objectives ranges from partially important (2.85 in a 1 to 7 scale) for sensory awareness up to 1.82 for teamwork. In this context, students seem to rate learning objectives addressing procedural and personal skills as more important than other basic learning objectives. Preliminary studies show a striking difference in the importance of ethically correct behavior between men and women. Some of the new learning objectives for the working world 4.0 proposed by the CrossLab joint project are rated as very important and others as rather less important.

It must be taken into account that all students were asked about all learning objectives, which means that the statements of a person with a consistently negative or positive attitude towards the learning objectives have an impact on the result.

**Table 2.** Descriptive statistics of learning outcomes of [6] and [8]

|       | Mean value | Median | Standard deviation | Max | Min |
|-------|-----------|--------|--------------------|-----|-----|
| LO1a  | 2.260     | 2      | 1.130              | 5   | 1   |
| LO1b  | 2.030     | 2      | 0.920              | 5   | 1   |
| LO2   | 2.100     | 2      | 0.920              | 5   | 1   |
| LO3   | 2.140     | 2      | 1.120              | 6   | 1   |
| LO4   | 1.920     | 2      | 0.900              | 4   | 1   |
| LO5   | 2.030     | 2      | 1.000              | 5   | 1   |
| LO6   | 2.040     | 2      | 0.990              | 5   | 1   |
| LO7   | 2.570     | 2      | 1.120              | 6   | 1   |

(*continued*)

**Table 2.** (*continued*)

|  | Mean value | Median | Standard deviation | Max | Min |
|---|---|---|---|---|---|
| LO8 | 2.460 | 2 | 1.180 | 6 | 1 |
| LO9 | 2.070 | 2 | 1.220 | 6 | 1 |
| LO10 | 2.080 | 2 | 1.010 | 6 | 1 |
| LO11 | 1.820 | 2 | 0.950 | 4 | 1 |
| LO12 | 2.270 | 2 | 1.300 | 6 | 1 |
| LO13 | 2.850 | 3 | 1.390 | 6 | 1 |
| LO14 | 2.350 | 2 | 1.110 | 5 | 1 |
| LO15 | 1.990 | 2 | 0.890 | 5 | 1 |
| LO16 | 2.230 | 2 | 1.150 | 6 | 1 |
| LO17 | 2.470 | 2 | 1.070 | 5 | 1 |
| LO18 | 2.200 | 2 | 1.090 | 5 | 1 |
| LO19 | 2.420 | 2 | 1.020 , | 5 | 1 |
| LO20 | 2.150 | 2 | 0.910 | 5 | 1 |

## 3.2 Examination of the Scale of all Learning Objectives

Due to the high number of variables, an exploratory factor analysis is carried out to reduce the dimensions of the 21 learning objectives examined. This involves looking for groups of learning objectives that are linearly related to other learning objectives and at the same time less related to other groups of other learning objectives. The aim is to group the learning objectives and crystallize commonalities and factors. First, a correlation matrix of all learning objectives is examined in order to identify any correlations that occur. This shows that there are many positive correlations, often in the range of 0.5 or greater. This suggests fundamental correlations, which is why a factor analysis is appropriate at this point.

The Scree test, the Kaiser criterion and the variance are used to create the factor analysis. In the Scree test, the eigenvalues of the factors are plotted against their index and these data points are connected according to the index sequence of the learning objectives. All factors to the left of the point with the largest gradient are then selected in the scree plot shown in Fig. 1.

It can be seen that the largest gradient occurs after the first factor. According to the Scree test, one factor is therefore appropriate. The Kaiser criterion states that all factors whose eigenvalues are greater than or equal to 1 are taken into account. This means that all factors that contribute at least as much to the overall explanation as a single learning objective itself are taken into account. This is shown by the red line in Fig. 1. According to the Kaiser criterion, there are two relevant factors.

The third option is to look at the variance. This involves looking at how much variance the factors cover. Each learning objective contributes to the variance. If all learning objectives are considered, the variance is 100%. Accordingly, the variance is

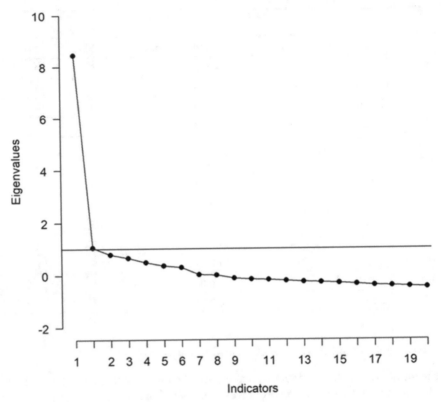

**Fig. 1.** Scree plot of perceived importance of laboratory learning objectives

greater when many factors are considered than when fewer factors are considered. The results of the factor analysis are now considered. The factor loadings are in the interval from −1 to +1. The significance of the factor loading on the factor is as follows: large values, positive as well as negative values, have an impact on the factor. If a loading has a value lower than 0.3, it is unimportant for this work and can be neglected. All loadings between 0.3 and 0.5 can be classified as not particularly important but not as irrelevant as well. Any factor loadings greater than 0.5 is considered as relevant due to the factor analysis. In Table 3 the factor loadings for two factors are shown.

Some learning objectives can be clearly assigned to a factor. Learning objectives 1a, 2, 5 and 6 can be clearly assigned to factor 2. Learning objectives 3, 4, 7, 8 and 9 can also be assigned to factor 2. A large proportion of these are assigned to factor 2, but a small proportion is also assigned to factor 1. Learning objectives 10, 11, 14, 15 and 20 can be clearly assigned to factor 1. Learning objectives 18 and 19 behave in the opposite way to learning objectives 3, 4, 7, 8 and 9. This leaves learning objectives 1b, 12, 13, 16 and 17, which cannot be assigned to any factor. This means that these learning objectives are included in factor 1 and factor 2 in almost equal proportions. When looking at the learning objectives, factor 1 can be classified as soft skills and factor 2 as hard skills. The mean values of all LO belonging to factor 1 and factor 2 are quite similar.

**Table 3.** Load of all learning objectives for two factors

|         | F2    | F1    |
|---------|-------|-------|
| LO 1a   | 0.600 | 0.020 |
| LO 1b   | 0.380 | 0.300 |
| LO 2    | 0.630 | 0.210 |
| LO 3    | 0.610 | 0.400 |
| LO 4    | 0.620 | 0.320 |
| LO 5    | 0.740 | 0.250 |
| LO 6    | 0.550 | 0.260 |
| LO 7    | 0.570 | 0.320 |
| LO 8    | 0.620 | 0.400 |
| LO 9    | 0.600 | 0.380 |
| LO 10   | 0.240 | 0.550 |
| LO 11   | 0.170 | 0.530 |
| LO 12   | 0.400 | 0.500 |
| LO 13   | 0.420 | 0.470 |
| LO 14   | 0.120 | 0.770 |
| LO 15   | 0.280 | 0.720 |
| LO 16   | 0.550 | 0.470 |
| LO 17   | 0.450 | 0.430 |
| LO 18   | 0.390 | 0.720 |
| LO 19   | 0.320 | 0.550 |
| LO 20   | 0.230 | 0.650 |

In order to be able to summarize the learning objectives into a scale, the reliability of this scale in the sample is crucial. The Cronbach's alpha is calculated to determine the reliability. The Cronbach's alpha indicates how strongly the individual learning objectives correlate with the overall scale. With low alphas, less than 0.7 [15], the individual learning objectives hardly correlate with the scale, which would mean that the individual learning objectives have nothing to do with each other. Large values for alpha, from 0.7 [15] mean exactly the opposite. The values for alpha range from 0 to 1. A Cronbach's alpha close to 1 shows that the learning objectives are strongly related. This means that the same questions were asked too often from a statistical point of view. A Cronbach's alpha in the range of 0.7 and 0.9 would therefore be a meaningful scale. This would ensure a good consistency test and not too frequent duplication of questions. When calculating the Cronbach's alpha, we obtain an alpha of 0.93. This indicates a reliable scale with probable duplication of some questions, i.e. students do not rate the importance of these learning objectives differently.

### 3.3  Gender Influence

This section examines whether the learning objective and gender have a significant influence on the importance of the individual learning objectives. For this purpose, the p-value was calculated for both influencing factors examined. The p-values of a F-test for "learning objective" and "gender" are listed in the Table 4.

**Table 4.** F-test of the individual learning objectives

| Model term | p.value |
|---|---|
| Learning objective | $1.081 \times 10^{-20}$ |
| Gender | 0.931 |

When looking at the results of the F-test for the significance of the "learning objective" and "gender" in the evaluation of the learning objectives, it can be seen that only the "learning objective" has a significant effect with a p-value of $1.081 \times 10^{-20}$. This means that gender has no influence on the importance of the individual learning objectives, only the learning objective itself. The greatest differences are in LO13 sensory awareness, which is rated with 3.2 by man and 2.6 by women. Male students rate LO15 (working style) with 1.8 slightly better than women with 2.2 but female students rate CPS proficiency with 2.2 better than male students with 2.6. Obviously there are small differences.

### 3.4  Qualitative Results

The answers of the students to the first question Q1 map to the 13 fundamental learning objectives by [6] except: i.) work in vocational environment ii.) organization of groups iii.) dealing with unknown problems and developing own solution strategies, and iv.) interdisciplinary cooperation. There is no difference between male and female students. The answers to the second question are presented in Table 5.

## 4  Discussion

The results show that the students assess the importance of different learning objectives differently but that both the fundamental learning objectives of Feisel and Rosa [6] and those of the joint project CrossLab [8] are relevant from the students' point of view. Some learning objectives were seen as important on average, some as significantly less important like sensory awareness, particularly by male students. In particular, learning objectives of hard skills were rated as less important than soft skills. Thus, in the planning of experiments and the preparation of materials, the necessity arises to motivate some learning objectives in particular, in order to increase the motivation of the students in the execution of the laboratory experiments. This is especially true when experiments only address learning objectives like sensory awareness that is perceived as less important.

**Table 5.** Answers to open question Q2 and frequency.

| Answers | Frequency | |
|---|---|---|
| | female | male |
| Independence and problem solving with limited information | 9 | 12 |
| Application of inventory analysis | | 3 |
| Interdisciplinary work | | 2 |
| Awareness (environment, responsibility, economy) | 5 | 4 |
| Math-reality relationship | | 2 |
| Digital literacy (digital twins, software,…) | | 7 |
| Home office work | | 1 |
| Curiosity | | 2 |
| Resilience | 3 | 2 |
| Social competence | 2 | 3 |
| International cooperation / communication | 1 | |
| Leadership with personnel responsibility | 1 | 1 |

This additionally relates to gender differences in perception. Investigating the learning objectives that students perceive as lacking will either allow for an extension of the learning objectives or focus on the need to explain existing learning objectives more transparently in order to increase student motivation and satisfaction in this way as well.

According to the qualitative results the most important learning objectives are implemented in the 13 fundamental learning objectives of [6]. Both questions show that self-organized dealing with unknown problems seems to be important, which is also identified in literature for Working World 4.0 [16–18]. Every fifth student mentioned this in Q2. Students are also aware that interdisciplinary and international cooperation is a skill they should develop, which is also one of the future skills of work 4.0 as well as awareness of responsibilities [19]. Interestingly some students mention to work in a vocational environment as also industry wants this to be a learning objective [20] and what may be addressed using work integrated learning or a real world scenario in laboratory teaching [16].

But there are also striking differences: while 7 male students independently mentioned "data literacy", not a single women thinks mentioned this as a proper skill, as well as the improvement of existing inventory or the ability to get the key points how to describe real world phenomena with math, which corresponds to the highest cognitive level of the SOLO taxonomy (extended abstract understanding) [21].

## 5  Conclusion

All laboratory objectives are perceived to be important. There seems to be a difference between more conventional objectives and more Working World 4.0-related objectives. The objectives data analysis, working style and teamwork are perceived to be most important, sensory awareness is the least important. A comparison of the stated learning objectives with those for Working World 4.0 of [8] shows greater similarities. However, the factor analysis shows that this does not affect more than one factor in terms of perceived importance. Accordingly, these learning objectives could be condensed further. The learning objectives were condensed into two factors, soft skills and hard skills. For the teachers, there is no change in the addressing and motivation of the learning objectives with regard to men and women except small differences in sensory awareness, CPS proficiency, and working. In general, the gender has no significant effect on the importance of the learning objectives.

**Acknowledgments.** The work presented in this paper is part of the research project "CrossLab-Flexibel kombinierbare Cross-Reality Labore in der Hochschullehre: zukunftsfähige Kompetenzentwicklung für ein Lernen und Arbeiten 4.0" funded by Stiftung Innovation in der Hochschullehre (funding code: FBM2020-VA-182-3-01130).

## References

1. Biggs, J.B., Tang, C., Kennedy, G.: Teaching for quality learning at university. What the student does, 5th edn. McGraw-Hill/Open University Press, Maidenhead (2022)
2. Biggs, J.B.: Enhancing teaching through constructive alignment. High. Educ. **32**, 347–364 (1996). https://doi.org/10.1007/BF00138871
3. Meixner, J., Müller, K.: Angewandter Konstruktivismus. Ein Handbuch für die Bildungspraxis in Schule und Beruf. Shaker Verlag, Aachen (2004)
4. Lawlor, K.B.: Smart goals: How the application of smart goals can contribute to achievement of student learning outcomes. In: Developments in Business Simulation and Experiential Learning: Proceedings of the Annual ABSEL Conference, vol. 39 (2012)
5. Bjerke, M.B., Renger, R.: Being smart about writing SMART objectives. Eval. Program Plann. **61**, 125–127 (2017). https://doi.org/10.1016/j.evalprogplan.2016.12.009
6. Feisel, L.D., Rosa, A.J.: The role of the laboratory in undergraduate engineering education. J. Eng. Educ. **94**, 121–130 (2005). https://doi.org/10.1002/j.2168-9830.2005.tb00833.x
7. Aubel, I., et al.: Adaptable digital labs – motivation and vision of the crosslab project. In: 2022 IEEE German Education Conference (GeCon) (2022). https://doi.org/10.1109/GeCon55699.2022.9942759
8. Boettcher, K., Aubel, I., Ortelt, T., Terkowsky, C., Soll, M.: Did you check it? Checklist for Redesigning a Laboratory Experiment in Engineering Education addressing Competencies of Learning and Working 4.0, REV2023 – Remote Engineering & Virtual Instrumentation, Thessaloniki, Greece, Lecture Notes in Networks and Systems, in press (2023)
9. Terkowsky, C., Schade, M., Boettcher, K., Ortelt, T.: Once the child has fallen into the well, it is usually too late – An action research approach to formative evaluation of laboratory manuals and practices, REV2023 – Remote Engineering & Virtual Instrumentation, Thessaloniki. Lecture Notes in Networks and Systems, in press, Greece (2023)

10. Boettcher, K., Terkowsky, C., Schade, M., Brandner, D., Grünendahl, S., Pasaliu, B.: Developing a real-world scenario to foster learning and working 4.0 – on using a digital twin of a jet pump experiment in process engineering laboratory education. Eur. J. Eng. Educ. **48**(5), 949–971 (2023). https://doi.org/10.1080/03043797.2023.2182184
11. Boettcher, K., Schade, M., Terkowsky, C., Grünendahl, S.: Ein real-world-szenario laborversuch am digitalen zwilling in der virtuellen realität auf basis des constructive alignment. In: Tagungsband Ingenieurpädagogische Wissensgesellschaft, Jahrestagung 2022 (S. Seitennummer/en). Regionaltagung, Technische Universität Dortmund 2022. Technische Bildung für eine Nachhaltige Entwicklung (2023). ISBN 978-3-9818728-6-6
12. Visible Learning Homepage. https://www.visiblelearningmetax.com/Influences. Last accessed 4 Dec 2023
13. Ryan, R.M., Deci, E.L.: Self-determination theory and the facilitation of intrinsic motivation, social development, and well-being. Am. Psychol. **55**(1), 68–78 (2000). https://doi.org/10.1037/0003-066X.55.1.68
14. Tsai, W., Tai, W.: Perceived importance as a mediator of the relationship between training assignment and training motivation. Pers. Rev. **32**(2), 151–163 (2003). https://doi.org/10.1108/00483480310460199
15. Schmitt, M.: Uses and abuses of coefficient alpha. Psychol. Assess. **8**, 350–353 (1996)
16. Schade, M., Terkowsky, C., Boettcher, K., Ortelt, T.R.: Work in Progress: Cobbler, stick to your last! On providing engineers Constructive Alignment. In: ICL Conference Madrid (2023)
17. Lemaître, D. (ed.) Innovation, entrepreneurship, and management series. Training Engineers for Innovation. ISTE, Ltd.; John Wiley & Sons, Inc, London, UK, Hoboken, NJ
18. Felder, R.M., Brent, R.: Designing and teaching courses to satisfy the ABET engineering criteria. J. Eng. Educ. **92**(1), 7–25 (2016)
19. Gottburgsen, A., Wannemacher, K., Wernz, J., Willige, J.: VDI-Studie: Ingenieurausbildung Digitale Transformation (2019)
20. Soll, M., Boettcher, K.: Expected Learning Outcomes by Industry for Laboratories at Universities. In: 2022 IEEE German Education Conference (GeCon), pp. 1–6 (2022)
21. Biggs, J.B., Collis, K.F.: Evaluating the quality of learning: The SOLO taxonomy (structure of the observed learning outcome). Educational psychology series. Academic Press (1982). ISBN: 0120975505

# Automated Code Readability Feedback on Student Awareness

Oscar Karnalim$^{(\boxtimes)}$ [ID], Sendy Ferdian Sujadi [ID], and Rossevine Artha Nathasya [ID]

Universitas Kristen Maranatha, Suria Sumantri Street No 65, 40164 Bandung, Indonesia
oscar.karnalim@it.maranatha.edu

**Abstract.** Despite its importance in industry, writing readable code is often overlooked in academia. We developed CQIS, an automated checker that automatically provides feedback about code readability. Each time a program is submitted, a code readability report will be generated, and its unique link will be sent to the author. The report highlights parts of the submitted program that can be made clearer. The use of the checker needs to be evaluated, whether students are aware about code readability due to the checker. There is also a need to know which code readability aspects that students are least aware of and need further attention. Consequently, we evaluate the checker for one academic year (six courses with 222 students), and directly ask students about their awareness of code readability. Students seem to be generally more aware of code readability aspects, especially about having clear comments and having only useful variables. Students occasionally check the quality of their code after the code works at functional level. Few misconceptions might occur on higher programming level.

**Keywords:** Code Readability · Immediate Feedback · Programming · Engineering Education

## 1 Introduction

Software maintenance is a crucial task in software development and can be labor-intensive [1]. For efficient maintenance, it is important to document the software and make the code readable with high quality [2].

Despite its importance in industry, writing readable code is often overlooked in academia [3]. Some instructors consider it in their assessments but since the portion is relatively small and students still need to complete other tasks, code readability is less prioritized. Students can be encouraged to use automated checkers for code readability, like checkstyle [4] or flake8 [5]. However, not all the reported issues are relevant for novices, and they are usually delivered with many technical terminologies.

In response to that, several automated checkers dedicated to academia have been developed. They are dedicated to report code readability issues that are relevant to novices. Some of them have the messages simplified so that they are easier to understand. However, they cannot be easily integrated into many existing teaching environments. They either work as standalone applications [6, 7], are embedded to specific integrated development applications (IDE) [8, 9], and/or rely on historical data [10].

© The Author(s), under exclusive license to Springer Nature Switzerland AG 2024
M. E. Auer et al. (Eds.): STE 2024, LNNS 1027, pp. 56–66, 2024.
https://doi.org/10.1007/978-3-031-61891-8_5

We addressed the concerns by developing an automated checker and integrating the checker into an assessment submission system [11]. Each time a program is submitted, a code readability report will be generated, and its unique link will be sent to the author. The report highlights parts of the submitted program that can be made clearer. The author can voluntarily check the report and resubmit the work if needed. Such a mechanism can be applied to many existing teaching environments given that it does not affect the flow of teaching and completing assessments.

This study evaluates the checker for two academic semesters. Our checker was employed into six courses with a total of 222 students. At the end of each course, we asked the students about their agreement toward 14 code readability aspects for Java and Python, the two common programming languages for novices [12]. We also asked the students about when they fix code readability issues during the assessments. Based on 168 responses, we summarized the findings.

While our study is focused on a particular tool (CQIS), we believe the findings can still be useful for educators interested in applying automated code readability feedback tool. The findings can also be useful for the future development of similar tools.

## 2   Related Work

A number of automated checkers about code readability have been developed. Many of them are general-purpose. Checkstyle [4] for example, is a Java automated checker that can report over 196 code readability issues. Flake8 [5] is an automated checker for Python that is built on other three Python tools: pycodestyle, pyflakes, and mccabe. CppDepend [13] is a commercial automated checker for C/C++, covering many aspects from type usages to object-oriented paradigm. DeepScan [14] is another commercial automated checker but for JavaScript. It covers not only code readability but also potential runtime errors.

Some automated checkers are dedicated to novices in academia. Keuning et al. [7] for example, presented Refactor Tutor, a standalone tool where students can learn how to improve the readability of their code via step-by-step instructions. The tool currently focuses on Python code readability in program statements, expressions, branching, and looping.

Ala-Mutka et al. [6] presented Style++, an automated checker for C++ programming language. The tool reports non-declarative identifier names, unhelpful comments, and large code size. Students are expected to voluntarily use the tool while completing the assessments.

Ureel II and Wallace [9] developed WebTA, an automated checker for Java. The tool works as an IDE that constantly provides feedback about issues on localized instance variables, repetitive resource instantiation, pseudo-implementation, misplaced code, and dead code.

Blau and Moss [8] developed FrenchPress, an automated checker for Java programs. Unique to the tool, it is embedded to an IDE (i.e., Eclipse) as a plug-in. Four code readability issues are reported: field misuse, public modifiers misuse, loop control misuse, and boolean misuse.

Choudhury et al. [10] wrote AutoStyle, an automated checker where instructors can employ their own code readability rules based on student historical data. The tool is embedded to an assessment submission system, reporting detected issues for each submission.

Several studies observed student submissions regarding code readability issues. De Ruvo et al. [15] found sixteen program flow issues on introductory programming courses, starting to self-assigning variables (e.g., 'a = a') to dead branches in if-else statements. Keuning et al., [16] found that students often wrote bad and/or inefficient code on a large BlueJ data set. Similar finding was observed by Aivaloglou & Hermans [17]; they found that at least a quarter of programs in their observed Scratch repository have unused code. Karnalim et al. [18] found that students tend to write code as short as possible and that can sometimes harm the code readability.

We developed an automated checker named Code Quality Issue Reporter (CQIS) [11] and integrated that on an assessment submission system [19]. CQIS reports 52 code readability issues: 32 of them are dedicated to Java while the rest are for Python. Examples of the reported issues are unhelpful comments, non-declarative identifiers, inconsistent naming style, overly deep nested program statements, overly complex expressions, and locally undefined variables. A complete list of the reported issues can be read in the corresponding publication [11].

For each assessment of courses with CQIS, students are allowed to resubmit their work so long as the submission time is no later than the submission deadline. Sometimes, resubmission upon the deadline is permitted with mark deduction. Per submission, the student will get a report listing all code readability issues. Each issue is highlighted on their corresponding code location and explained in human language. The report is written in HTML and its link is sent to the student via university email.

## 3   Method

This study evaluates our approach on student awareness about code readability. The approach was employed for two academic semesters: the second semester of 2021/2022 and the first semester of 2022/2023. For all assessments, students were expected to upload their work on our assessment submission system and a code readability report would be generated per submission.

Involved courses with their numbers of students can be seen in Table 1. All courses have programming assessments that should be completed in less than one week. In total, there are 222 students enrolled in six courses.

Introductory programming was a mandatory first-year course about basic concepts of programming (e.g., sequential programs, branching, looping, and functions). Students were expected to complete one lab assessment (2 h) and one homework assessment (submitted before next week's lab session). Each assessment has three tasks: trivial, medium, and challenge. The course was offered to Information Technology (IT) students in the second semester of 2021/2022 and to Information System (IS) students on the first semester of 2022/2023. The former expected the solutions written in Java while the latter expected the solutions written in both Java and Python.

Data structure was a course offered to second-year IT students in the first semester of 2022/2023. It covered common data structures including linked lists and trees. Students were expected to implement a data structure in Python and to use it for a particular case each week.

**Table 1.** Involved Courses with Their Numbers of Students and Survey Responses

| Course | Students | Responses |
|---|---|---|
| Introductory programming IT | 55 | 45 |
| Introductory programming IS | 38 | 25 |
| Data structure IT | 34 | 29 |
| Object-oriented programming IS | 43 | 30 |
| Machine intelligence IT | 33 | 28 |
| Business application programming IS | 19 | 11 |
| **Total** | 222 | 168 |

Object-oriented programming was offered to second-year IS students, and it covered object-oriented concepts in Java including classes, objects, inheritance, and polymorphism. Per week, students were expected to write a program based on given specifications.

Machine intelligence covers artificial intelligence concepts and how it is implemented with third-party libraries in Python. It was exclusively offered to IT students (third-year or higher). Weekly assessments about how to solve a particular case were given.

Business application programming was the successor of the object-oriented programming course for IS students. It was about developing applications with graphical user interface and database. The course had weekly assessments about simple application developments.

At the end of each course offering, a voluntary survey was conducted. The questions can be seen in Table 2 while the number of responses can be seen in Table 1.

Q01-Q14 ask about student perspective regarding aspects of good code readability. The aspects are summarized from those stressed by checkstyle and flake8, validated by several computing instructors from our faculty. Per question, students need to show their agreement in 5-points Likert scale (1 = strongly disagree, 3 = neutral, 5 = strongly agree). While responses to these questions were likely to "agree", students might respond "neutral" if they were unaware about said code readability aspect. They might respond with "disagree" or "strongly disagree" if they misunderstood the practice.

Q15 asks how frequently students consider code readability while completing their assessments and Q16 asks reasons for the response. Q15 has five options: "not at all", "seldom", "occasionally". "frequently", and "always". Q16 on the contrary, is an open-ended question.

Q17 specifically asks when students consider code readability while completing assessments. The options are: "every time", "while writing the code", "after the code works", and "never".

The analysis was conducted at two granularity levels: overall and programming levels. Overall analysis considers all responses at once. Programming level analysis separates the responses based on the courses' programming levels. Introductory programming IT and IS are grouped as CS1 courses as they are first offered. Data structure

**Table 2.** Survey questions

| ID | Question |
|---|---|
| Q01 | Comments should be clear and easy to understand |
| Q02 | Comments should exist for each program block |
| Q03 | Names of identifiers (variables, functions, etc.) should be clear and easy to understand |
| Q04 | Names of identifiers (variables, functions, etc.) should have consistent transition style; for instance, all identifiers use camel case – i.e., thisIsVariable instead of underscore – i.e., this_is_variable |
| Q05 | Each line of code should not be too long |
| Q06 | Each line of code should have only one program statement |
| Q07 | Modules should be imported when used |
| Q08 | Variables declared and assigned with values should be limited to those actually used |
| Q09 | One-line branching should be avoided; for instance, $count = (count == N) ? 0: count + 1;$ |
| Q10 | Boolean operators in an expression or a condition for branching/looping should not be too many |
| Q11 | Syntax blocks (e.g., branching or looping) should not have empty components |
| Q12 | A function body should not have too much code |
| Q13 | Nested branching, looping, or try-catch should not be too deep |
| Q14 | Braces, brackets, and semicolons should be adequately used |
| Q15 | How frequent do you consider readability while completing your assessments? |
| Q16 | Explain your response to Q15 |
| Q17 | When do you consider code readability while completing your assessments? |

IT and object-oriented programming IS are grouped as CS2 courses as they are offered right after CS1 courses. Machine intelligence IT and business application programming IS are considered as advanced courses since they are offered for third-year students who have completed both CS1 and CS2 courses.

For each analysis, responses of Q01–Q14 and Q15 were analyzed quantitatively based on their average Likert scores. Q16 would be summarized and discussed to complement Q15 findings. Q17 was analyzed quantitatively based on the proportion of responses for each option. Any comments from the instructors were considered to explain most of the phenomena.

We were also interested in checking whether awareness of code readability was improved on higher programming level. We compared average agreement score across programming levels for both the whole code readability aspects (Q01–Q14) and each aspect. We also checked whether consideration of code readability was affected by programming level (Q15–Q17).

# 4  Results and Discussion

This section summarizes our findings and discusses them. We first present our findings at the overall level, followed by those for each programming level (CS1, CS2, and advanced) and trend across programming levels. We end the section with a discussion and some recommendations.

## 4.1  Overall Findings

Students tended to agree with our code readability statements; all of them scored no less than 3 (neutral) on average. Q01, Q03, and Q08 were three questions which average response score is higher than 4. Q01 was about the readability of comments and the issue might have been reported frequently. Students sometimes wrote unhelpful comments due to bad management of time. Some of them assumed that their unhelpful comments were still understandable by them later. Further, CQIS encouraged students to reread comments with words not listed in its dictionary.

Q03 was about the readability of identifier names. Some students run out of ideas for naming identifiers, resulting in meaningless names. CQIS reminded them about the matter by parsing the identifier names based on the transition style, and pointing out words that are not listed in the dictionary. However, since students without CQIS were somewhat aware about the matter, CQIS improvement might not be that substantial.

Q08 was about declaring and assigning only useful variables. Students sometimes had more variables than what they actually needed. CQIS reminded them to remove unused variables.

Q07, Q12, and Q13 were three questions with the least agreed statements. Q07 was about unused module imports; Q12 was about too much code in a function; and Q13 was about overly deep nested programming constructs. They rarely occurred on student submissions, especially at early programming levels (CS1 and CS2). Instructors are expected to focus on these aspects while informing students about code readability.

Students occasionally considered code readability while completing their assessments; their average response rate for Q15 is 3.8 (near to "frequently"). Half of them (50%) considered code readability after their code works at functional level. Nearly a third (29%) considered the matter while writing the code and 7% students always considered the matter (while both writing the code and after the code works). It is interesting to see that 15% of students did not consider such an aspect at all. This is perhaps because either these students did not see the importance of code readability, limited assessment marks were allocated to code readability, or these students relied too much on automated formatting from their programming workspace.

Most students (153 of 168) provided their reasons for Q15 as responses in Q16. Ninety responses were positive. Forty-six responses were to prevent readability issues in the future. They might reread and reuse the code. Some of the responses stated that getting used to writing readable code could prepare them for future collaboration as professionals. Twenty-five responses were about keeping the code simple; some parts could be more effective and efficient. Nineteen responses were to acknowledge that code readability is an important aspect in writing code.

Sixty-three responses were about why students did not fully consider code readability. Thirty-one of them considered code readability if it was part of assessment requirements. They focused on short-term benefits of applying code readability. Twenty-five of them considered code readability if it was seen achievable. They either forgot some code readability aspects or did not have time to apply that due to bad time management. Seven responses argued that code readability was relative and depended on the programming style.

### 4.2   CS1 Level Findings

Students generally agreed about the code readability statements. Q01, Q03, and Q08 were still three questions with the most agreed statements, though Q03 and Q08 experienced a slight reduction. The reported issues (readability of comments, readability of identifier names, and using only useful variables) were generally common to occur at CS1 level.

Q07, Q12, and Q13 were also consistent to be questions with the least agreed statements. However, Q07 had the least agreement score for CS1 instead of Q12. This is expected as CS1 programming seldom uses external modules (or libraries).

Student frequency of consideration of code readability is slightly lower than that at overall level (3.6). For introductory level, being able to write readable code was not a main objective. The courses were more focused on encouraging students to be able to write working code.

Based on responses summarized from Q16 (63 responses), more than half (35) were not positive. Eighteen responses stated that code readability would be considered when it was part of the marking. Sixteen responses stated that code readability would be considered if such consideration was achievable: students still had time to improve code readability before the submission due date and they understood which code readability aspects were in need for consideration. One response argued that code readability was subjective.

Twenty-eight responses were positive. Twenty students were aware that having good readability on code can prevent readability issues (13) or can result in simpler code (7). Others (8) only acknowledged that consideration of code readability is needed.

The frequency of considering code readability (Q17) was comparable to that at overall level. Most of the students considered code readability (89%), but the consideration usually happened after the code works (51%).

### 4.3   CS2 Level Findings

Q01, Q03, and Q08 were questions with the most agreed statements. Unique to CS2 level, Q08 was the most agreed instead of Q01. CS2 level was where students wrote large programming solutions for the first time, and it is expected that they occasionally forgot to use variables that had been already declared and assigned. CQIS could remind them about that.

Q06, Q10, and Q12 were the least agreed by CS2 students. Q06 was about ensuring that each code line should contain only one program statement. Some CS2 students appeared to believe that having two or more program statements in one line is acceptable for short statements (e.g., assigning three variables representing coordinates: x, y, and z). Although having two or more program statements in one line resulted in fewer code lines, it might be less helpful for other programmers, especially those who are weak in programming.

Q10 was about having a reasonable number of boolean operators in expressions or conditions while Q12 was about having reasonable amount of code in one function. These issues were seldom reported by CQIS as they were unlikely to happen at CS2 level. The number of boolean operators and the amount of code per function were manageable by default. Although CS2 assessments were more advanced than those of CS1, they still imposed several restrictions to prevent over-complexity. Our data structure and object-oriented programming assessments for examples, sometimes provided a class structure to follow.

CS2 students had a similar level of consideration of code readability (Q15) to that at overall level (3.8). It was higher than CS1 level of awareness since students became more mature in programming. This was supported by the fact that more than two thirds of the reasons for Q15 (collected from Q16) were positive: 38 of 54. They believed that writing good code was somehow important for software development: it could prevent readability issues and could keep the code simple.

Sixteen responses were not positive. Seven believed that consideration of code readability was only worth when it was asked on the assessment document. Five believed that code readability could be considered if they had time. Four believed code readability was relative.

Most CS2 students (90%) considered code readability. This was slightly higher than that of CS1 (89%). A larger proportion of students considered code readability while writing the code (27% to 32%) or after the code works (51% to 53%). This consequently reduced the proportion of students considering code readability on all occasions (10% to 5%).

## 4.4 Advanced Level Findings

For advanced programming level, Q01 and Q03 were still questions with the highest agreement rate. These students were already aware that comments and identifier names should be clear and easy to understand. CQIS could also help them to identify unhelpful comments and less meaningful identifier names.

Exclusive to the advanced level, Q04 was also a question with the most agreed statements. It was about consistent transition style for identifier names (using either camel case or underscore as the separator). The importance of such consistency became more prevalent at advanced level.

Q05, Q09, and Q12 were three questions with the least agreed statements. Q12 was the only question that is consistent for the remaining levels: CS1 and CS2. It was about not having too much code in a function. Although advanced level assessments had fewer constraints, the issue was seldom reported.

Q05 was about not having too much code in one line. The issue seldom occurred at advanced level since the courses encouraged the use of programming workspace (IDE) that automatically reformats code.

Q09 was about avoiding one-line branching. Such a branching was seldom used at advanced level due to difficulties in catering complex expressions and long branching bodies.

Students considered code readability slightly more frequent than that at CS2 level (3.8 to 3.9). This was expected since in advanced level courses, code readability was considered as part of the marking. When asked about the reasons (Q16), twenty-four responses were positive. More than half of the positive responses (13) were about preventing readability issues. Students and/or their colleagues might revisit the code. Five positive responses were about keeping the code simple and clear for effectiveness and efficiency. Six positive responses implicitly addressed the importance of considering code readability.

Twelve responses were not positive. Ten students only considered code readability if such a thing was considered as part of the marking (6) and/or there was still some time left to improve readability of the code (4). Two students believed code readability was part of programming style and it was subjective.

Although students considered code readability slightly more frequent than those of CS2 and CS1, such an improvement occurred in a smaller proportion of students (72%). Some students completely ignored code readability while completing their assessments on advanced level. They used IDE that automatically fixed major readability issues via reformatting. Further, some identifier names were already defined by the instructors. In addition, some identifiers were automatically generated.

### 4.5 Relationship Between Code Readability Awareness and Programming Level

Code readability awareness was generally improved for higher programming level. The average agreement rate for Q01-Q14 was increased from 3.7 (CS1) to 3.8 (CS2) and 4 (advanced), though it was not substantial. However, when analyzed per question, the improvement was only applied on seven of 14 questions: Q01, Q03, Q04, Q07, Q11, Q13, and Q14. Other code readability aspects might be reduced on higher programming level, as described in Sect. 4.2 to Sect. 4.4.

As the programming level became higher, students considered code readability slightly more frequently. Q15 agreement score was slightly improved from 3.7 (CS1) to 3.8 (CS2) and 3.9 (advanced). The slight improvement was partly since some students started to realize the importance of writing readable code (Q16). However, on advanced level, few students might ignore code readability as they believed the matter would be fixed automatically by their IDE. Further, some identifier names were not freely defined by the students.

### 4.6 Discussion and Recommendations

Students with CQIS were generally aware of the clarity of comments (Q01), the clarity of identifier names (Q03), and the need to have only useful variables (Q08). They could be more aware of consistent transition in identifier names (Q05) on advanced level. Students without CQIS were also quite aware of the clarity of writing identifier names but their awareness seemed to be lower (not reaching 'agree').

Students with CQIS were least aware of not having unused module imports (Q07), too much code in a function (Q12), and overly deep nested programming constructs (Q13). They might also be less aware of not having too much code in one line (Q05), not having too many program statements in one line (Q06), and not having one-line branching (Q09). They either seldom got such issues reported or thought these were acceptable in their circumstances.

Students occasionally checked for code readability and the frequency became higher on students with more programming experience. Students considered code readability as they believed it was important in software development, especially to prevent readability issues and to keep the code simple. Some students only considered the matter if they had time or code readability is part of the marking. Few of them believed that code readability was part of programming style, and it was thus relative.

Many students checked the readability of their code after the code works. On advanced level, few of them ignore code readability, assuming that their IDE could fix major readability issues.

In response to the findings, we have a number of recommendations. First, the use of code readability feedback tool for teaching programming is encouraged. CQIS in our case can help students to be aware of some code quality aspects. The benefits can be more substantial at a higher programming level. Second, instructors need to remind students about code quality aspects that are not reported by the code readability feedback tool. Some aspects might seldom occur in student submissions. In our case, they are about not to have unused module imports, overly deep nested programming constructs, and too much code or program statements in one line or a function. Third, there is a need to address student misconceptions that are developed as a result of having better programming experience. Few students relied too much on IDE to fix code readability issues while such IDE could only fix few. They might also think that having some unreadable code was acceptable as the code was still readable for them. Fourth, given that some students only considered code readability when it was part of the marking and/or they had time to do it, it is recommended to allocate some marks in code readability and to allocate extra time for improving quality of the submitted code.

## 5   Conclusion and Future Work

We evaluated the impact of employing an automated checker for code readability (CQIS in our case) on student awareness of code readability for a year. The checker seemed to generally help students understand code readability, especially on having clear comments and having only useful variables. Students occasionally checked the quality of their code after they had ensured that the code worked at functional level. Few misconceptions might occur on higher programming level.

For future work, we plan to occasionally remind students about code readability aspects that either are seldom reported by the tool or are commonly misunderstood. We are also interested in comparing student awareness of code readability with CQIS to that of students without CQIS from multiple perspectives: surveys and identified code readability issues. There is also a need to replicate the study to other institutions.

# References

1. Kaur, S., Singh, P.: How does object-oriented code refactoring influence software quality? Research landscape and challenges. J. Syst. Softw. **157**, 110394 (2019). https://doi.org/10.1016/J.JSS.2019.110394
2. Fakhoury, S., Ma, Y., Arnaoudova, V., Adesope, O.: The effect of poor source code lexicon and readability on developers' cognitive load | IEEE Conference Publication | IEEE Xplore. In: IEEE/ACM 26th International Conference on Program Comprehension (ICPC). IEEE (2018)
3. Kirk, D., Crow, T., Luxton-Reilly, A., Tempero, E.: On assuring learning about code quality. In: 22nd Australasian Computing Education Conference, pp. 86–94 (2020). https://doi.org/10.1145/3373165.3373175
4. checkstyle — Checkstyle 10.7.0, https://checkstyle.sourceforge.io/. Last accessed 1 Feb 2023
5. Flake8: Your Tool For Style Guide Enforcement – flake8 6.0.0 documentation. https://flake8.pycqa.org/en/latest/. Last accessed 1 Feb 2023
6. Ala-Mutka, K., Uimonen, T., Jarvinen, H.-M.: Supporting students in c++ programming courses with automatic program style assessment. J. Inform. Technol. Educ.: Res. **3**, 245–262 (2004). https://doi.org/10.28945/300
7. Keuning, H., Heeren, B., Jeuring, J.: A tutoring system to learn code refactoring. In: 52nd ACM Technical Symposium on Computer Science Education (2021)
8. Blau, H., Moss, J.E.B.: FrenchPress gives students automated feedback on Java program flaws. In: ACM Conference on Innovation and Technology in Computer Science Education (2015)
9. Ureel II, L.C., Wallace, C.: Automated critique of early programming antipatterns. In: 50th ACM Technical Symposium on Computer Science Education (2019)
10. Choudhury, R.R., Yin, H., Fox, A.: Scale-driven automatic hint generation for coding style. In: International Conference on Intelligent Tutoring Systems (2016)
11. Karnalim, O., Simon, Chivers, W., Panca, B.S.: Automated reporting of code quality issues in student submissions. In: IFIP World Conference on Computers in Education, 685, (2023)
12. Simon, Mason, R., Crick, T., Davenport, J.H., Murphy, E.: Language choice in introductory programming courses at Australasian and UK universities. In: The 49th ACM Technical Symposium on Computer Science Education, pp. 852–857. ACM Press, Baltimore (2018). https://doi.org/10.1145/3159450.3159547
13. CppDepend : C/C++ Static Analysis and Code Quality Tool, https://www.cppdepend.com/. Last accessed 1 Feb 2023
14. How to ensure JavaScript code quality | DeepScan, https://deepscan.io/. Last accessed 1 Feb 2023
15. De Ruvo, G., Tempero, E., Luxton-Reilly, A., Rowe, G.B., Giacaman, N.: Understanding semantic style by analysing student code. In: 20th Australasian Computing Education Conference (2018)
16. Keuning, H., Heeren, B., Jeuring, J.: Code quality issues in student programs. In: ACM Conference on Innovation and Technology in Computer Science Education (2017)
17. Aivaloglou, E., Hermans, F.: How kids code and how we know: an exploratory study on the Scratch repository. In: ACM Conference on International Computing Education Research (2016)
18. Karnalim, O., Simon, Chivers, W.: Work-In-Progress: Code Quality Issues of Computing Undergraduates. In: IEEE Global Engineering Education Conference (2022)
19. Karnalim, O., Simon, Chivers, W., Panca, B.S.: Educating students about programming plagiarism and collusion via formative feedback. In: ACM Transactions on Computing Education, vol. 22 (2022)

# An Easy-to-Use Experimental Platform for Advanced Control Teaching in Wuhan University

Xingwei Zhou[1], Jiuzheng Su[1(✉)], Wenshan Hu[1], Guo-Ping Liu[2], and Zhongcheng Lei[1]

[1] Wuhan University, Luojia Hill, Wuhan, China
jiuzheng.su@whu.edu.cn
[2] Southern University of Science and Technology, No. 1088 Xueyuan Avenue, Shenzhen, China

**Abstract.** In the field of control engineering education, applying theoretical knowledge to practical scenarios is essential, allowing students to reinforce their understanding of the curricula. Based on feedback from control engineering students, it is evident that implementing and validating advanced algorithms in physical test rigs poses challenges for students due to their limited knowledge of embedded development. In light of these challenges, an easy-to-use experimental platform covering the entire control process (e.g., graphical algorithm building, simulation of mathematical models for advanced test rigs, and monitoring of these test rigs) is crucial for control engineering education. Therefore, this paper proposes an easy-to-use experimental platform for verifying advanced algorithms in physical test rigs. This allows students to focus solely on designing control algorithms without being burdened by the complexities of developing the experimental setup. An experimental case with this platform is conducted to illustrate its usage and advantages, particularly in the context of control education.

**Keywords:** experimental platform · advanced control teaching · engineering education · Matlab

## 1 Introduction

Experiment serves as a practical platform for students to apply theoretical knowledge acquired in the classroom to real-world scenarios, establishing a robust groundwork for future engineering endeavors [1]. Experimental teaching plays a pivotal role in engineering majors like control engineering, offering students a comprehensive grasp of control theory and the ability to tackle real-world issues through hands-on practice [2–4].

However, in reality, students often find themselves needing to invest more time in crafting intricate code and configuring diverse hardware environments, rather than delving into the understanding of control algorithms. This is a challenge currently faced in control engineering and other similar engineering disciplines during experimental teaching. For the convenience of students, many schools have established customized physical experimental platforms [5–9]. A low-cost, portable DC motor control teaching experiment platform for control engineering education is investigated in [6]. An experimental testbed that translates control systems theory into hands-on experience by implementing interactive control of the orientation of a quadcopter is proposed in [8]. However, these platforms mainly focus on supporting a limited number of simple algorithms and seldom offer complex control algorithm capabilities. Additionally, students may need to invest a significant amount of effort in adapting to these platforms, incurring a considerable transition cost [10,11].

In order to improve students' concentration on learning complex control algorithms and cater to their usage preferences, this paper introduces an easy-to-use experimental platform based on Matlab & Simulink. The platform aims to simplify the process by eliminating the need for intricate code logic writing and environment configuration, while providing a diverse range of sophisticated control algorithm designs.

The platform consists of a PC with Matlab & Simulink for graphical algorithm design, a microcontroller-based control box for communication, and a physical test rig. Students can utilize existing Matlab & Simulink blocks to design advanced control algorithms without complex hardware development. The PC connects to the control box to send commands and receive sensor data from the test rig. Custom pre-built blocks handle low-level functions like reading encoder pulses, setting motor speed, stopping motor, etc. Students simply drag-and-drop these to build their control system. Thanks to the introduction of Matlab, which not only simplifies the programming approach, but also seamlessly supports the extensive toolbox provided by Matlab and simplifies the process of writing complex algorithms, students can more easily focus on understanding complex control algorithms through physical experiments rather than spending a lot of time on environment configuration. The hardware communication module provided by the platform can well deliver the control information sent by Matlab and send the return data to Matlab in real time, realizing hardware-in-the-loop control. This enables students to realize the physical device control experiments even if they don't know any hardware communication protocols, such as serial port, RS485 and so on. In addition, in order to enrich the teaching extension content, the platform not only provides basic control experiments, but also provides a series of advanced control algorithms model, for students who have the ability to learn to provide further learning content.

## 2    Design and Implementation of the Easy-to-Use Experimental Platform

Students who major in control engineering are familiar with Matlab & Simulink where graphical algorithm design is quite easy and convenient for students to use [12–15]. On account of such concept, the easy-to-use experimental platform proposed in this paper is designed based on Matlab & Simulink. Students can design their advanced control algorithms with Matlab & Simulink which provides a wealth of modules that can almost implement any advanced algorithm's design.

This section is divided to three subsections to introduce the design and implementation of the experimental platform in detail, which are physical hardware setup, specific Simulink module design and data transmission, respectively.

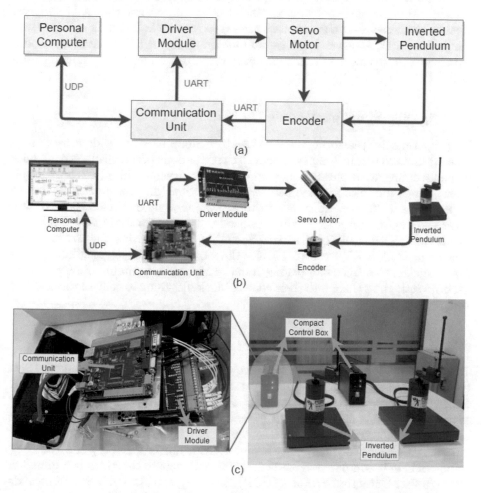

**Fig. 1.** The structure of the easy-to-use experimental platform. (a) System schematic diagram; (b) detailed physical item connections; (c) physical hardware setup.

## 2.1 Physical Hardware Setup

In terms of the physical hardware setup, it consists of a compact control box and physical test rigs, as Fig. 1 shows. A communication unit, a power module and a driver module are neatly installed inside the control box. A STM32 micro-controller plays the role of the communication unit, transmitting data between Matlab & Simulink and physical test rigs. Just as its name implies, the power module supplies power for whole physical hardware, such as the communication unit and the driver module. While the driver module takes responsibilities to drive the physical test rigs to move. A rotational inverted pendulum is chosen as the test rig in the easy-to-use experimental platform. These components are housed within a sealed control box, which features only three external interfaces: a cable interface for data transmission between Matlab & Simulink and the communication unit, a power interface for supplying power to all hardware, and a data line interface for data exchange between the test rig and the communication unit.

## 2.2   Specific Simulink Module Design

The experimental platform utilizes Matlab & Simulink to provide a user interface for users to conduct experiments, so that a serious of indispensable modules communicating with physical hardware must be designed in advance by developers. These modules are designed with 's-functions' in Matlab & Simulink to implement specific functions. For instance, the module of 'ReadChannelPulse' takes the responsibility to obtain the encoder data from the rotational inverted pendulum, the module of 'SetMotorSpeed' is used to drive the motor in set speed, the module of 'SetStop' disables the motor, etc. These essential modules are designed in advance and packaged in the Matlab & Simulink library, so that users directly drag them into their control block diagrams to control the physical test rigs.

## 2.3   Data Transmission

Data transmission consists of two parts, one between Matlab & Simulink (Personal Computer) and the communication unit (STM32 microcontroller) and another one between the communication unit and the driver module, as illustrated in Fig. 1(a) and Fig. 1(b). During the experimentation, the driver module obtains the motor encoder's data and transmit it to the communication unit through Universal Asynchronous Receiver/Transmitter (UART) and the communication unit will parse the data and transmit that to the Matlab & Simulink through User Datagram Protocol (UDP). Similarly, the Matlab & Simulink sends the motor's set speed to the communication unit through UDP and the communication unit will parse the data and transmit that to the driver module driving motor to move at a specified speed through UART.

In short, a PC, a compact control box and a physical test rig constitute the experimental platform, where the Matlab & Simulink is employed as the frontend in the PC to design DIY algorithms to control the test rig and a series of supporting blocks are developed to send and request experimental data to physical hardware setup.

# 3  Application of the Easy-to-Use Experimental Platform in the Advanced Control Teaching

The inverted pendulum system exhibits characteristics such as nonlinearity, high order, multivariable, and strong coupling, making it a typical representative of unstable systems [17]. The inverted pendulum system can be abstracted as a fundamental model for various dynamic equilibrium objects and serves as an ideal platform for learning and understanding automatic control theories [18]. Many key control concepts, such as system stability and robustness, can be intuitively reflected through the control of inverted pendulum systems. Beyond its educational utility, the characteristics of the inverted pendulum control system have led many researchers in modern control theory to consider it a classic subject of study [19]. They continually explore new control strategies from the study of inverted pendulums and apply them to various high-tech fields, including aviation, aerospace, navigation, and intelligent robotics. Therefore, using the inverted pendulum as the test rig in teaching enables students to acquire a deep understanding of various modern and advanced control algorithms. In this section, an advanced experiment is conducted to exhibit the application of this easy-to-use experimental platform in the control engineering education.

## 3.1  Curriculum Set with the Experimental Platform

A series of control experiments are designed to support bridge between the practice and both classic and advanced theories for students, which is illustrated in Fig. 2. The experiments available at present are listed in the following, covering from classic control theories to modern control theories, and even to intelligent ones.

1. Classic Control Experiments
   - Root Locus Correction Experiment
   - Frequency Response Correction Experiment
   - Single-Loop PID Control Experiment
   - Dual-Loop PID Control Experiment
2. Modern Control Experiments
   - State-Space Pole Placement Control Experiment
   - Linear Quadratic Regulator (LQR) Control Experiment
   - LQR Control (Energy Swing-Up) Experiment
   - Sliding Mode Variable Structure Control Experiment
   - LMI-Based H∞ Robust Control (Energy Swing-Up) Experiment
   - Active Disturbance Rejection Control (ADRC) (Energy Swing-Up) Experiment
3. Intelligent Control Experiments
   - Fuzzy Logic Control Experiment
   - Fuzzy Logic Control (Energy Swing-Up) Experiment
   - Variable Universe Adaptive Fuzzy Control Experiment
   - Fuzzy Adaptive PID Control Experiment

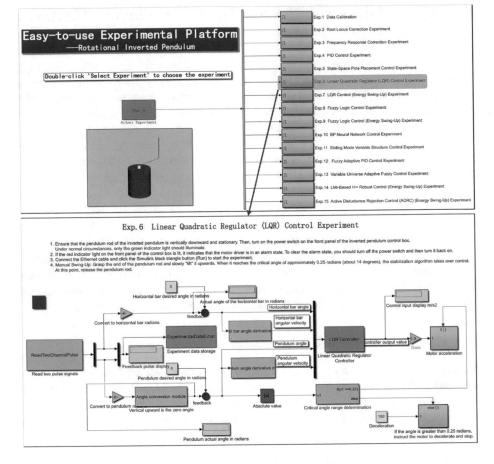

**Fig. 2.** Wide-range control experiments provided for teaching.

– Backpropagation Neural Network Control Experiment

According to past theoretical teaching experience, students usually struggle to bridge the gap between theory and practice [16]. Even if they can comprehend theoretical knowledge, students may still find it challenging to apply that knowledge in practical contexts. However, this experimental platform addresses this gap, allowing students to validate the theoretical knowledge they have acquired from textbooks in a tangible, hands-on way.

## 3.2    Experimental Case

In the course of the advanced (or modern) control theory or others, students usually conduct experiments in the form of simulation, seldom executing experiments in the physical test rigs, which causes the gap between theory and practice [17]. The algorithm of Linear Quadratic Regulator (LQR) taken as the example to explain the usage of the experimental platform and indicate the connection between the theory and practice [18].

In this experiment, students are required to establish the mathematical model of the inverted pendulum at first with the knowledge learning in the modern control theory class. A series of parameters, such as pendulum length and horizontal bar length, will be provided to the students, and they will, in turn, use the principles of the inverted pendulum to establish its state-space equations and transfer functions. After this, students need to use the calculated parameters as the parameters of the Matlab & Simulink control model, and then adjust the inputs to observe the state of motion of the inverted pendulum and the output results.

For single-stage rotating inverted pendulum, the system input is the rotational angular acceleration, which is typically taken as a unit step signal. The state variables are assumed to be as shown in Eq. (1), where $\theta_1$ represents the angle of the horizontal position and $\theta_2$ represents the angle of the vertical position.

$$x = \begin{bmatrix} \theta_1 & \dot{\theta}_1 & \theta_2 & \dot{\theta}_2 \end{bmatrix} \tag{1}$$

For short, the state-space model of the linearized single-stage rotating inverted pendulum is given as Eq. (2), where $A$ and $B$ are the system and control matrices, respectively, which are constant matrices determined by the length, mass, and initial position of the inverted pendulum arm.

$$\dot{x} = Ax + Bu \tag{2}$$

According to [19], it is useful to find a control vector $u$ that minimizes the quadratic cost function given in Eq. (3), where $Q$ is a nonnegative definite matrix and $R$ is a positive definite matrix. By solving the associated algebraic equation, this leads to the optimal control law $u = -Kx$ as shown in Eq. (4), where $K$ is determined by matrices $A, B, Q, R$.

$$J = \int_0^\infty (x \cdot Qx + u \cdot Ru)dt \tag{3}$$

$$u = -Kx \tag{4}$$

After mathematical calculations, the students obtain the mathematical model of the inverted pendulum, and the LQR algorithm is based on the state-space equations of the inverted pendulum for control.

At this point, problem-driven learning enables students to realize how knowledge from textbooks is applied. Next, they need to build an LQR control algorithm block diagram using Simulink, which further requires students to apply theory in the context of practical control. In the LQR algorithm, students need to calculate the feedback gain matrix first and then apply the calculated parameters to the algorithm. They should observe the operational state of the inverted pendulum. In addition, as shown in Fig. 3, the experimental data can be well preserved. Additionally, they should continuously adjust the values of $u, K$ and observe the impact of parameter changes on the behavior of the inverted pendulum. As can be seen from Fig. 3, with the proper parameters, smaller changes in the horizontal position do not affect the state of the inverted pendulum (set value of position changes from −20° to 20° and from 20° to −40°), and the inverted pendulum quickly returns to the vertical state even with larger disturbances (set value of position changes from −40° to 40°).

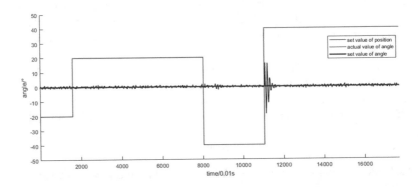

**Fig. 3.** The experiment record of LQR-controlled inverted pendulum.

In this experiment, only the easy-to-use experimental platform is provided to the students. They have to flexibly apply the knowledge they have learned from textbooks, starting from mathematical model calculations, algorithm development, and optimization of parameters. All of these steps are problem-driven. Students use the experimental platform to validate their mathematical models, algorithms, and parameters, which undoubtedly deepens their understanding of theoretical knowledge and hones their ability to solve real-world problems.

## 4    Conclusion

An easy-to-use experimental platform proposed in this paper is designed and implemented for the propose of control engineering teaching and offering students a tool to conduct advanced algorithm verification, which effectively bridges the gap between theoretical knowledge and practical application. With the physical experimental platform, students can not only verify algorithms, especially advanced and complex ones, through simulation, but can also visually access algorithms effect through physical test rigs [3]. This boosts practical skills and intuition for real-world applications.

**Acknowledgement.** This work was supported in part by National Natural Science Foundation of China under Grant 62073247, Grant 62188101 and Grant 62173255.

# References

1. Wang, L., Yu, Z., Zhang, D., Qin, G.: Research on multi-cycle CPU design method of computer organization principle experiment. In: 2018 13th International Conference on Computer Science & Education (ICCSE), Colombo, Sri Lanka, pp. 1–6 (2018). https://doi.org/10.1109/ICCSE.2018.8468694
2. Xiaoxia, G., Deshan, Z., Yan, W., Hongwei, S., Xin, L., Peng, J.: Construction and application of multi-dimensional teaching model of medical morphology experiment. In: 2015 7th International Conference on Information Technology in Medicine and Education (ITME), Huangshan, China, pp. 403–407 (2015). https://doi.org/10.1109/ITME.2015.145
3. Li, L., Chen, Y., Li, Z., Li, D., Li, F., Huang, H.: Online virtual experiment teaching platform for database technology and application. In: 2018 13th International Conference on Computer Science & Education (ICCSE), Colombo, Sri Lanka, pp. 1–5 (2018). https://doi.org/10.1109/ICCSE.2018.8468849
4. Liu, M., Chen, W., Yang, L.: Optimization of experimental teaching system based on ACSI model. In: 2022 International Conference on Information System, Computing and Educational Technology (ICISCET), Montreal, QC, Canada, pp. 106–109 (2022). https://doi.org/10.1109/ICISCET56785.2022.00035
5. Ojeda Misses, M.A., Jiménez, N.J.: Development of a platform with real-time performance for electrical circuits education. IEEE Lat. Am. Trans. **19**(12), 2147–2155 (2021). https://doi.org/10.1109/TLA.2021.9480158
6. Wang, S., Zhang, F., Tang, Q., Zhang, X., Zhao, R.: A take-home motor control teaching experiment platform for control engineering-related courses. IEEE Trans. Educ. **65**(2), 115–123 (2022). https://doi.org/10.1109/TE.2021.3094981
7. Aloulou, A., Boubaker, O.: Enhancing technical skills of control engineering students in robotics by using common software tools and developing experimental platforms. In: International Conference on Education and e-Learning Innovations, Sousse, Tunisia, pp. 1–5 (2012). https://doi.org/10.1109/ICEELI.2012.6360601
8. Toapanta, C., Villafuerte, J., Cruz, P.J.: 3DoF multi-rotor experimental testbed for teaching control systems. In: IEEE Third Ecuador Technical Chapters Meeting (ETCM), Cuenca, Ecuador, pp. 1–6 (2018). https://doi.org/10.1109/ETCM.2018.8580337
9. Lu, D., Liang, S., Wang, M., Wang, Y., Han, D., Cheng, Y.: Intelligent manufacturing education and tutoring experimental platform studies based on reconfigurable module. In: 2023 2nd International Joint Conference on Information and Communication Engineering (JCICE), Chengdu, China, pp. 179–183 (2023). https://doi.org/10.1109/JCICE59059.2023.00044
10. Jun, Z., Haoran, W., Zhanghuang, G., Hao, X.: Design and implementation of experimental platform for multi-agent system. In: 2023 9th International Conference on Control Science and Systems Engineering (ICCSSE), Shenzhen, China, pp. 196–201 (2023). https://doi.org/10.1109/ICCSSE59359.2023.10245091
11. Guo, H., Ge, J., Wang, Y., Wang, Y., Yang, X., Fu, P.: Experimental teaching design of mechanical specialty based on multidisciplinary knowledge fusio. In: 2019 IEEE 11th International Conference on Engineering Education (ICEED), Kanazawa, Japan, pp. 16–19 (2019). https://doi.org/10.1109/ICEED47294.2019.8994909
12. Chengwei, Z., Yihua, L., Liu, C., Yang, S.: Research on teaching reform of "Motor and Drive" based on Matlab simulation. In: 2021 2nd International Conference on Education, Knowledge and Information Management (ICEKIM), Xiamen, China, pp. 272–276 (2021). https://doi.org/10.1109/ICEKIM52309.2021.00066

13. Da-you, H., et al.: Teaching research of signal and system based on MATLAB. In: 2019 IEEE 3rd Advanced Information Management, Communicates, Electronic and Automation Control Conference (IMCEC), Chongqing, China, pp. 397–400 (2019). https://doi.org/10.1109/IMCEC46724.2019.8984042
14. Meilong, C., Xiehua, Y., Shaozi, L., Fengfei, K.: Design of graphic interactive experimental platform based on MATLAB. In: 2022 12th International Conference on Information Technology in Medicine and Education (ITME), Xiamen, China, pp. 729–733 (2022). https://doi.org/10.1109/ITME56794.2022.00153
15. Bentounsi, A., Djeghloud, H., Benalla, H., Birem, T., Amiar, H.: Computer-aided teaching using MATLAB/Simulink for enhancing an IM course with laboratory tests. IEEE Trans. Educ. **54**(3), 479–491 (2011). https://doi.org/10.1109/TE.2010.2085046
16. Brunhaver, S.R., Korte, R.F., Barley, S.R., Sheppard, S.D.: Bridging the gaps between engineering education and practice, NBER Chapters. In: US Engineering in a Global Economy, pp. 129–163. National Bureau of Economic Research, Inc. (2018)
17. Sugaya, J., Ohba, Y., Kanmachi, T.: Simulation of standing upright control of an inverted pendulum using inertia rotor and the swing type inverted pendulum for engineering education. In: 2017 9th International Conference on Information Technology and Electrical Engineering (ICITEE), Phuket, Thailand, pp. 1–6 (2017). https://doi.org/10.1109/ICITEED.2017.8250436
18. Lee, H.W.: Performance the balance of circular inverted pendulum by using LQR controlled theory. In: 2017 IEEE International Conference on Consumer Electronics - Taiwan (ICCE-TW), Taipei, Taiwan, pp. 415–416 (2017). https://doi.org/10.1109/ICCE-China.2017.7991172
19. Ogata, K.: Modern control engineering fifth edition (2010)

# LLM Integration in Workbook Design for Teaching Coding Subjects

Magnus Westerlund and Andrey Shcherbakov[✉]

Arcada University of Applied Sciences, 00560 Helsinki, Finland
{magnus.westerlund,andrey.shcherbakov}@arcada.fi

**Abstract.** This work in progress paper explores the integration of Large Language Models (LLMs) into educational support tools, particularly in studying basics of Natural Language Processing. It investigates how generative LLMs can align with Bloom's Taxonomy to support various learning levels, and how Prompt Engineering (PE) can improve communication with educational tools backed by LLMs. The approach involves planing integration LLMs into digital workbooks, employing techniques like Chain of Thoughts (CoT) and the Socratic method. These methods are utilized to provide tailored, interactive learning experiences that stimulate analytical thinking and deeper understanding of new concepts as well as providing instant response for coding exercises. Preliminary findings suggest that LLMs could significantly enhance the learning experience in coding education. The study suggests that PE combined with CoT and Socratic method could effectively addresses conversational problems with LLMs, enhancing learner engagement and understanding. This integration of LLMs facilitates a more dynamic form of learning, centered around dialogical exploration and practical experimentation, which aligns with the evolving objectives of lifelong learning. The ability to adapt and personalize knowledge acquisition to meet these ever-changing educational goals is a key advantage of this approach. The paper emphasizes the necessity for continuous research focused on the tailored integration of generative LLMs into educational tools. This involves employing PE techniques and leveraging a curated knowledge base to ensure that the LLMs' responses are both relevant and contextually appropriate for educational purposes.

**Keywords:** LLM · Educational Supportive Tool · Teaching Coding · Prompt Engineering

## 1 Introduction

Educators recognize the value of re-imagining supportive materials to enhance student aid tools. The emergence of assistive technology based on Large Language Models (LLMs) promises innovative pedagogical methodologies that

---

M. Westerlund and A. Shcherbakov — Both authors contributed equally to this work.

© The Author(s), under exclusive license to Springer Nature Switzerland AG 2024
M. E. Auer et al. (Eds.): STE 2024, LNNS 1027, pp. 77–85, 2024.
https://doi.org/10.1007/978-3-031-61891-8_7

engage the learner and facilitate personalized and experiential pedagogical approaches beyond the traditional classroom environment. This manuscript delineates the preliminary findings derived from experimentation and juxtaposes these insights with established didactic formalism. In [1] they state the proposition of deploying LLMs as an aim to enhance metacognitive aptitudes within educational domains. Integrating LLMs into educational support tools, such as digital workbooks, educators are empowered to architect bespoke learning experiences that cultivate metacognitive reflection and foster critical thinking. In alignment with Bloom's Taxonomy, LLMs exhibit proficiency at supporting the student within each of the levels. For example, at the "analyzing" tier by aiding in summarization, explication, and comparison of scholarly works while also extending support at the "creating" tier by enabling scenario simulation and experimentation through a code interpreter (i.e., running the code in a virtual sandbox environment). This amalgamation of artificial intelligence and pedagogical methods aspires to incessantly nurture the growth of critical thinking skills, thereby resonating with the higher-order cognitive levels as delineated in Bloom's Taxonomy, see Table 1.

**Table 1.** Bloom's Revised Taxonomy Levels and LLM Relevance [2]

| Level in Bloom's Taxonomy | Description | Example Questions | LLM Application |
|---|---|---|---|
| Remembering | Retrieving knowledge | What is the definition of...? | Data retrieval, fact-checking |
| Understanding | Explaining ideas or concepts | Explain how...? | Summarizing, paraphrasing, clarifying |
| Applying | Using information in new situations | How is...applied in...? | Problem-solving, real-world application scenarios |
| Analyzing | Breaking down information | What are the causes of...? | Analyzing, comparing, and contrasting |
| Evaluating | Justifying a decision or course of action | Why is...better than...? | Evaluating arguments, assessing scenarios |
| Creating | Generate code or architectural perspectives | What would happen if...? | Scenario simulation, predictive analytics |

This paper is structured as follows. Section 2 briefly introduces prior work in supportive tooling for coding studies. In Sect. 3, we discuss the considered design methods for defining prompt sequences and perspectives. Section 4 introduces the experimental setup, and in the final section, we conclude.

# 2  Prior Work

## 2.1  Prompt Engineering

Prompt Engineering (PE) is an emerging field crucial for effectively interacting with LLMs. It involves the practice of designing, refining, and implementing prompts to guide the output of these models. The task of PE often involves constructing meta-prompts that guide LLMs to perform automatic PE and refining [3,4]. This is a task that requires experimentation and discovering patterns in the LLM's response generation. The work involves developing a step-by-step reasoning template and context specification to enhance the relevance of LLMs output. Prompt patterns provide a structured and repeatable approach to interacting with LLMs. These patterns and techniques are applied to solve common problems in conversations with LLMs, like generating irrelevant, inaccurate, or contextually inappropriate responses.

Jules White et al. [5] provide a comprehensive overview of the techniques used in PE to address these issues. Their catalog presents PE techniques in a pattern form, specifically designed to tackle frequent problems in LLM conversations and improve the output. The paper categorizes prompt patterns into various types to address distinct aspects of LLM conversations. These include Input Semantics, which focus on clarifying the user's intent; Output Customization, Error Identification, Prompt Improvement, and Interaction. However, it's important to acknowledge the rapidly evolving nature of this field, making continual adaptation essential for staying current with advancements in PE.

Khanmigo[1], an educational tool, utilizes PE to craft tailored content, feedback, and tone for each learner, aligning with their skill level, interests, and preferences. This method integrates data from the Khan Academy library and individual learner profiles to personalize tutoring sessions. Khanmigo leverages this approach to meet students at their current level of understanding, offering immediate feedback and guiding them towards self-explanation. Moreover, it enriches the learning experience by infusing responses with personality, humor, and emojis, fostering engagement and motivation. This approach also includes offering encouragement and praise, recognizing learners' efforts and achievements. The use of PE in Khanmigo exemplifies how AI can be tailored to enhance educational experiences, ensuring that interactions are not just informative but also engaging and supportive of the individual learning journey.

## 2.2  Online Supportive Tools

Workbooks are online documents that contain interactive exercises, quizzes, videos and other multimedia elements that aim to enhance the learning experience and outcomes of online learners. Workbooks can be used to supplement online courses, provide self-paced learning opportunities, or offer personalized learning paths.

---

[1] https://blog.khanacademy.org/learner-khanmigo/.

Auto-Grading Systems (AGS) are online tools that automatically evaluate the learners' responses to code assignments, tests or quizzes; they can provide immediate and consistent feedback to learners, as well as save time and resources for instructors. AGS can also motivate learners to improve their performance and monitor their progress. AGS can be based on various methods, such as automated testing, multiple-choice questions, fill-in-the-blanks, short answer questions, or even natural language processing and machine learning techniques. The use of AGS for more complicated coding exercises is still problematic and needs teaching assistance's intervention [6].

### 2.3    Instant Verification and Feedback

Feedback has motivational aspects, and can produce feelings akin to rewards. A study involving seventh-grade math students in a randomized controlled design found that receiving immediate corrective feedback while doing homework might be more effective than delayed feedback. The effectiveness of immediate feedback, however, appears to be dependent on various factors, including the learner's prior knowledge and the context in which learning occurs [7].

A study involving novice high-school programmers [8] demonstrated that the Adaptive Immediate Feedback (AIF) system[2] that offers students real-time information on their progress, code correctness, and potential errors, alongside encouragement during programming, significantly increased students' intentions to persist in computer science. It was found that students using AIF were more engaged compared to those in a control group. Students found the AIF system enjoyable and helpful, reporting increased focus and engagement. This indicates that AIF can enhance learning in software engineering education.

## 3    Design Methods

### 3.1    Chain of Thought

Chain-of-Thought (CoT) reasoning has been proposed as a prompting protocol for improving the reasoning capabilities of LLM interaction [9]. CoT prompting is a hierarchically structured way of reasoning that is intended for the LLM to deduce what the human expects in return. Depending on the LLM in use, the CoT reasoning can have different success, and it is assumed that particularly smaller models may have a bigger improvement by using CoT.

### 3.2    Persona-Based Counterarguments

Persona-based design is a widely recognized method across various fields. Attempts have been made to systematize this approach [10], making it highly

---

[2] https://isnap.csc.ncsu.edu/home/public/project/aif/.

relevant and effective for integrating into the Socratic method of response generation by LLMs. Personas can be prompt-engineered to shape responses, adding a layer of adaptability to educational workbooks.

The Socratic method, which emphasizes guiding learners through pertinent questions that stimulate information-seeking, aligns well with the conversational capabilities of LLMs, which can be transformed into various personas, each undertaking specific roles such as Presenter, Prompt Editor, Fact Retriever, Critic, Coding Copilot, etc. Recent research highlights the benefits of employing such agents [11], and it would be beneficial to utilize this technique in generative LLM-based educational tools.

However, challenges exist, including limitations in effective prompt length, which may restrict the depth of persona definitions. With the context window of LLMs expanding, currently reaching up to 128k tokens for GPT-4-turbo (ca. 300 text pages), these challenges are expected to diminish over time. Additionally, the impact of shaping techniques on model responses, while prompting, requires more systematic experimentation and further research is necessary to fully understand and optimize the influence of persona-based shaping on the responses of LLM models.

## 4   Experimental Design

This section presents an innovative approach with aim to integrate LLMs into the educational process focusing on teaching elementary concepts in NLP to graduate students. It will present description of the use case, methodology of implementation, metrics and data collection, main concerns, and non-functional requirements.

### 4.1   Description of the Use Case

Traditional NLP methods remain relevant and useful, especially in areas where precision, efficiency, interpretability, and domain specificity are key. As NLP continues to evolve, a hybrid approach that capitalizes on the strengths of both traditional methods and Language Models (LLMs) can yield more robust solutions. Our primary objective is to utilize LLMs to enhance the learning experience for students studying traditional NLP. This involves imparting theoretical knowledge and translating it into practical skills. To achieve this, we emphasize interactive case studies and hands-on experimentation with NLP methods. By automating the Socratic method to tailor dialogical paths, we aim to foster analytical thinking and a deeper understanding of text processing. CoT reasoning serves as the method for amplifying the performance of both LLMs and students. Additionally, introducing students to the Persona-based counterargument approach alongside the Socratic method is expected to cultivate analytical skills while augmenting core factual knowledge.

## 4.2    Methodology of Implementation

The aim is to implement the best practices developed for educational didactics in higher education with an emphasis on meaningful experiences and student ownership of learning. The methodology for integrating LLMs into the curriculum involves several steps. Firstly, we aim to embed LLM support, for introducing elementary NLP concepts. This begins with identifying the modules where LLM integration can significantly enhance learning. By instructing the generative model to tailor content according to the student's existing knowledge base and to adapt to the complexity of content, we can configure the model to provide a choice between different levels of abstraction in the presentation. For example, in a module on sentiment analysis, we use LLMs to present different approaches to sentiment analysis of text data. For instance, LLMs could be used for initial explorations of a topic, followed by in-depth discussions led by the instructor: a flipped classroom approach where students interact with LLM-generated content as preparatory work.

Interactive tutorials can further be developed where LLMs guide students through NLP concepts using examples and exercises that students can follow along and replicate. We plan to implement problem-based learning activities, such as form validation or parsing structured text, which can be reliably handled by rule-based systems that follow explicit instructions. These activities will be adapted to student progress by analyzing the student-model conversation using an LLM for evaluation. An example of a few initial steps is provided in Fig. 1. A dialogical assessment framework will also be developed to create LLM-generated test challenges and compare student solutions with model-generated answers.

Customized learning pathways will be another focus. This includes implementing Socratic dialogues facilitated by LLMs to promote analytical thinking and a deeper understanding of lightweight NLP techniques (like regex, basic tokenization, or simpler models), which are practical for many engineering tasks. The learning content will be selected and stored in a knowledge repository. A repository of LLM-generated content, including case studies, example code, and theoretical explanations tailored to the curriculum, will be created. Interactive exercises will also be developed for students by the LLM, like having the model generate datasets for analysis or providing instant feedback on coding exercises.

## 4.3    Knowledge Base

To significantly improve the instructional capabilities of generative models in NLP, the following resources should be integrated, by fine-tuning, into the knowledge base: NLP academic textbooks and papers, ML and DL tutorials, case studies, and resources on AI and NLP ethics. Adding such resources can significantly enhance the generative model's ability to focus on facts, and guide students in understanding and applying NLP concepts and techniques effectively.

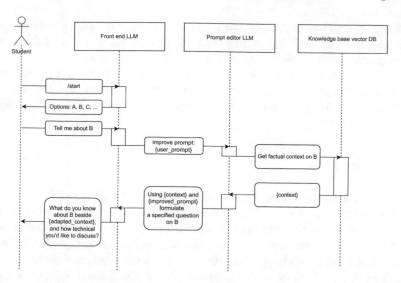

**Fig. 1.** An example of the first steps in dialogical, knowledge base backed student-model communication.

## 4.4   Evaluation Metrics and Data Collection Methods

The evaluation of this methodology will be based on several metrics. Learning outcomes will be assessed by means of automated quizzes produced by the generative model to measure the improvement in theoretical knowledge. Engagement levels will be monitored and logged for statistical analysis to gauge student interaction with LLM-generated content and activities. The effectiveness of problem-based learning activities will also be evaluated by measuring the time students spend on solving training challenges. Personalization effectiveness will be measured by registering the number of additional clarifying inquiries to the instructing LLM to see how well the LLM-driven environment adapts to individual learning needs.

Data collection methods will include pre and post-assessment surveys to gauge knowledge gain and student satisfaction. Usage data analysis will track interaction with the instructional environment. Performance tracking will monitor progress on problem-solving activities and assessments. Feedback will be collected through semi-structured interviews to gather qualitative feedback from students on their learning experiences.

## 4.5   Main Concerns

There are several main concerns that need to be addressed in this design. One is the potential overemphasis on technology, which might come at the expense of pedagogical principles. The balance between technology and traditional teaching methods needs to be carefully managed.

More research about risks linked to ethical considerations, concerning equity, social and physical well-being and communication skills should be addressed, including over-reliance on LLMs and data privacy concerns.

While the design mentions encouraging analytical thinking, it doesn't clearly outline how student autonomy will be fostered in an environment potentially dominated by automated responses and content and what this might bring in must be discussed.

# 5   Conclusions

Integration of LLMs in educational workbooks, utilizing Chain of Thought (CoT) and the Socratic method, offers transformative pedagogical benefits. These methodologies steer the PE process, ensuring that learning is rooted in dialogical exploration and practical experimentation, which are crucial for lifelong learning and critical thinking. Educational supportive tools backed by LLMs not only adapt to knowledge acquisition and evolving educational objectives but also enrich it with learner engagement and may support a deeper understanding of novel concepts and analytical thinking. The effectiveness of the LLMs in enhancing learning is assessed through various means: automated quizzes, monitoring of engagement levels, and analysis of time spent on training challenges. This evaluation extends to gauging the effectiveness of personalization.

# References

1. Khan, U.A., Alamäki, A.: Harnessing AI to boost metacognitive learning in education (2023). http://urn.fi/URN:NBN:fi-fe20230825108259. ESignals Pro
2. Anderson, L.W., Krathwohl, D.R.: A Taxonomy for Learning, Teaching, and Assessing: A Revision of Bloom's Taxonomy of Educational Objectives, Complete. Addison Wesley Longman Inc. (2001)
3. Ye, Q., Axmed, M., Pryzant, R., Khani, F.: Prompt engineering a prompt engineer (2023). https://doi.org/10.48550/arXiv.2311.05661
4. Zhou, Y., et al.: Large language models are human-level prompt engineers (2023). https://doi.org/10.48550/arXiv.2211.01910
5. White, J., et al.: A prompt pattern catalog to enhance prompt engineering with ChatGPT (2023). https://doi.org/10.48550/arXiv.2302.11382
6. Sinambela, F.K., Simaremare, M.E.S., Sidabutar, E.F.D., Gultom, R.P.: An integrated automatic-grading and quality measure for assessing programming assignment. JATISI 8(3), 1508–1514 (2021). https://doi.org/10.35957/jatisi.v8i3.1271
7. Tricomi, E., DePasque, S.: The role of feedback in learning and motivation. Adv. Motiv. Achiev. 19, 175–202 (2016). https://doi.org/10.1108/S0749-742320160000019015
8. Marwan, S., Akram, B., Barnes, T., Price, T.W.: Adaptive immediate feedback for block-based programming: design and evaluation. IEEE Trans. Learn. Technol. 15(3), 406–420 (2022). https://doi.org/10.1109/TLT.2022.3180984
9. Wei, J., et al.: Chain-of-thought prompting elicits reasoning in large language models (2022). https://doi.org/10.48550/arXiv.2201.11903

10. Salminen, J., Guan, K.W., Jung, S.G., Jansen, B.: Use cases for design personas: a systematic review and new frontiers. CHI 2022, New Orleans, USA (2022). https://doi.org/10.1145/3491102.3517589
11. Qiao, B., et al.: Taskweaver: a code-first agent framework (2023). https://doi.org/10.48550/arXiv.2311.17541

# A Research-Led Contribution of Engineering Education for a Sustainable Future

Brit-Maren Block$^{(\boxtimes)}$ (iD) and Marie Gillian Guerne

Leuphana University, Universitätsallee 1, 21335 Lüneburg, Germany
brit-maren.block@leuphana.de

**Abstract.** In order to optimally support engineering students and prepare them for future challenges, the integration of education for sustainable development (ESD) with a linking of technical and sustainability-oriented topics within the curriculum is essential. This paper contributes in two ways. On the one hand, a systematic literature analysis will focus on existing modules integrate sustainability aspects into engineering education and revealed framework conditions for successful implementation. Framework conditions such as the learning objectives of a course, the possible main topics, the didactic methods used, the challenges and obstacles to implementation and examples of good practice were identified. On the other hand, determined aspects were used developing an exemplary teaching-learning concept and putting it into practice. The proposed concept consists of a tripartite module structure, including a content level, a methodological-didactic concept and an individual process level for anchoring sustainability aspects in the teaching of engineering and strengthening student awareness.

**Keywords:** Sustainable Learning · Engineering Education Research · Engineering Education of the Future

## 1 Introduction

Engineering plays an important role in the development of innovations [1]. In 2008, the National Academy of Engineering identified the major challenges for engineering in the 21st century. The 14 challenges include improving global health, expanding renewable energy, access to clean water or eliminating carbon emissions [1]. Future engineers must develop solutions that enable technology to be designed in a way that is economically, environmentally and socially sustainable. At the same time, the solutions should help meet the basic needs of all people while leaving existing supplies untouched [2]. This illustrates the increasingly important influence that engineering will have on shaping the future. The education of students must be adapted to this future role. This includes, for example, the teaching of competencies in the areas of communication skills, entrepreneurial skills, teamwork, creativity, lifelong learning and problem-solving skills [3, 4]. In order to optimally support engineering students and prepare them for future challenges, the integration of education for sustainable development (ESD) with a linking of technical and sustainability-oriented topics within the curriculum is essential.

M. E. Auer et al. (Eds.): STE 2024, LNNS 1027, pp. 86–97, 2024.
https://doi.org/10.1007/978-3-031-61891-8_8

This paper contributes in two ways to this discourse. On the one hand, a systematic literature analysis will show which concepts exist that attempt to integrate sustainability aspects into engineering education. On the other hand, a contribution is made to teaching practice by developing an exemplary teaching-learning concept and putting it into practice. The elective module was implemented in the Master's degree program "Management & Engineering" in the winter semester 2022/2023 at Leuphana University. The learning objectives, the concept and initial student feedback are presented in this paper.

## 2 Research Questions

This paper is based on two central research questions, that in the sense of design-based work, will contribute both to the research landscape of engineering sciences and to teaching practice [5]:

(1) What existing concepts and approaches exist for integrating sustainability into engineering education?
(2) What does a comprehensive concept for the integration of sustainability topics in engineering look like? How can this be implemented and what framework conditions must be considered during implementation?

For research question (1), an introduction and overview of the integration of sustainability in engineering is given. A systematic literature review looks at concepts that currently already integrate sustainability topics in engineering curricula. It is expected that sustainability has so far found little integration within the engineering curriculum. Furthermore, it is assumed that the economic dimension of sustainability is addressed more than the ecological and social dimension.

To answer the research question (2), the identified framework conditions will be addressed. The aim is to develop an interdisciplinary concept that enables the integration of sustainability topics into the engineering sciences in order to promote sustainable development. To this end, existing concepts will be reviewed and implemented.

## 3 Methodological Approach

This chapter deals with the methodological approach. The first subchapter describes the procedure for the systematic literature review of existing concepts for integrating sustainability into the engineering sciences. The development of a concept for pedagogical practice is based on the DBR approach and is described in Sect. 3.2.

### 3.1 Design-Based Research Approach

Design-based research (DBR) is a methodological approach that aims to generate research-based insights into both the theoretical understanding of teaching and learning and pedagogical practice [5, 6]. The approach consists of several stages that typically extend over several cycles. For this paper, the generic model according to McKenney and Reeves was chosen, which aims at the systematic and simultaneous generation of new knowledge and solution development for practical and complex educational problems within it and was therefore selected as a compatible model for the planned research [7]. The model is shown in Fig. 1 and illustrates the three stages of the process model for DBR: analysis, design and evaluation. The three stages are interdependent and require the implementation of the respective solution approaches. Results are delivered in two ways, both for theoretical understanding and for intervention in teaching practice. Feedback loops enable continuous improvement and adaptation [7].

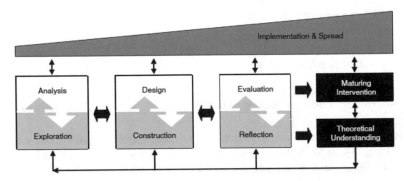

**Fig. 1.** Model for the implementation of a DBR approach, illustration based on [5]

In the **Analysis phase**, the problem is analysed and defined [7, 8]. In the case of this paper, the systematic literature analysis reveales that sustainability aspects are insufficiently addressed in engineering teaching. The aim is to develop a solution concept that is strongly focussed on the future needs of students. In the **Design phase** of the DBR approach, a solution is designed. A concept for a pedagogical intervention or learning environment is created and must be **implemented** and tested in a real environment [8, 9]. The **Evaluation phase** involves evaluating whether the developed solution proves to be effective. In this study students were regularly asked for feedback and a quantitative before-and-after survey was conducted. If necessary, the teaching concept can and must be revised in real time [7, 10]. Finally, the experiences of the teaching experiment are documented and reflected upon. Successful measures and improvement measures are identified and the solution concept is revised [8]. The research cycle, in which teaching materials are designed, tested, evaluated, reflected on and optimised, is repeated in order to arrive at a satisfactory solution to the problem defined at the beginning in an iterative process. The process ends when a solution to the initial problem has been found and a differentiated, theoretical understanding has been gained [11].

### 3.2 Systematic Literature Review

The aim of a systematic search is to analyse all empirical data that meet the specified inclusion criteria in order to answer the corresponding research question [12]. The process in this paper is based on the Preferred Reporting Items for Systematic reviews and Meta-Analyses (PRISMA) guidelines [12].

The first step is to describe the background and objectives of the systematic analysis [12]. The study is intended to answer both research questions. For research question (1), an overview of the current academic literature on the topic of sustainability education in the research landscape of engineering studies is to be compiled. As a basis for answering research question (2), concepts will be identified that are used in the engineering curriculum to teach students competences in the field of sustainability.

A protocol is drawn up in which the analysis is documented. Search terms and suitable databases are determined and the definition of inclusion and exclusion criteria is carried out [12]. The inclusion and exclusion criteria is documented in Table 1.

**Table 1.** Inclusion and exclusion criteria, table based on [12]

|  | Inclusion criteria | Exclusion criteria |
|---|---|---|
| Language | English, German | Other languages |
| Document type | Articles from academic journals, conference reports | Books, book chapters, corrections, editorials, introductions, forewords, letters, opinions, dissertations |
| Time period | 1995 until today | Publications before 1995 |
| Availability | Full text must be accessible | Access only to the title /abstract |
| Subject area | The title/abstract must indicate that the article is about teaching and learning concepts that integrate sustainability into engineering education. | It is not clear from the title/abstract that the article is about teaching-learning concepts that attempt to integrate sustainability into engineering education. |

Search terms are defined to match the research question, making it possible to find articles on the inclusion criteria defined above. The search terms that are considered useful for the underlying research question are "engineering education", "sustainab*" (for "sustainability" and "sustainable"), "concept" and "climate change". The keywords were linked with the Boolean operator "AND" and "OR", which means that only articles in which the keywords come together are displayed in the results [13]. The database Web of Science was used for this systematic literature search. Table 2 describes the search strategy including the results.

To facilitate comprehension, the selection of included sources is shown in a flow chart according to the PRISMA method in Fig. 2. The 211 results of the final search strategy and seven relevant articles from previous searches are indicated at the top. Two duplications were removed. Subsequently, various screenings were carried out, which checked the inclusion criteria in the title, abstract and finally in the full text. Reasons for exclusion are documented. In the present process, 75 articles were excluded on the basis of the title, a further 115 after reviewing the abstract and finally 15 articles were excluded when the full texts were checked. The most common reasons for excluding articles are that the article did not fit the topic. For example, excluded articles do not present a teaching concept, but analyse the need to integrate sustainability. In the area of the target level, articles are designed for other target groups, such as children or pupils.

**Table 2.** Search history of the systematic literature search (Searched until 05.05.2023, 11:45 a.m.), table based on [12]

| ID | Search | Results |
|---|---|---|
| #1 | "Engineering Education" | 18.296 |
| #2 | Sustainability | 300.550 |
| #3 | Sustainable | 622.186 |
| #4 | Sustainab* | 862.382 |
| #5 | Concept | 1.042.956 |
| #6 | "Climate Change" | 380.293 |
| #7 | #1 AND #5 AND (#4 OR #6) | 211 |

#7 exported for screening; search until 05.05.2023; https://www.webofscience.com/wos/woscc/citation-report/0c34298b-702e-45db-9438-9a4efa6df2cd-878498e2

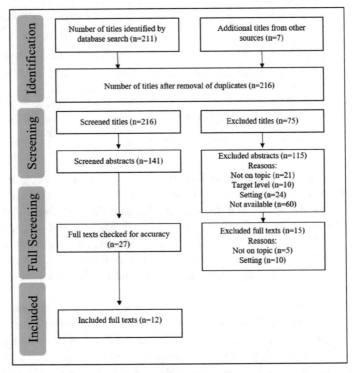

**Fig. 2.** Research process using the PRISMA method, illustration based on [12]

There were also articles whose settings are too specific for replication to be possible. For example, specific concepts are presented for regions in Africa or Saudi Arabia. A total of twelve articles fulfil the inclusion criteria. The content of selected studies is analysed in the following chapter.

## 4 Results of the Analysis of Concepts for Sustainable Technology Education

The aim of this analysis is to review existing concepts that integrate sustainability into the teaching of engineering in order to derive ideas and experiences for an elective module at Leuphana University. As described in Sect. 3.2, after a screening of the initial 218 articles, ultimately twelve sources are included in the literature analysis. The most important findings of the twelve articles are summarised in six inductively formed categories.

These five categories are:

(1) the learning objectives of a course,
(2) the possible main topics,
(3) didactic methods that are used,
(4) challenges and obstacles to implementation,
(5) best practice examples.

The **learning objectives of the courses** are defined very differently and range from general to very specific. They include, for example, the description, discussion and evaluation of key concepts of sustainability, sustainable development and sustainable engineering, the application of sustainable engineering principles to the design of engineering systems, and the evaluation of the sustainability of technologies. Working effectively as a member of a planning team and communicating effectively through reports and presentations are also part of the learning objectives [14, 15]. Other important aspects include the development of a critical and holistic perspective, problem-solving and decision-making issues involving sustainability criteria, ethics and social responsibility, interdisciplinary projects between different areas and courses, and the involvement of other social stakeholders in higher education [16, 17].

The analysis focussed on a variety of **main topics** such as environmental policy, ethics, law, management, industrial ecology, greener production, resource efficiency, life cycle assessment, eco-design, environmental impact assessment, sustainable urban drainage management and sewage sludge management [14, 15, 18, 19]. Table 3 summarises suggested topics for a basic course.

**Didactic methods** Used in the identified articles are often active forms of learning. One of these methods is problem-based learning (PBL). This is a pedagogical methodology based on students learning while tackling or trying to solve real life problems in groups and under supervision [20]. PBL is an ideal opportunity for students to gain experience in the areas of problem solving, project management, interpersonal skills in the form of written and oral communication, productive teamwork, work-integrated education and leadership skills, which are also used in Chau's (2007) concept [18]. Baier (2013) focuses on study-centred learning, which also appears in the approach of Rodríguez-Chueca et al. (2019) [21, 22]. Rodríguez-Chueca et al. (2019) increases learning success

**Table 3.** Main topics for a basic course on the integration of sustainability

| Derived Main Topics | |
|---|---|
| Historical Background and Overview | - Definitions of sustainability<br>- Challenges of international sustainable development<br>- Evidence of global climate change<br>- Limits to growth |
| Sustainability Concepts | - 3 pillar model<br>- Strong/Weak sustainability<br>- Ecological footprint<br>- Circular economy |
| Instruments for environmental assessment | - Life cycle analysis<br>- Eco-balancing<br>- Environmental impact assessment<br>- Sustainability assessment |
| Sustainable technologies | - Solar, wind or hydropower<br>- Electric vehicles and carsharing systems<br>- Precision farming/organic farming |

through a flipped classroom approach by promoting autonomous learning among students [22]. In addition, skills such as critical thinking, creativity and collaborative work are strengthened. Huang et al. (2020) shows that a design-based constructivist approach is suitable is well suited for teaching sustainability skills [16]. Students are encouraged to solve real world challenges and reflect on the learning process through the application of design activities. In general, it can be said that methods should be chosen that are practice-orientated, actively involve the students and allow them to adopt different perspectives. They should promote critical thinking, for example through plenty of room for discussion, and the autonomy of the students.

When implementing a module, **challenges and obstacles** can be identified both on the student side and from the lecturers' perspective. From the lecturers' perspective, obstacles that have frequently emerged in the literature are the lack of flexibility of the course content, the large amount of time required for restructuring, a lack of materials and interdisciplinarity [23, 24]. There are concerns that content would have to be cut elsewhere to integrate corresponding topics, which would mean the loss of other essential content. would mean the loss of other essential content. However, the most frequently mentioned aspect is the resource or knowledge of the teacher themselves [16, 18, 24]. Challenges were also identified on the student side. For example, Huang et al. (2020) consider that active forms of learning may be less suitable for more passive students [16]. Rodríguez-Chueca et al. (2019) fears differences in students' basic knowledge at the beginning of the course and little engagement with the knowledge imparted [22]. This can lead to difficulties when working in a group later on. Specifically in the area of interest, identified problems include a lack of motivation in class, low participation in debates or discussions and irregular attendance.

Finally, **best practices** for the integration of sustainability into technical degree programmes will be summarised. Lu (2015) identified five recommendations for successful sustainability integration [25]. These include,

- incorporating student feedback to improve and adapt courses,
- tie into current events to discuss the history of the environmental movement and the development of environmental science/engineering practices,
- use relevant examples to better understand sustainability concepts,
- integrating research into the classroom to make lectures more interesting, and
- inviting external speakers or experts.

In addition, interdisciplinary projects have been identified as good methods for conveying a comprehensive understanding of sustainability [15, 24]. Active learning methods, such as project work, case studies and group work, can also improve the understanding of sustainability in technical degree programs [23]. Furthermore, field experience can also lead to great learning progress and increase motivation [24]. The Blue Engineering concept can be emphasised as a constructivist approach to sustainable engineering education, which attempts to bridge the gap between sustainability and engineering [2, 21].

## 5 Development of a Concept for Sustainable Technology Education

Based on the results of Sect. 4, a concept for sustainable technology education was developed. A Master's elective module was designed and implemented in the winter semester 2022/2023. The learning objectives and the concept developed are explained below. The Blue Engineering module, was chosen as the basis for the elective module [2].

### 5.1 Learning Objectives

The defined learning objectives for the planned module were categorised at the levels of knowledge, understanding and application and are displayed in Table 4.

### 5.2 Development of a Concept for Sustainable Technology Education

In order to anchor sustainability aspects in the teaching of engineering and to raise students' awareness, particularly in the area of social and ecological sustainability, this paper proposes the concept of a tripartite module structure, which attempts to which attempts to combine and integrate different didactic approaches.

The developed concept for sustainable competences in engineering is shown in Fig 3. The tripartite structure at the centre consists of (1) **a content level**, (2) **a methodological-didactic concept** and (3) an **individual process level** and is based on the concept of the didactic triangle [26]. However, the elements of teachers, learners and content were modified for the present concept by integrating constructivist learning approaches. The

**Table 4.** Derived learning objectives for the new concept

| Learning Objectives | |
| --- | --- |
| Knowledge | - Imparting knowledge about the connections between social, economic, ecological, local and/or global problems.<br>- Transfer of the key aspects of sustainability in engineering. |
| Understanding | - Understand and evaluate the interrelationship between engineering, the individual, nature and society through interdisciplinary learning.<br>- Understand societal and ethical obligations of engineers to recognise potential challenges, risks and impacts.<br>- Understand the natural and social systems in our society and the problems caused by humans and technology in these systems.<br>- Develop personal perspectives through holistic and critical thinking. |
| Application | - Contribute to the design of the seminar through self-directed learning.<br>- Promote responsible, forward-looking action.<br>- Promote social interaction and collaboration to develop shared perspectives and alternative courses of action.<br>- Promoting creativity and innovation.<br>- Promoting future and vision orientation.<br>- Acquisition of the twelve design competencies of ESD in the areas of technical and methodological competence, social competence and self-competence. |

constructivist learning approach changes the role of the teacher, who instead acts as a learning facilitator in order to promote individual learning processes.

At the **content level**, the technical basics in the field of sustainability science and didactics are taught first. Open access building blocks [2] are included in the module. By participating in the first thematic teaching-learning modules, students should begin to understand the interrelationships. Through the selection and implementation of an existing blue engineering module and the subsequent in-depth study of a topic of their own choice, participants should acquire a technical foundation. A variety of learning methods should be used to ensure a successful **methodological and didactic concept**. Personal perspectives and points of view should be developed and enable participants to take action. The acquisition of technical and methodological competence, social competence and self-competence should take place on an **individual level**.

The accompanying survey showed that the students are very satisfied with the course and their increased knowledge. The module motivates them to engage with and reflect upon the topic.

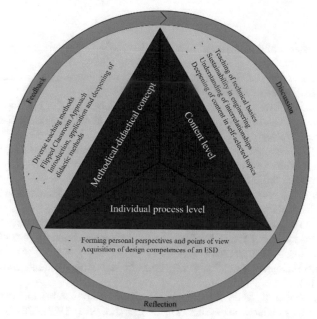

**Fig. 3.** Concept for sustainable competences in engineering: A tripartite module structure, own illustration

## 6 Summary

In order to answer the research questions of what comprehensive concepts for the integration of sustainability topics into engineering could look like, a systematic literature review was used to identify framework conditions for successful implementation. Based on those theoretical findings a corresponding concept was developed and implemented at Leuphana University as an elective module on sustainable engineering in educational practice. A tripartite module structure is proposed for anchoring sustainability aspects in the teaching of engineering and strengthening student awareness. The use of active learning methods, such as student-centered learning, problem- and project-based learning or the flipped classroom method are recommended for comprehensive concepts. The course received positive feedback from the participants and enabled an increase in knowledge.

For a more comprehensive evaluation to identify further concepts, other databases such as Scopus, ResearchGate or IEEE Xplore Digital Library must be included in a literature analysis in addition to the Web of Science database. Another limitation is that the elective module is only carried out once. It is necessary to repeat and adapt the module and to evaluate the skills developed by the students. Effectiveness analyses must be used to evaluate the skills developed.

# References

1. National Academy of Engineering: Grand Challenges for Engineering: http://www.engineeringchallenges.org/challenges.aspx. Last accessed 1 Jul 2023
2. TU Berlin: Bausteine als Lehr-/Lerneinheiten, http://www.blue-engineering.org/wiki/Baukasten:Startseite. Last accessed 3 May 2023
3. Brunhaver, S.R., Korte, R.F., Barley, S.R., Sheppard, S.D.: Bridging the Gaps between Engineering Education and Practice, U.S. Engineering in a Global Economy. The University of Chicago Press, Chicago and London (2018)
4. Alcorta de Bronstein, A., Lampe, S., Halberstadt, J.: Fostering future engineers as transformational agents: integrating sustainability and entrepreneurship in engineering education. Procedia Comput. Sci. **219**, 957-962 (2023)
5. McKenney, S.E., Reeves, T.C.: Systematic review of design-based research progress: is a little knowledge a dangerous thing? Educ. Res. **42**(2), 97–100 (2013)
6. Kelly, A.E.: Design-based research in engineering education. In: Johri, A., Olds, B.M. (eds.) Cambridge Handbook of Engineering Education Research, pp. 497–518. Cambridge University Press, New York (2013)
7. McKenney, S.E., Reeves, T.C.: Conducting Educational Design Research, 2nd edn. Routledge, Milton (2018)
8. Scott, E.E., Wenderoth, M.P., Doherty, J.H.: Design-based research: a methodology to extend and enrich biology education research. CBE—Life Sci. Educ. 19(3), (2020).
9. Reiser, B.J., Smith, B.K., Tabak, I., Steinmuller, F., Sandoval, W.A., Leone, A.J.: BGuILE: strategic and conceptual scaffolds for scientific inquiry in biology classrooms. In: Carver, S.M., Klahr, D. (eds.) Cognition and instruction: Twenty-Five Years of Progress, pp. 263–305. Lawrence Erlbaum Associates, Inc., Mahwah, NJ (2001)
10. Collins, C.J., Hanges, P.J., Locke, E.A.: The relationship of achievement motivation to entrepreneurial behavior: a meta-analysis. Hum. Perform. **17**(1), 95–117 (2004)
11. Fishman, B.J., Penuel, W.R., Allen, A.-R., Cheng, B.H., Sabelli, N.: Design-based implementation research: an emerging model for transforming the relationship of research and practice. Yearb. National Soc. Study Educ. **112**, 136–156 (2013)
12. Moher, D., Liberati, A., Tetzlaff, J., Altman, D.G.: PRISMA Group: preferred reporting items for systematic reviews and meta-analyses: the PRISMA statement. Ann. Intern. Med. **151**(4), 264–269 (2009)
13. Witteriede, H.: Serie Wissenschaftliches Arbeiten – Teil 3 Systematische Literaturrecherche. www.katho-nrw.de/h.witteriede/swa. Last accessed 1 May 2023
14. Aurandt, J.L., Butler, E.C.: Sustainability education: approaches for incorporating sustainability into the undergraduate curriculum. J. Profess. Issues Eng. Educ. Pract. **137**(2), 102–106 (2011)
15. Bhandari, A., Ong, S.K., Steward, B.L.: Student learning in a multidisciplinary sustainable engineering course. J. Profess. Issues Eng. Educ. Pract. **137**(2), 86–93 (2011)
16. Huang, Z., Peng, A., Yang, T., Deng, S., He, Y.: A design-based learning approach for fostering sustainability competency in engineering education. Sustainability **12**(7), 2958 (2020)
17. Sánchez-Carracedo, F., et al.: Tools for embedding and assessing sustainable development goals in engineering education. Sustainability **13**, 12154 (2021)
18. Chau, K.W.: Incorporation of sustainability concepts into a civil engineering curriculum. J. Profess. Issues Eng. Educ. Pract. **133**(3), 188–191 (2007)
19. Bielefeldt, A.R.: Incorporating a sustainability module into first-year courses for civil and environmental engineering students. J. Profess. Issues Eng. Educ. Pract. **137**(2), 78–85 (2011)
20. Kolmos, A., Fink, F.K., Krogh, L.: The Aalborg PBL Model: Progress, Diversity and Challenges. Aalborg University Press, Aalborg (2004)

21. Baier, A.: Student-driven courses on the social and ecological responsibilities of engineers. Sci. Eng. Ethics **19**(4), 1469–1472 (2013)
22. Rodríguez-Chueca, J., Molina-García, A., García-Aranda, C., Pérez, J., Rodríguez, E.: Understanding sustainability and the circular economy through flipped classroom and challenge-based learning: an innovative experience in engineering education in Spain. Env. Educ. Res. 1–15 (2019).
23. Børsen, T., et al.: Initiatives, experiences and best practices for teaching social and ecological responsibility in ethics education for science and engineering students. Eur. J. Eng. Educ. **46**(4), 1–24 (2020)
24. Ngo, T.T., Chase, B.: Students' attitude toward sustainability and humanitarian engineering education using project-based and international field learning pedagogies. Int. J. Sustain. High. Educ. **22**(2), 254–273 (2020)
25. Lu, M.: Integrating sustainability into the introduction of environmental engineering. J. Prof. Issues Eng. Educ. Pract. **141**(2) (2015)
26. Einecke, G.: Was ist Fachdidaktik? http://www.fachdidaktikeinecke.de/1_Unterrichtsplanung/was_ist_fachdidaktik.htm. Last accessed 11 Nov 2022

# IoT Laboratories Based on Reconfigurable Platform with Analog Coprocessor

Elena-Cătălina Gherman-Dolhăscu[1]([✉]), Horia Modran[1], Cornel Samoilă[1,3], and Doru Ursuțiu[1,2]

[1] CVTC "Transilvania" University of Brașov, Brasov, Romania
{elena.dolhascu,horia.modran,csam,ursutiu}@unitbv.ro
[2] AOSR – Academy of Romanian Scientists, Bucharest, Romania
[3] ASTR – Technical Sciences Academy of Romania, Bucharest, Romania

**Abstract.** New challenges and opportunities have arisen in the field of engineering education and research with the advent of Internet of Things (IoT). Due to the fast technological progress in the use of IoT, a need has been identified for engineering students to learn IoT concepts easily, with focus on hands-on activity [1]. Using a reconfigurable platform in studying IoT concepts offers advantages regarding scalability, adaptability, costs, small size, power consumption and above all, having the possibility to host various applications. The motivation behind this paper emerged from the increasing importance of IoT in various sectors and the need to bridge the gap between theoretical knowledge and practical application. The main goal of this paper is to present a comprehensive perspective of IoT laboratories using reconfigurable platform with PSoC Analog Coprocessor, highlighting their design and implementation and their potential in enhancing the hands-on learning experiences for students. The reconfigurability nature of the platform is explored, highlighting its customizability to meet the educational needs. Illustrating the implementation IoT laboratories based on reconfigurable platform with Analog Coprocessor in engineering education, showcasing the enhanced learning outcomes and engaging experience by students and improvement of their understanding of IoT concepts.

**Keywords:** IoT · Reconfigurable · PSoC · Analog Coprocessor

## 1 Introduction

The emergence of the Internet of Things (IoT) has introduced fresh challenges and prospects in the realm of engineering education and research. The integration of IoT in education enhances the teaching and learning procedures, leading to improved student outcomes that align more closely with current industrial requirements. This holds significant importance, especially in the field of engineering education. The rapid technological advancements in IoT underscore the necessity for engineering students to grasp its concepts efficiently, emphasizing practical engagement [2].

Many educators, including both teachers and lecturers, have dedicated time and effort to curate specialized training materials designed specifically for IoT learning.

M. E. Auer et al. (Eds.): STE 2024, LNNS 1027, pp. 98–105, 2024.
https://doi.org/10.1007/978-3-031-61891-8_9

These tailored resources aim to enhance the understanding and skills of students in the realm of IoT education [3].

The significance of hands-on experience, as well as the fostering of students' creativity and initiative, is underscored anew through the introduction of the concepts of Pocket Lab and IoT in the paper [4].

As we already noted in the paper [5], the students were delighted with the PSoC Analog Coprocessor Pioneer Kit from Infineon, formerly Cypress, both for its ability to easily implement sensor interfaces and because it has "everything they need" at a convenient price, as well as for the fact that it allowed them an invaluable full hands-on experience.

Leveraging a reconfigurable platform for studying IoT not only provides benefits in terms of scalability, adaptability, cost-effectiveness, compact size, and low power consumption but also affords the flexibility to host a diverse range of applications. The impetus for undertaking this paper stems from the escalating significance of the IoT across diverse sectors and the imperative to narrow the divide between theoretical understanding and practical implementation. In response to this burgeoning need, the primary objective of this paper is to offer a thorough insight into IoT laboratories that leverage reconfigurable platforms featuring the PSoC Analog Coprocessor.

Delving into the intricacies of IoT laboratories, this investigation places a spotlight on the fundamental role assumed by the PSoC Analog Coprocessor in facilitating seamless interaction with a diversity of sensors. One key aspect under scrutiny is the inherent reconfigurability of the platform, revealing its capacity to adapt and conform to specific educational requirements.

This discourse delves into the implementation of IoT laboratories within the realm of engineering education, specifically focusing on the utilization of reconfigurable platforms featuring Analog Coprocessors.

The article explores the concept of a mini laboratory designed for students, utilizing the PSoC Analog Coprocessor Pioneer Kit from Infineon Technologies as a foundational element. The evolution of the laboratory experience commences with the utilization of sensors on the kit board, progresses to the incorporation of additional sensors beyond the kit, and ultimately integrates with the Laboratory Virtual Instrument Engineering Workbench (LabVIEW) software component. This incremental approach allows students to gain a comprehensive understanding of IoT concepts in a gradual and accessible manner.

## 2 Exploring PSoC Analog Coprocessor Pioneer Kit and PSoC Creator for IoT Projects

### 2.1 PSoC Analog Coprocessor Pioneer Kit and PSoC Creator Overview

PSoC Analog Coprocessor Pioneer Board (see Fig. 1) is a development platform offered by Infineon Technologies designed to facilitate the exploration and development of applications leveraging the PSoC Analog Coprocessor. The kit includes the necessary hardware and software components to kickstart projects in the analog and sensor processing domain.

**Fig. 1.** PSoC Analog Corprocessor Pioneer board

The PSoC Analog Coprocessor Pioneer board comprises essential components, including the PSoC Analog Coprocessor, the PSoC 5LP for programming and debugging, and five distinct sensors: PIR Motion Sensor, Thermistor, Humidity Sensor, Ambient Light Sensor, and Proximity Inductive Sensor [6]. This kit facilitates the straightforward integration of sensor interfaces, offering students a flexible and comprehensible system for hands-on learning and experimentation.

PSoC Creator, designed by Infineon Technology, stands as an advanced Integrated Development Environment (IDE) meticulously designed for the creation of applications using PSoC devices. Presenting a user-friendly drag-and-drop interface, PSoC Creator empowers students to design their systems graphically with intuitive ease. This graphical environment not only simplifies the integration of PSoC components but also expedites the entire development process. A rich catalog of pre-built components simplifies the inclusion of functionality, such as analog and digital peripherals, communication interfaces, and more. Eliminating the need for extensive manual coding, PSoC Creator dynamically generates low-level code based on the graphical schematic [6]. This approach allows students to concentrate on system-level design and functionality, as the IDE adeptly handles the generation of optimized code. Further enhancing the development experience, the IDE is equipped with robust debugging tools. These tools, encompassing real-time monitoring and interactive debugging features, prove invaluable in aiding students to efficiently identify and resolve issues during the development process.

### 2.2  IoT Projects with PSoC Analog Coprocessor Pioneer Kit and PSoC Creator

The programming process with PSoC Creator unfolds through a series of easily navigable and straightforward steps. These steps are designed to provide a user-friendly experience, ensuring that both novice and experienced programmers can effortlessly engage with the platform. From initiating the development environment to configuring components, the workflow is intuitively structured. This approach not only simplifies the programming journey but also promotes efficiency in creating applications for PSoC devices.

Also, the PSoC Analog Coprocessor Pioneer Kit simplifies the learning process by providing dedicated examples for all five sensors included. This feature enhances the user's comprehension of the kit's functionality, offering practical insights into the operation of each sensor. The examples provided by [7] offer a thorough breakdown of the project, encompassing key elements such as schematics, pins configuration, and the generated code (see Fig. 2).

This comprehensive information serves as a valuable resource, facilitating a deeper understanding and successful implementation of the projects by providing insights into both hardware and software aspects.

| | Name | Port | Pin | Lock |
|---|---|---|---|---|
| ☐ | \EZI2C:scl\ | P4[0] ⌄ | 37 ⌄ | ☑ |
| ☐ | \EZI2C:sda\ | P4[1] ⌄ | 38 ⌄ | ☑ |
| ☐ | Pin_LED | P1[4] ⌄ | 4 ⌄ | ☑ |
| ☐ | TIA_IN | P2[4] ⌄ | 14 ⌄ | ☑ |
| ☐ | TIA_OUT | P2[3] ⌄ | 13 ⌄ | ☑ |

**Fig. 2.** Ambient Light Sensing pins configuration

This flexibility allows developers to not only modify existing code but also seamlessly integrate new components into the schematics, offering a dynamic and customizable approach to PSoC device programming.

The extensibility of the language, coupled with the ability to incorporate additional components, provides a versatile environment for crafting tailored solutions within PSoC Creator.

## 3  Augmenting Sensing Component of PSoC Analog Coprocessor Pioneer Kit and Further Considerations

### 3.1  PSoC Analog Coprocessor Pioneer Kit and Other Sensors

Advancing to the subsequent phase of the project, we introduced an external module to the kit (KY-024), specifically incorporating the magnetic sensor module, commonly known as the Hall module. (see Fig. 3) This strategic addition extends the sensor capabilities of the kit, opening new possibilities for data collection and enhancing the overall functionality of the project.

**Fig. 3.** PSoC Analog Coprocessor Pioneer Board and Hall Module

The module serves the purpose of sensing the magnetic field and responding when it is detected. It incorporates a potentiometer allowing the adjustment of sensor sensitivity. Additionally, the module offers two outputs – one analog and one digital.

The sensor operates by detecting the presence of a magnetic field and subsequently showcasing its analog value. This functionality allows users to glean real-time information about the intensity or strength of the detected magnetic field, providing a straightforward and tangible output for further analysis or application in the system.

In emphasizing the adaptability of the Analog Coprocessor platform, we created two illustrative applications using the Hall module and the CY8CKIT-048 PSoC Analog Coprocessor board. In the first application, the board's LED illuminates in red when the sensor detects a magnetic field and switches off when no magnetic field is detected. To implement the project, we initially established the hardware connections as follows: We connected the VCC and GND of the KY-024 module to the respective power and ground pins on the PSoC Analog Coprocessor Pioneer board. The digital output (DO) pin of the KY-024 module was linked to the digital input port P2 [7] on the PSoC device, which corresponds to pin 17 of the PSoC Analog Coprocessor. Additionally, we use the RGB LED, connected to pin 4 of the PSoC Analog Coprocessor, to indicate the state changes.

Following the hardware setup, we proceeded to configure the project in PSoC Creator. Upon launching the software, we developed the Hall KY-024 module schematics, we incorporated a Digital Input Pin (Din), which is connected to the DO pin of the KY-24 and a Digital Output Pin (Pin_Red) which is connected to the RGB LED from the PSoC Analog Coprocessor Pioneer Board.

Subsequently, in the main.c source file, we implemented the code to define the behavior of the system. After completing the coding process, we proceeded to build the project and programmed it onto the PSoC Analog Coprocessor for execution.

This practical demonstration underscores the platform's flexibility and showcases its capability to dynamically respond to varying environmental conditions, offering a tangible representation of its reconfigurable nature.

The objective of this secondary application is to emphasize the AO of the KY-024 module. (see Fig. 4).

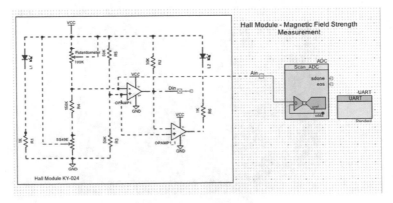

**Fig. 4.** Hall Module Project 2 Schematics

This is achieved by measuring the strength of the magnetic field in the vicinity and presenting the results through an emulator terminal, such as Putty in this context.

In Fig. 5, the code for the program is displayed, while Fig. 6 showcases a screenshot capturing the results in the terminal emulator, presenting the recorded values for magnetic field strength.

```
12  #include "project.h"
13  #include "stdio.h"
14  #include "cyapicallbacks.h"
15  #include "cytypes.h"
16  #include <cydevice_trm.h>
17  #include <cyfitter.h>
18  int main(void)
19  {
20      CyGlobalIntEnable; /* Enable global interrupts. */
21      UART_Start(); /* Start UART communication */
22      ADC_Start(); /* Start ADC conversion */
23
24      for (;;)
25      {
26          /* Read the analog signal from the AO pin */
27          float Ain = ADC_GetResult32(0);
28          /* Convert analog output to magnetic field strength (adjust based on sensor characteristics) */
29          float magneticFieldStrength = Ain*(5.0/4095.0);
30          /* Send the magnetic field strength over UART */
31          char uartBuffer[50];
32          sprintf(uartBuffer, "Magnetic Field Strength: %.2f Gauss\r\n", magneticFieldStrength);
33          UART_UartPutString(uartBuffer);
34          CyDelay(1000); /* Delay for 1 second (adjust as needed) */
35      }
36  }
```

**Fig. 5.** Hall module – Magnetic Field Strength Measurement Application – Code

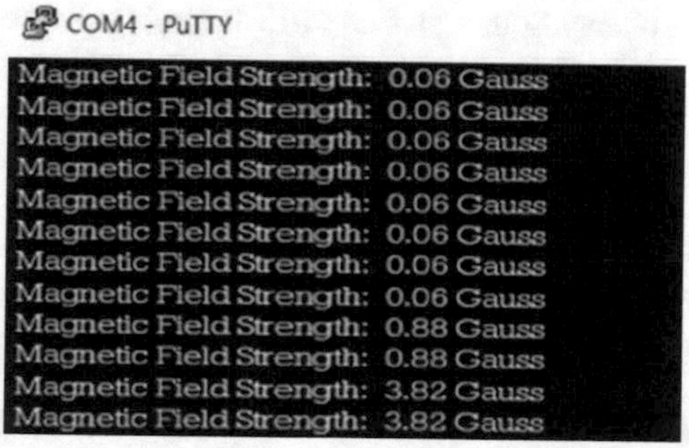

**Fig. 6.** Hall module – Magnetic Field Strength Measurement Application – Results

## 3.2 Further Developments with PSoC Analog Coprocessor Pioneer Kit, LabVIEW and NI ELVIS

The seamless compatibility and effective collaboration between the PSoC and Lab-VIEW ecosystems have demonstrated their capability for swift prototyping of embedded designs [8].

As further developments, we are considering the integration of the PSoC Analog Coprocessor Pioneer Board, LabVIEW, and NI ELVIS (National Instruments Educational Laboratory Virtual Instrumentation Suite) in order to introduce a comprehensive solution for hands-on learning, experimentation, and prototyping in the field of analog signal processing and embedded systems.

The PSoC Analog Coprocessor Pioneer Board serves as the core hardware platform, providing analog signal processing capabilities, versatile peripherals, and flexibility for various applications, while NI ELVIS is a multifunctional platform that includes both hardware and software components, with a variety of I/O interfaces, allowing seamless integration with LabVIEW and other educational modules. LabVIEW represents the software component, that acts as the graphical programming environment, enabling users to design, simulate, and deploy applications seamlessly. Its user-friendly interface makes it suitable for both beginners and experienced engineers. Utilizing the PSoC board to interface with sensors, to acquire analog data, and process it in real-time and LabVIEW to facilitate easy visualization and analysis of acquired data, becomes an ideal combination for sensor-based applications.

## 4   Conclusions

Employing IoT laboratories that leverage reconfigurable platforms featuring the PSoC Analog Coprocessor has proven to significantly enhance learning outcomes and student engagement. The successful integration of these advanced technologies facilitates a hands-on learning approach, which, in turn, cultivates a more profound understanding of IoT concepts among students. This system plays a pivotal role in smoothing the journey for students as they transition from theoretical understanding to the practical application of knowledge in their day-to-day experiences. It stands out as an invaluable tool, facilitating and enhancing this crucial shift in their educational journey. Distinguished as an invaluable tool, it not only eases, but also enhances this critical shift in their educational journey, enriching their overall learning experience.

By employing these reconfigurable platforms, equipped with Analog Coprocessors, students not only gain hands-on exposure but also witness an improvement in their comprehension of IoT concepts. This exploration aims to underscore the transformative impact of integrating such technology into educational settings, highlighting its role in fostering a more dynamic, interactive, and effective learning journey for students in the realm of IoT.

The interactive nature of these laboratories, coupled with the capabilities of the PSoC Analog Coprocessor, ensures that students are actively involved in practical applications, leading to a more robust grasp of theoretical principles. This approach not only elevates the educational experience but also aligns with contemporary teaching methodologies that prioritize experiential learning to prepare students for the dynamic challenges within the IoT environment.

## References

1. Zainuddin, A.A., et al.: Trends and challenges of internet-of-things in the educational domain. Malaysian J. Sci. Adv. Tech. (2021). https://doi.org/10.56532/mjsat.v1i3.17
2. Asad, M.M., Naz, A., Shaikh, A., Alrizq, M., Akram, M., Alghamdi, A.: Investigating the impact of IoT-based smart laboratories on students' academic performance in higher education. Univ. Access Inf. Soc. (2022). https://doi.org/10.1007/s10209-022-00944-1
3. Nurulhuda, A.R., et al.: Development of educational kit for IoT online learning. Int. J. Technol. Inn. Hum. 1(1), 26–32 (2020)
4. Cvjetkovic, V.M.: Pocket labs supported IoT teaching. Int. J. Eng. Ped. 8(2), 32–48 (2018). https://doi.org/10.3991/ijep.v8i2.8129

5. Ursutiu, D., Samoila, C., Neagu, A., Florea, A., Chiricioiu, A.: "Software reconfigurable hardware" in IoT student training. In: Auer, M.E., Tsiatsos, T. (eds.) The Challenges of the Digital Transformation in Education: Proceedings of the 21st International Conference on Interactive Collaborative Learning (ICL2018) – Volume 1, pp. 410–416. Springer International Publishing, Cham (2020). https://doi.org/10.1007/978-3-030-11932-4_39
6. Getting Started with PSoC Analog Coprocessor: https://www.infineon.com/dgdl/Infineon-AN211293_Getting_Started_with_PSoC_Analog_Coprocessor-ApplicationNotes-v02_00-EN.pdf?fileId=8ac78c8c7cdc391c017d0d23d16b61b2. Last accessed 21 Dec 2023
7. PSoC Analog Coprocessor Pioneer Kit Guide: https://www.infineon.com/dgdl/Infineon-CY8CKIT-048_PSOC_ANALOG_COPROCESSOR_PIONEER_KIT_GUIDE-UserManual-v01_00-EN.pdf?fileId=8ac78c8c7d0d8da4017d0efc783e12a6. Last accessed 21 Dec 2023
8. Chamunorwa, T., Ursutiu, D., Samoila, C., Modran, H.A.: Embedded system learning platform for developing economies. In: Auer, M.E., Rüütmann, T. (eds.) Educating Engineers for Future Industrial Revolutions: Proceedings of the 23rd International Conference on Interactive Collaborative Learning (ICL2020), vol. 2, pp. 617–627. Springer International Publishing, Cham (2021). https://doi.org/10.1007/978-3-030-68201-9_60

# Open-Source Software and Digital Sovereignty
## A Technical Case Study on Alternatives to Mainstream Tools

Dennis Biström[1]([⊠]), Kristoffer Kuvaja Adolfsson[2], and Matteo Stocchetti[2]

[1] School of Engineering, Culture and Wellbeing, Arcada University of Applied Sciences, Jan-Magnus Janssons plats 1, 00560 Helsinki, Finland
dennis.bistrom@arcada.fi
[2] Graduate School and Research, Arcada University of Applied Sciences, Jan-Magnus Janssons plats 1, 00560 Helsinki, Finland
{kuvajaan,matteo.stocchetti}@arcada.fi

**Abstract.** EU's vision of digital sovereignty conflicts with tools and platforms used in modern working life. We are dependent on software controlled by global technology giants.

The dependence on commonly used services and platforms provided by Microsoft, Google or Amazon must be removed to achieve digital sovereignty. Open-source software has been readily available for a long time, but studies looking into excluding all proprietary software from working life are scarce. This paper investigates the feasibility of replacing proprietary software as a step towards digital sovereignty.

The ability to self-host services is key to ensuring independence, and the ability to verify software code is key to ensuring integrity. Therefore open-source code and transparency are transient properties for sovereign software. We present problems by mapping the problematic software used by teachers, researchers and student assistants at Arcada UAS with a varied set of requirements on the digital tools of their daily lives. We research alternatives and deploy a suite of software compatible with EU policies on digital sovereignty.

We rely on many services offered by tech giants in our professional lives. There are several open-source software suites that can be self-hosted and that have permissive licenses for business use. Extensive technical knowledge is required for the deployment of software. Most daily tasks can be accomplished by using sovereign software, although not all needs of the test group were met.

Companies can deploy sovereign software for most daily tasks, reducing dependency on tech giants. However, software adoption was low. Arcadas IT-support does not host, enforce or endorse the usage of sovereign software. Public code repositories enable independent code verification, in contrast to proprietary software. Some software cannot be easily replaced, and digital habits are hard to change.

**Keywords:** Digital sovereignty · open-source · transparency · accountability · privacy

© The Author(s), under exclusive license to Springer Nature Switzerland AG 2024
M. E. Auer et al. (Eds.): STE 2024, LNNS 1027, pp. 106–113, 2024.
https://doi.org/10.1007/978-3-031-61891-8_10

# 1   Introduction

## 1.1   Sovereignty and Control

In the debate about digital sovereignty, the role of free and open-source alternatives and the concept of digital commons supporting them is sometimes neglected. The paper is an attempt to fill this gap, as the paper aims to evaluate the role and contributions of free and open-source software in the deployment of a digital infrastructure more compatible with democratic sovereignty.

Our starting point is that corporate control of the digital infrastructure is detrimental to democratic sovereignty as this infrastructure is a crucial element of political sovereignty. While the current debate on digital sovereignty engages with conceptual and legal issues [4, 5, 7, 9, 11, 15, 16, 18, 19][4, p. 1274], this paper focuses on the technological dimension.

Free and open-source software is equally accessible to everyone and can also be developed collectively, democratically. An infrastructure built upon open-source hardware and software is not under corporate control but can be hosted independently. From the perspective of technology, closed-source software is also problematic due to a lack of transparency. Privacy and security can be ensured through an audit of a software's source code. For closed-source software, the code is not available for audits.

In line with the concerns about the "enormous power of the corporate actors... owning vital social structures ... that makes them one of the biggest challenges to the concept of democratic sovereignty" [7], we argue that free and open-source software supports a digital infrastructure that allows for more control for the citizens, thus being more suitable for democratic digital sovereignty. On theoretical grounds, open-source software has the potential to bring about the 'subversive rationalisation' and democratisation of technology and its development, in line with the normative ambitions of the critical theory of technology.

On more practical grounds, an evaluation of existing technological alternatives is key to establishing the material conditions of democratic digital sovereignty. The importance of digital sovereignty involves us all and cannot be overrated since, as Floridi argued:

"The fight for digital sovereignty may remind one of the Investiture Controversy, the medieval conflict between the Church/Pope and the State/Emperor in Europe over the ability to choose and install bishops and more generally over secular and spiritual power. (...) Today, the fight is not over secular and spiritual power but over corporate and political power over the digital, yet the roots of this clash are very old. But most importantly that medieval debate reminds us that whoever will win the fight for digital sovereignty will determine the lives of all people on both sides of the digital divide, exactly like the Investiture Controversy affected all people, no matter whether religious or not. This is why (...) digital sovereignty is not just a matter of interest for some specialists. It is already affecting everybody. And this is why it is essential to design it as well as possible, together." [7, p. 377]

## 1.2    Approach

This paper's argument relies on the critical theory of technology, examining the effects of technology within society. We agree that technology is not neutral but embedded with values, and we build upon three ideas associated with the critical theory as presented in, among others, the work of Andrew Feenberg [6]:

**Social Constructivism vs Technological Determinism** - We argue that society and technology co-construct each other and critique technological determinism, the idea that technology develops independently and shapes our society.

**Subversive Rationalisation** - We seek to use technology for democratic and socially responsible purposes and state that technological systems should be made to serve broad interests such as empowering individuals and communities and increasing social well-being. We stand opposed to traditional views on technology, seeing it only as a tool to maximise output and minimise costs in the interest of corporations or states.

**The Democratisation of Technology** - We argue for a broader range of stakeholders to be included in the decision-making processes of technological development. The concept is to create a more equitable landscape to share the benefits of technological progress.

## 2    Method

The biggest tech giants today are Microsoft, Google, Amazon, Meta (Facebook), Apple and Netflix (NASDAQ). At Arcada UAS, there is no requirement to use Meta or Netflix accounts or products. Google, Apple and Amazon accounts are in use by teachers and students, but they are opt-in. A mandatory Microsoft account is used for gaining access to the Microsoft 365 cloud productivity suite as Microsoft Teams and Outlook are everyday tools. The natural focus of this paper is, therefore the Microsoft 365 product family of cloud productivity services.

Through implementing, testing and evaluating alternative software, we can assess the potential contribution of the chosen alternatives on the democratisation of technology as well as on subversive rationalisation. By targeting open-source alternative software in use by personnel and students at Arcada university of applied sciences, we strive to apply the ideas of free and open-source software to the field of higher education.

### 2.1    Assessment Criteria

The objective of the paper is to examine open-source alternatives to software currently in the hands of a few large corporations and to evaluate the role and contribution of these open-source alternatives to the larger problem statement. For evaluating the alternative software, we have outlined the following criteria for assessment:

**Accessibility & Usability** - Availability of the software (platforms) and ease of use for individuals with varying levels of digital literacy.

**Sustainability & Support** - Evaluate the strength of community behind the active development.

**Compatibility & Privacy** - How well the alternative integrates with existing systems and assess level of privacy.

## 2.2   Mapping

We have mapped the software and hardware in use by teachers, researchers and student assistants at Arcada UAS. The software and hardware were classified according to a set of functional requirements on the digital tools in the participants' daily lives (e.g. file storage, document editing). We collected a list of software according to what was installed on the colleagues' phones and computers and classified the software according to use cases (e.g. backing up smartphone pictures in the cloud). We also recorded the brands and models of phones and laptops of the participants (Fig. 1).

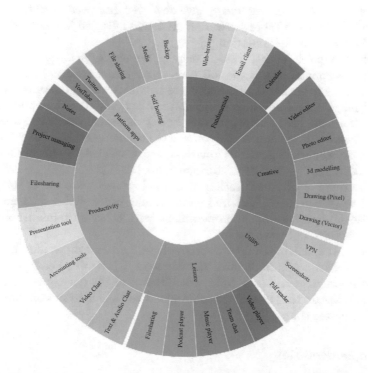

**Fig. 1.** *Interactive* sunburst of software and purpose areas, made by authors using D3.js [3].

## 2.3   Software Review

To ensure software independence and integrity the user needs to have the ability to self-host services and verify its code. Therefore transparency and open-source code are transient properties for sovereign software. A software review was conducted on cloud native open-source alternatives to the Microsoft 365 productivity suite compatible with EUs policies on digital sovereignty. The software suite was deployed and reviewed according to the outlined assessment criteria.

The software fulfilling the functional purposes of the Microsoft 365 suite as well as our criteria for open-source and self-hostable software limited our deployment choice to the following three collaborative suites: Nextcloud Hub, OnlyOffice Workspace and Collabora Online. One additional alternative was identified (CryptPad), but the option was excluded due to the software not offering the same type of modularity as the others.

## 3   Results

### 3.1   Transparency and Collaboration

The benefits of open-source software have been well discussed since the beginning of the GNU (1983) and Libre software movements initiated by Richard Stallman and the Free Software Foundation [8].

In software development, transparency or the lack of it impacts the security, privacy and accountability of the software. Open-source software can be audited by third parties to ensure that the software complies with modern security standards. Transparency enables us to evaluate privacy by examining what data is collected from the user and where and how it is stored and sent. Most open-source software utilise version control, where each code revision is connected to a user (a developer). Since code revisions are traceable to specific users, this ensures accountability in case of infringements [1].

On the topic of democratisation of technology, software under open-source licenses, like MIT, Apache, GPL and more, enables peer collaboration and derivative works. Integration and interoperability with other systems are also made possible due to code transparency.

However, the benefits of open-source software are not automatic but depend on how the software is implemented and managed. Active contributors, good governance practices and long-term commitment of the developers is a requirement, and software that is not well maintained can suffer from vulnerabilities be it open- or closed source [2].

### 3.2   Implementation

The objective of the paper was to examine open-source alternatives to software currently in the hands of a few large corporations. A Nextcloud Hub instance was hosted on a local machine, and later moved to a cloud server at the Finnish datacentre CSC. Nextcloud offers a comprehensive suite of collaboration features for secure file sharing and hosting as well as communication and documentation, some of our most frequent features used in Microsoft Teams. Nextcloud Hub also provides an array of apps and plugins to extend the functionality of the software suite. Some key features include:

- Secure file sharing and management tools.
- Collaborative editing with tools like Collabora Online or ONLYOFFICE.
- Integrated chat and video calls, user group management.
- Search functionality for file names, tags, comments, and content.
- Mobile app features like multi-account support, auto-upload of photos/videos and image editing.
- LDAP/Active Directory integration for user management and SSO [14].

After trials of both ONLYOFFICE and Collabora Online in our implementation of the Nextcloud platform, the latter was chosen as the provider of the office functionality.

### 3.3 Assessment

The software review assessment criteria are both technical and user centric. Accessibility for example is achieved through the software being available for different platforms (Linux, Mac OS), hardware requirements (legacy, modern) and capital capabilities (free, cost). But accessibility also refers to ease of use, text legibility, and customisation.

The modularity of Nextcloud Hub makes it suitable for replacing the wide range of features of the Microsoft 365 cloud suite. However, Microsoft 365 cloud is just one software suite, and the replacement of the suite is not a solution to our reliance on proprietary software.

**Accessibility & Usability.** The Nextcloud Hub is accessible in all major browsers (Chrome, Firefox, Bing) and therefore available for all major desktop operative systems (Linux, OSX, Windows). The Nextcloud hub is accessible on mobile phones, but the user interface is somewhat limited depending on the size of the display. All features are available when turning on desktop mode on mobile browsers with the drawback of the interface and text being small, impacting usability. On mobile phones Nextcloud provides companion apps for most features of the cloud [10]. Two users of the test group reported problems with the simple method of the dark colour theme of the platform.

Based on reviews on Gartner, Nextcloud Hub has a clean and consistent interface that is easy to use and administrate, the web interface works well with most modern browsers, and the mobile companion apps offer a seamless experience across different devices.

**Sustainability & Support.** The Nextcloud community forum is active, has over 100 000 registered users and over 200 000 topics. The Nextcloud Github organisation has over 2000 repositories and over 1000 collaborators. The main repository nextcloud-server has over 24 000 stars and around 4000 followers. For comparison, the world's most popular open-source database management server MySQL-server has over 9000 stars and around 1000 followers [12].

**Compatibility & Privacy.** The Nextcloud Hub uses the Collabora or the ONLYOFFICE software suite to provide the document editing properties familiar from the Microsoft 365 suite. Our platform used Collabora for the office functionality. Both Collabora and ONLYOFFICE claim to have high compatibility with Microsoft Office formats, such as docx, xlsx and pptx [17,20]. The compatibility is not perfect, and some formatting or features may not be preserved or displayed correctly in either of the office suites. Collabora is based on LibreOffice while ONLYOFFICE is an independently developed product by Ascensio System [13].

Heinrich Long reviewed the privacy and security of the Nextcloud Hub in 2021. He found that security and privacy levels in Nextcloud vary based on the self-hosted configuration. Nextcloud also offer hosting solutions where the security of the data and therefore the privacy of the user depends on the hosting partner. Nextcloud's privacy policies are user-friendly, but the impact of third-party plugins and hosting services must be evaluated individually [10].

## 4  Conclusion

We rely on proprietary services offered by technology giants in our daily lives. The digital infrastructure is part of political sovereignty and corporate control of it is detrimental to democratic sovereignty. An evaluation of existing technological alternatives is key to establishing the material conditions of democratic digital sovereignty.

Free and open-source software offers advantages to proprietary software. The advantages have been presented by the FOSS movement and include benefits of increased transparency, collective code review and equal access to software. We found that the use of free and open-source software in society is in line with the subversive rationalisation of technology predicated by the critical theory of technology. A more widespread use of this software supports the democratisation of technological development to the extent that it challenges the influence of few large tech companies.

The implementation of the alternative office suite proved successful in replicating important functionality of the Microsoft 365 suite and the assessment of the role of open-source software for digital sovereignty was completed. However, we found that extensive technical knowledge is often required for the deployment and maintenance of the software, this impacts the citizens ability to democratically make decisions about the development of the software.

There are several actively and collectively developed, free and open-source software suites that can be self-hosted by individuals and communities, with permissive licenses for business use. Companies can deploy sovereign software for most users' daily tasks and reduce dependency on tech giants.

# References

1. Ana, J.S.: Why transparency is critical to your open source project's security (2021). https://opensource.com/article/21/6/security-transparency
2. del Bianco, V., Lavazza, L., Morasca, S., Taibi, D.: Quality of open source software: the QualiPSo trustworthiness model. In: Boldyreff, C., Crowston, K., Lundell, B., Wasserman, A.I. (eds.) Open Source Ecosystems: Diverse Communities Interacting. IFIP Advances in Information and Communication Technology, vol. 299, pp. 199–212. Springer, Berlin (2009). https://doi.org/10.1007/978-3-642-02032-2_18
3. Bostock, M.: D3js (2018). https://github.com/d3/d3
4. Broeders, D., Cristiano, F., Kaminska, M.: In search of digital sovereignty and strategic autonomy. J. Common Mark. Stud. **61**(5), 1261–1280 (2023)
5. Couture, S., Toupin, S.: What does the notion of 'sovereignty' mean when referring to the digital? New Media Soc. **21**(10), 2305–2322 (2019). https://doi.org/10.1177/1461444819865984
6. Feenberg, A.: Critical Theory of Technology. Oxford University Press, New York (1991)
7. Floridi, L.: The fight for digital sovereignty: what it is, and why it matters, especially for the EU. Philos. Technol. **33**(3), 369–378 (2020). https://doi.org/10.1007/s1334-020-01462-x, https://doi.org/10.1007/s13347-020-00423-6
8. Foundation, F.S.: Working together for free software - our initiatives (2023). https://www.fsf.org/
9. Heller, K.J.: In defense of pure sovereignty in cyberspace. International Law Studies (2021). https://digitalcommons.wcl.american.edu/cgi/viewcontent.cgi?article=1432&context=ilstudies
10. L., H.: Self-hosted cloud storage with drawbacks (2021). https://restoreprivacy.com/cloud-storage/reviews/nextcloud/
11. Mueller, M.L.: Against sovereignty in cyberspace. Int. Stud. Rev. **22**(4), 779–801 (2020). https://doi.org/10.1093/isr/viz044
12. MySQL: Mysql and nextcloud (2023). https://github.com/mysql and https://github.com/nextcloud
13. Nextcloud: Collabora vs onlyoffice (2022). https://help.nextcloud.com/t/collabora-vs-onlyoffice/134994/2
14. Nextcloud: Nextcloud features that put you in control (2023). https://nextcloud.com/features/
15. Ávila Pinto, R.: Digital sovereignty or digital colonialism? New tensions of privacy, security and national policies. Internet Democracy **15**(27), 15–27 (2018)
16. Pohle, J., Thiel, T.: Digital sovereignty. Internet Policy Rev. **9**(4) (2020). https://doi.org/10.14763/2020.4.1532
17. Productivity, C.: Collabora online your own private office suite in the cloud (2023). https://www.collaboraoffice.com/collabora-online/. Accessed 15 Jan 2024
18. Roberts, H., Cowls, J., Casolari, F., Morley, J., Taddeo, M., Floridi, L.: Safeguarding European values with digital sovereignty an analysis of statements and policies. Internet Policy Rev. **10**(3) (2021). https://doi.org/10.14763/2021.3.1575
19. Robles-Carrillo, M.: Sovereignty vs. digital sovereignty. J. Digit. Technol. Law **1**(3), 673–690 (2023). https://doi.org/10.21202/jdtl.2023.29
20. SIA, A.S.: Onlyoffice, supported formats of spreadsheets. https://helpcenter.onlyoffice.com/ONLYOFFICE-Editors/ONLYOFFICE-Spreadsheet-Editor/HelpfulHints/SupportedFormats.aspx. Accessed 15 Jan 2024

# Demand for Future Skills: Education on AI in Comprehensive Digital Business Development, Big Data Analytics, and Ubiquitous Approach to Data in Business

Martin Zagar[1]([⊠]), Jasminka Samardzija[1], Ana Havelka Mestrovic[1], Muhieddin Amer[1], and Jinane Mounsef[2]

[1] RIT Croatia, Dubrovnik, Croatia
`martin.zagar@rit.edu`
[2] RIT Dubai, Dubai, UAE

**Abstract.** The goal of this paper is to identify which future skills will be in demand and to offer the student's program the competencies needed for those. Future job skills will require simulations of different options in Digital business development and an update on education on future business skills that will tackle raising AI Impacts on Academia, Industry, and Society and address a ubiquitous approach to different global needs. These will also shape some emerging jobs like Specialist in AI and ML and Digital transformation specialist. Together with some technical skills such as Categorization of data; Analytical queries; and Development of predictive and analytical models, we also identified additional skills needed for such kinds of jobs. Based on the initial research outcomes, it was possible to design and initially evaluate our program that will address the desired set of skills/competencies in a way of eight courses that provide a comprehensive education on Big data and AI techniques, including problem-solving, data analytics, natural language processing, recommender systems, image analysis, and Blockchain technologies.

**Keywords:** Program on Future Business Skills · Problem-Solving Techniques in AI · AI and Productivity · AI and Employment · Interdisciplinary Education

## 1 Introduction

For organizations, the development of new business models and competitive advantages through the integration of artificial intelligence (AI) in business and IT strategies holds considerable promise. Human beings have the capability to resolve most of their issues or problems by applying fast, intuitive judgments, instead of using the conscious, step-by-step assumption that the early AI researchers were using with their previous models. Consider this example - suppose you are traveling by car and unfortunately, you get a flat tire. The step-by-step deduction would demand you to call a mechanic and search for a solution. Whereas, working according to intuitive judgment would enable you to change

M. E. Auer et al. (Eds.): STE 2024, LNNS 1027, pp. 114–121, 2024.
https://doi.org/10.1007/978-3-031-61891-8_11

the tire yourself. AI has made a lot of progress at replicating "sub-symbolic" problem solving: which are basically the personalized approaches that emphasize the importance of sensorimotor skills to higher-order reasoning. The neural net that is involved in attempts to simulate the structures inside the brain will further give rise to specialized skills such as statistical approaches being used in AI. The main goal is to look exactly just like humans. Hence, nowadays it becomes even more important how humans develop and implement different skills to use AI more often in various fields.

As analyzed by (Georgieff and Hyee, 2021), over the period 2012–2019, employment grew in nearly all occupations. Overall, there appears to be no clear relationship between AI exposure and employment growth. However, in occupations where computer use is high, greater exposure to AI is linked to higher employment growth. The increase in labor productivity and output counteracts the direct displacement effect of automation through AI for workers with good digital skills, who may find it easier to use AI effectively and shift to non-automatable, higher-value-added tasks within their occupations. The opposite could be true for workers with poor digital skills, who may not be able to interact efficiently with AI and thus reap all potential benefits of the technology.

From the financial perspective, as stated in (Thormundsson, 2022), in 2022 revenues from the global artificial intelligence market were expected to reach 433 billion U.S. dollars. The global AI market is forecast to see rapid growth in the coming years, reaching more than half a trillion U.S. dollars by 2023 showing enormous potential from an educational and applicative perspective.

Statistics from (Accenture, 2017) show the impact AI will potentially have on the real gross value added (GVA) of various industries worldwide in 2035. By 2035, with AI absorbed into the economy, the manufacturing industry could potentially increase their gross value added by 12.2 trillion U.S. dollars, 3.8 trillion U.S. dollars more than the baseline, non-AI scenario. With this global potential in mind, we started to design our international multidisciplinary program on future AI applications in businesses that would meet the demand for future skills we identified. The rest of the paper will show our design approach and key points we wanted to meet for specific program and course learning outcomes.

## 2  Impact and Needs

Businesses must use several distinctive technologies such as AI to build adaptive transformation and sense-and-respond capabilities that will spur innovation, enhance customer service and experience, and promote improved performance. For many organizations, applying AI to general business workflows and processes could represent a new stage in the process of digital transformation and one that has the potential to be disruptive in all the wrong ways if it's not managed well. AI needs to be integrated into the business culture – and as part of that, it needs to be trusted by the people who must work with it and will be affected by it. That of course includes staff trusting it to help them work better and smarter, and to do more with less, not simply to put them out of a job. Those concepts are taking a long time to trickle down into end-user organizations. According to (Microsoft, 2019), the state of AI in the UK mentioned that 96% of people said they have never been consulted by their superiors on the introduction of AI and conversely, 83%

of business leaders said their employees have never asked about AI. This thing matters because of the transformational and cultural aspects. Industrialization and technological development have increased and now it is time to move back to humans and soft skills. Without the human factor, there will be no successful implementation of AI.

With this paper, we would like to offer the answer to the fear of people that we will become robots and that artificial intelligence will take over creativity and would like to point out the productivity influence of artificial intelligence and the impact on the increase in human capital connected with sophisticated problem-solving skills. We would also like to discover which skills should have the young individual of the future to remain mentally healthy and productive – the goal is to achieve a better work-life balance than before AI. Also, the goal is to emphasize the positive influence on the person, company, and national productivity which would increase the standard of living measured by GDP per capita, increase wages due to the increase in the demand for labor force with developed human capital, and enable individuals to have more leisure time.

Keeping in mind all the above mentioned, we propose to develop new courses and short program for students, that could also be delivered as a life-long learning program. This program is the first one that offers this type of education and we also merged economy, engineering, and psychology into one interdisciplinary program. In order to analyze the above/mentioned goals we designed the survey to identify objectives related to Problem-Solving Techniques in AI; AI and productivity and AI and employment.

## 2.1 Survey Description

Here we will shortly describe the important parts of validated survey content, with examples of questions, since in this paper we are not focusing on the survey itself, but on the program we designed and offered, while using this part just to show the reasons why we addressed specific topics:

- Technology-based services (questions like: "Would you allow a computer to help diagnose and treat a medical problem?", answering on a Likert scale of $1 =$ very undesirable $- 5 =$ very desirable)
- Using technology (focusing on how often users are using technologies like Autonomous Vehicles, Facial Recognition, or Recommender Systems in Health)
- Future of work requirements/expectations (focusing on how often users are using different types of virtual AI assistants, like Virtual Financial Assistants, Virtual Teaching Assistants, or Virtual Medical Assistants)
- Optimism in using AI (questions like: "You prefer to use the most advanced technology available." or "Technology makes you more efficient in your occupation.", answering on a Likert scale of $1 =$ strongly disagree$- 5 =$ strongly agree)
- Innovativeness in using AI (questions like: "You prefer to use the most advanced technology." or "You find you have fewer problems than other people in making technology work for you.", answering on a Likert scale of $1 =$ strongly disagree$- 5 =$ strongly agree)
- Discomfort in using AI (questions like: "Sometimes, you think that technology systems are not designed for use by ordinary people." or "Many new technologies have health or safety risks that are not discovered until after people have used them.", answering on Likert scale $1 =$ strongly disagree$- 5 =$ strongly agree)

- Insecurity in using AI (questions like: "You do not consider it safe to give out a credit card number over a computer." or "If you provide information to a machine or over the Internet, you can never be sure it really gets to the right place.", answering on Likert scale 1 = strongly disagree– 5 = strongly agree)

We run the survey on the introduction of Big Data Analytics and Problem-solving courses as a part of our Master's program. The number of respondents was N = 37. From Fig. 1, it can be seen that the vast majority of the respondents (for all statements above 80%) strongly agree with our statements on the AI application.

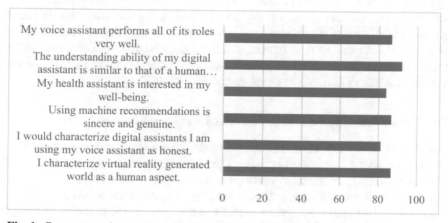

**Fig. 1.** Percentage of respondents that strongly agree with the statements from the survey.

# 3  Problem-Solving Techniques in AI

When addressing future technologies like Big Data analytics, artificial intelligence, distributed ledger technologies (Blockchain), e-commerce and online platforms, augmented and virtual reality, Internet of Things (IoT), and cloud computing as part of education on problem-solving techniques in artificial intelligence, we focused on to show how these techniques help in identifying, analyzing, and resolving complex problems and how they form the foundation of AI applications.

## 3.1  Big Data Analytics

Primarily, we oriented our educational program toward Big Data analytics, where we explained the concept of data analytics and its relevance in problem-solving. We describe how AI techniques can be employed to extract valuable insights from vast amounts of data and how this can aid decision-making and problem-solving processes. AI algorithms can analyze large datasets to identify patterns, correlations, and trends that may not be immediately apparent to humans. By recognizing these patterns, AI models can make predictions and generate insights that can inform decision-making. For example, in financial markets, AI algorithms can analyze historical data to predict future stock prices or market trends (Chopra and Sharma, 2021), helping investors make informed decisions.

AI techniques can automatically classify and categorize vast amounts of unstructured data, such as text documents, images, or videos (Wang et al., 2019). This enables efficient organization and retrieval of information, making it easier for business decision-makers to access relevant data quickly. For instance, AI-powered text classification algorithms can automatically categorize customer feedback into positive or negative sentiments, helping businesses understand customer satisfaction levels and identify areas for improvement. That's why we pointed in this area towards the data preprocessing and extracting some relevant features from the text, and model training to be able to perform text classification. We are aiming to cover Naive Bayes, Support Vector Machines (SVM), decision trees, random forests, and basic deep learning models, such as Convolutional Neural Networks (CNN) and Recurrent Neural Networks (RNN) like Long Short-Term Memory (LSTM) or Gated Recurrent Unit (GRU).

## 3.2   Recommender Systems Based on NLPs and Image Analysis

Following the data analytics, we also identified that Natural Language Processing (NLP) techniques enable AI systems to understand and process human language, both written and spoken, so by applying NLP techniques to textual data, AI models can extract key information, perform sentiment analysis, and generate summaries or insights (Allahyari et al., 2017). This capability can be useful in areas like customer feedback analysis, content moderation, and information extraction from documents.

In order to prove integrity and transparency in those data models, the best way is to implement some Blockchain technologies. Although Blockchain technologies in a core are not a part of the program, we provide education on Blockchain because it can be used to enhance data privacy and security in AI systems. By storing data in a decentralized environment, blockchain ensures that sensitive data remains private and cannot be altered without consent which is particularly useful in AI applications that involve handling sensitive personal information or confidential data. Blockchain also enables secure and transparent data sharing and collaboration among multiple parties. In AI, where large amounts of data are often required to train models effectively, blockchain can facilitate data sharing between different organizations or individuals without compromising data privacy. Smart contracts on the blockchain can define access rights, and data usage terms, and ensure fair compensation for data contributors, and thus can create decentralized AI marketplaces, connecting AI developers and consumers.

Finally, in this part, we also focus on image and video analysis. AI techniques like computer vision can analyze and interpret visual data from images and videos. This enables applications such as object recognition, image segmentation, and video tracking. In domains like healthcare, autonomous vehicles, or surveillance systems, AI-powered image analysis can aid in detecting abnormalities, identifying objects or individuals, and providing insights for decision-making.

# 4   Results

Based on the above-mentioned considerations and impacts we finally designed our program in a way of eight courses that provide a comprehensive education on Big data and AI techniques, including problem-solving, data analytics, natural language processing, recommender systems, image analysis, and blockchain technologies. The inclusion of the Blockchain Technologies and Applications course ensures that students gain an understanding of blockchain fundamentals and its applications in AI and business processes while analyzing the NLPs enabling text classification to categorize and organize unstructured textual data. This interdisciplinarity in the program also equips individuals with the necessary skills and knowledge to navigate the transformative potential of AI while addressing ethical considerations and emerging technologies. An overview of the courses is depicted in Table 1 - List of courses with Course learning outcomes.

**Table 1.** List of courses with course learning outcomes.

| Problem-Solving Techniques in Artificial Intelligence |
| --- |
| Identify the role of problem-solving techniques in AI applications |
| Analyze complex problems and develop strategies for their resolution using AI techniques |
| Apply AI algorithms to identify patterns, correlations, and trends in large datasets |
| Utilize AI models for prediction, decision-making, and generating insights |
| Evaluate and select appropriate AI techniques for different problem domains |
| **AI and Design** |
| Develop knowledge of various AI techniques and algorithms relevant to design processes |
| Assess skills to integrate AI tools and frameworks into the design workflow |
| Explain how AI can enhance creativity and innovation in design |
| Apply AI algorithms for optimization and problem-solving in design scenarios |
| Develop critical thinking skills to evaluate the effectiveness of AI-based design solutions |
| Compose collaboration and interdisciplinary skills by working on AI-driven design projects |
| **Problem-Solving Techniques in Artificial Intelligence** |
| Describe the concept of big data analytics and its relevance in problem-solving |
| Apply AI techniques to extract valuable insights from large datasets |
| Predict the impact of AI algorithms in decision-making processes |
| Identify patterns, correlations, and trends in data to inform business strategies |
| Analyze real-world case studies to show the effectiveness of Big data analytics |
| **Natural Language Processing and Large Language Models** |
| Create knowledge of natural language processing techniques for processing textual data |
| Discover key information from a text and perform sentiment analysis using AI models |
| Apply text classification algorithms to categorize and organize unstructured textual data |
| Evaluate the performance of text classification models and optimize them for accuracy |
| Detect the ethical considerations and challenges related to text analysis and classification |

*(continued)*

**Table 1.** (*continued*)

| Problem-Solving Techniques in Artificial Intelligence |
|---|
| **Recommender Systems and Personalized Recommendations** |
| Review the principles and algorithms behind recommender systems<br>Analyze user preferences and historical behavior to generate personalized recommendations<br>Explain the representation learning techniques used in recommender systems<br>Evaluate and rate the performance of recommender systems for accuracy and relevance<br>Inspect the ethical implications of personalized recommendations and privacy concerns |
| **Computer Vision** |
| Categorize computer vision techniques for analyzing and interpreting visual data<br>Apply AI algorithms for object recognition, image segmentation, and video tracking<br>Evaluate the impact of AI-powered image analysis in various domains, such as healthcare<br>Discuss the challenges and ethical considerations in the image and video analysis<br>Research real-world applications of AI in image and video analysis |
| **Blockchain Technologies and Applications** |
| Develop an understanding of the fundamentals of blockchain technology<br>Evaluate different types of blockchain networks and consensus mechanisms<br>Analyze the role of blockchain in enhancing data privacy, security, and trust<br>Practice about smart contracts and their applications in AI and business processes<br>Evaluate the potential of blockchain for decentralized AI marketplaces and data sharing |
| **AI Ethics and Responsible AI Practices** |
| Examine the ethical considerations and challenges associated with AI technologies<br>Discuss the importance of responsible AI practices, algorithmic bias, and fairness<br>Debate concepts like AI ethics, human-centric AI, and AI as human augmentation<br>Analyze scenarios to evaluate ethical dilemmas in AI apps<br>Develop strategies for implementing ethical and responsible AI practices in organizations |

## 5  Conclusion

According to (Chen et al., 2022) digital skills are skills that workers should acquire to work with artificial intelligence, (Yang, 2022) in his article claims that high-skilled workers have an advantage in working with artificial intelligence and to conclude with (Wang et al., 2022). Concluded that investing in human capital to get skills for working with artificial intelligence will be crucial to achieving productivity. The loss of blue-collar jobs has already happened due to the automation of manufacturing processes, but the increasing usage of artificial intelligence is also causing less need for white-collar jobs forcing people to continuously acquire new skills to stay competitive and employed. There are three new categories of jobs sustainers, trainers, and explainers, for the maintenance and support of artificial intelligence. Moreover, some of the new jobs created due to artificial intelligence are Digital Transformation Specialists, Human-Machine Interaction Designers, Data Analysts and Scientists, and Software and Applications Developers and this list is not finite. With our education program on future skills, we wanted to address skills demanded in the fourth industrial revolution: complex problem solving, critical thinking, creativity, data management, but also emotional intelligence, judgment and decision making, service orientation, and cognitive flexibility as a part of the soft skills. So far, our initial evaluation shows some outcomes that support our intention to design and address the demand for future skills in AI for comprehensive digital business development. To conclude, the skills that labor should possess will change and people will need to adapt and acquire new skills to stay competitive, and employed, to have a comparative advantage over AI alone.

# References

Accenture: Impact of artificial intelligence (AI) on real gross value added (GVA), by industry, by 2035 (in trillion U.S. dollars) [Graph]. In Statista (2017). Retrieved June 10 2023, from https://www.statista.com/statistics/757821/worldwide-artificial-intelligence-impact-on-gross-value-added-by-industry/

Allahyari, M., et al.: Text summarization techniques: a brief survey. Int. J. Adva. Comp. Sci. Appl. (IJACSA) **8**, 397–405 (2017). https://doi.org/10.14569/IJACSA.2017.081052

Aly, H.: Digital transformation, development and productivity in developing countries: Is artificial intelligence a curse or a blessing? Rev. Econo. Politi. Sci. **7**(4), 238–256 (2022). https://doi.org/10.1108/REPS-11-2019-0145

Chen, N., Li, Z., Tang, B.: Can digital skill protect against job displacement risk caused by artificial intelligence? empirical evidence from 701 detailed occupations. PLoS ONE **17**(11), e0277280–e0277280 (2022). https://doi.org/10.1371/journal.pone.0277280

Chopra, R., Sharma, G.D.: Application of artificial intelligence in stock market forecasting: a critique, review, and research agenda. J. Risk and Fin. Manage. **14**(11), 526 (2021). https://doi.org/10.3390/jrfm14110526

Georgieff, A., Hyee, R.: Artificial intelligence and employment: new cross-country evidence, OECD Social, Employment and Migration Working Papers No. 265 (2021)

Microsoft, U.K.: Accelerating competitive advantage with AI (2019). Retrieved June 10 2023, from: https://aka.ms/AcceleratingAI

Thormundsson, B.I.D.C., Statista: Artificial intelligence (AI) market revenues worldwide in 2020 and forecasts from 2021 to 2023 (in billion U.S. dollars) [Graph]. In: Statista (2022). Retrieved June 10 2023, from https://www.statista.com/statistics/694638/worldwide-cognitive-and-artificial-intelligence-revenues/

Wang, H., Zhang, Q., Lu, H., Won, D., Yoon, S.W.: 3D Medical image classification with depthwise separable networks. Procedia Manufacturing **39**, 349–356 (2019). https://doi.org/10.1016/j.promfg.2020.01.369

Wang, K.L., Sun, T.T., Xu, R.Y.: The impact of artificial intelligence on total factor productivity: Empirical evidence from China's manufacturing enterprises. Economic Change and Restructuring, 1–34 (2022). https://doi.org/10.1007/s10644-022-09467-4

West, D.M.: What happens if robots take the jobs? The impact of emerging technologies on employment and public policy. Center for Technology Innovation at Brookings, 1–20 (2016)

Yang, C.: How artificial intelligence technology affects productivity and employment: firm-level evidence from Taiwan. Res. Policy **51**(6), 104536 (2022). https://doi.org/10.1016/j.respol.2022.10453

# University 4.0: The Future of Higher Education in the Age of Technology

Abdallah Al-Zoubi[1] , María Isabel Pozzo[2]([envelope]) , Monica I. Ciolacu[3] ,
Gustavo R. Alves[4] , and Alexander A. Kist[5]

[1] Princess Sumaya University for Technology, Amman, Jordan
zoubi@psut.edu.jo
[2] National Technological University, Rosario, Argentina
pozzo@irice-conicet.gov.ar
[3] University of Passau, Passau, Germany
[4] Polytechnic of Porto, Porto, Portugal
[5] University of Southern Queensland, Toowoomba, Australia

**Abstract.** The fourth industrial revolution has had an unprecedented impact on humanity. The disruptive technologies driving the new revolution's pervasive change have the most significant impact on the younger generation. Universities should produce and promote the best use of such technologies. The next-generation university's form, function, and purpose must be clarified. The future university has already been given new names, including ecological, distributed cyborg, decentralized, and Zoom University, along with several models that have been imagined. The most important question, though, is whether universities will continue to operate under the same framework in the future or whether they will fundamentally alter the nature of higher education. To imagine the future university as described, this study aims to achieve two main objectives. First, to explore and discuss various models related to University 4.0 and the 4th Industrial Revolution. Second, to analyze the perceptions of Engineering Educators about the term 'University 4.0'. The data elicited from a non-probabilistic international sample of professionals in Engineering Education show partial features, among which 'digitalization' predominates. This tendency supports the idea of a continuation with partial changes (named University 4.0) rather than a profound alteration like higher education (named University 4.1) from their perspective. An overview of University 4.0 aims to spark a discussion about what lies ahead for future generations.

**Keywords:** University 4.0 · Fourth Industrial Revolution · Higher Education

## 1 Introduction

The convergence of new technologies in the fourth industrial revolution, combining physical, digital, and biological elements with disruptive technologies such as AI, IoT, and blockchain, is reshaping the world with excitement, uncertainty, and transformative impact on individuals, industries, and governments. This revolution includes adapting

M. E. Auer et al. (Eds.): STE 2024, LNNS 1027, pp. 122–134, 2024.
https://doi.org/10.1007/978-3-031-61891-8_12

to the new age of digitalization, tackling job losses, and improving skills [1]. Universities worldwide have the responsibility and the potential to produce graduates prepared to face the challenges of a new era using the Education 4.0 system, which encourages innovation and creativity [2]. This approach remains flexible to respond to the impact of Industry 4.0, where machines are interconnected and can communicate and collaborate independently during manufacturing and production processes [3]. The Education 4.0 framework envisions learning experiences that connect and enrich students' real-world experiences and includes a vision for the future of education where artificial and human intelligence work together for research and innovation in a symbiotic digital ecosystem [4]. Therefore, universities must rethink traditional face-to-face teaching models and integrate technology into education to create effective and intelligent connections between students and the latest technologies in a new learning environment.

One possibility for the university by 2040 is the mass adoption of emerging digital technologies, with the significant transition to blended classes that combine online learning components and MOOCs (massive open online course) with on-campus, in-person, and in-class lectures. Assessment of learning, based on individually oriented outcomes, micro-credentials, and graduation requirements, may significantly shift to customized learning [5]. Another possibility is that universities worldwide may experience a new paradigm shift with a completely different model from the present one. The "Future University" may physically transform from a brick campus onto a globally networked click, its primary mission may move from producing knowledge into creating innovation, the role of the professor may change from an information source sage to certified web activist and mentor, and the student from a passive learner to an active knowledge partner and creative designer [6].

Numerous calls for rethinking the design of the future university have been made in the literature. This university has been dubbed Fourth-Generation [7], Cyborg [8], Ecological [9], Entrepreneurial [10], Distributed [11], Singularity [12], Decentralized [13], Zoom [14], and the most general University 4.0 [15]. This paper seeks to initiate a discussion on the shape and type of the new generation of universities by imagining them as either a straightforward extension of the current traditional model or a transformation into a new model. To put it another way, the question is whether the university will completely change in the future, both in terms of its contribution to society's advancement and its outward appearance or if it will continue to change gradually over time as it has in the past. An overview of University 4.0 intends to stimulate the dialogue about the future of next generations.

## 2 Evolution of Education 4.0

In 2008, Harkins coined "Education 4.0" to describe the shift from knowledge-creating to innovation-encouraging education [16]. In his thesis, Harkins hypothesized that centuries of memorization practice, or Education 1.0, and Internet-enabled learning, or Education 2.0, make Education 3.0 possible, in which students are equipped to create knowledge rather than consume it. Students will soon have access to the resources they need to develop and produce innovations, or simply Education 4.0, which is the logical extension of knowledge creation. Harkins asserts that the first nations to adopt localized versions of

Education 4.0 support them with state-of-the-art technology and implement them from primary through higher education will become the leading creative economies in human capital development in the 21st century. He suggested that nations must leapfrog using innovative, time-money-saving techniques to advance beyond the current state-of-the-art in education and other fields.

Harkins modified a taxonomy his student John Moravec created while at the University of Minnesota to show that each of the four iterations of education is interactive but comparatively very different [17]. The traditional teacher-centered approach, represented by Education 1.0 in Harkins-Moravec's model, entails teachers disseminating knowledge to passive digital refugee students seen by the industry as line workers who must be trained but from whom little creation is expected. Learning occurs in a brick building where lectures are the primary form of instruction due to the emphasis on memorization and rote learning. Technology was only occasionally used to aid understanding. On the other hand, Education 2.0 denotes a time when learning is made possible by the Internet, where knowledge is freely accessible, and where instruction is given both by the teacher and by the students who are still digital immigrants. The university also has physical locations and an online presence where hybrid and fully online courses are offered, giving it a brick-and-click style. The teacher is also a qualified professional who collaborates with the students, parents, and other stakeholders to design engaging learning environments. Education 2.0 continued to adopt curriculum-based instruction and textbooks while emphasizing problem-solving and experiential learning, with teachers still playing a significant role in the educational process.

However, Education 3.0 is based on knowledge production, which happens everywhere in a digital world and "creative society"; teaching is a means of knowledge construction that can take many forms, including teacher-student-teacher and people-technology-people. The industry views graduates as knowledge-producing co-workers and entrepreneurs who can support the development of focused knowledge construction. Teachers are everywhere, backed by wireless devices that provide information and raw materials for knowledge production. Hardware and software are also inexpensive and used purposefully in this paradigm.

In Education 4.0, selective individual and team-driven embodiments in practice and targeted innovations are used to build the capacity for innovation. Positive innovation feedback loops amplify instruction in all living, learning, and working phases, everywhere and constantly. Universities are in a globally connected world where innovative supplemental and classroom-replacing tools are continually being developed. Teachers are everyone and about, and they are a creative source of production supported by perceptive software and social partners. The industry views graduates as innovation-producing co-workers and entrepreneurs who can sustain focused innovation construction because hardware and software are constantly being improved. Since almost all software is person-specific and unqualified expressions of familiarity and partnership, graduates are essential to the industry's innovation-producing workforce [18].

Several researchers adopted the model after Klaus Schwab coined the phrase "fourth industrial revolution" in a book with the same name in 2016 [19]. Still, they modified it to align with the term Industry 4.0, developed by the German government in 2011.

Researchers followed an era-division approach to define the four phases of education appropriately and with the distinct periods of the industrial revolutions (Fig. 1).

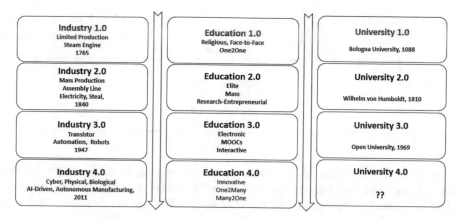

**Fig. 1.** Evolution of Education 4.0 related to Industry 4.0.

According to this model, Education 1.0 was classified as a pre-industrial phenomenon with a face-to-face and one-to-one modality that was primarily religious and that adhered to the model of Bologna University, the first university in the world founded in 1088. When Wilhelm von Humboldt added research to universities' duties in 1810, a new era in their development began. Henry Ford created the assembly line and mass production for the production of automobiles during a period known as Industry 2.0, and Andrew Carnegie and Michael Faraday made significant discoveries in electricity and steel production. Education 2.0 was subsequently categorized with a focus on elitist, for profit, and academically rigorous universities. The 1947 invention of the transistor sparked a series of innovations that resulted in the computer and the internet, featured industry automation, and saw the widespread use of electronic technologies in education. As a result, MOOCs and open universities were created, and Industry 3.0, Education 3.0, and University 3.0 emerged. With the convergence of several technologies that merged the physical, digital, and biological worlds, we live through the fourth industrial revolution, in which artificial intelligence rules the day and autonomous manufacturing is a reality. As a result, the educational system is responding to the Industry 4.0 phenomenon by developing a brand-new kind of education that can be identified by its innovative features and two-way many-to-one and one-to-many character, hence Education 4.0 and University 4.0.

However, the concepts of Education 3.0 and Education 4.0 make this model's categorization of the educational sequence different from the Harkins-Moravec model (Table 1). Most modern researchers disagree with Harkins' claim that Education 3.0 was knowledge production-based and instead claim that it is merely technology-based. The new model's Education 3.0 and 4.0 are equivalent to the Harkins-Moravec model's Education 2.0 and 3.0. Therefore, we propose Education 4.1 to correspond with Education 4.0 in the Harkins-Moravec model, thereby introducing a new term that accommodates all different conceptualizations of the evolution of education and eradicating confusion caused by using the various periods.

**Table 1.** Parallel between the New University 4.1 Model and the Harkins-Moravec Model.

| New Model | Harkins-Moravec Model |
| --- | --- |
| Education 1.0 | – |
| Education 2.0 | Education 1.0 |
| Education 3.0 | Education 2.0 |
| Education 4.0 | Education 3.0 |
| Education 4.1 | Education 4.0 |

To envision the University of the Future, say in 2040, one must abandon the notion that University 4.1 will only change in terms of its modes of instruction and learning. Along with changes in educational environments, other aspects of society will also change, including the purpose and organization of universities, their management and administration, the role of professors and students, and their relation to one another, the nature of research, and its connection to knowledge production, and the role of entrepreneurship and its relationship to innovation production. As a result, one should have a broad understanding of potential changes to the university's other pillars based on the knowledge triangle (education, research, and innovation) [20] and the triple-helix model's predictions for the university's interactions with the government and industry [21].

## 3   Learning in University 4.1

Education 4.0 refers to a new approach to education based on integrating advanced technologies and digital tools into the teaching and learning process. This concept is inspired by the Industry 4.0 movement, which seeks to create a "smart factory" that uses automation and data exchange to improve efficiency and productivity. Education 4.0 emphasizes using technology to support personalized learning, collaborative and project-based learning, and the development of critical thinking and problem-solving skills. This approach also emphasizes the importance of lifelong learning as technology and knowledge evolve rapidly. Education 4.1 builds on the concept of Education 4.0 but emphasizes integrating artificial intelligence (AI) and other advanced technologies into the education system. This includes using AI-based methods to support personalized learning, adaptive testing, and automated grading. Education 4.1 also emphasizes the development of digital literacy and computational thinking skills and integration of social and emotional learning into the curriculum. This approach seeks to prepare students for future challenges, where technology and data analysis will play an increasingly important role in society and the workplace. Overall, Education 4.0 and Education 4.1 represent a shift towards a more technology-driven and personalized approach to education, focusing on developing the skills and competencies needed for success in the 21st century.

# 4   Approach

A comprehensive literature review has been conducted to systematize various approaches and models to achieve the first objective. The results are presented in the previous section as a theoretical framework. To achieve the second objective, empirical evidence is elicited through two data collection methods, the first is an individual survey and the second is a group discussion followed by written production. Both methods were implemented with a non-probabilistic or non-randomized sample of Engineers and Engineering Educators attending an international event. Specifically, it was during the Workshop "University 4.0: Imagined Cyborg or Inevitable Reality?" held during the 20th International Conference on Remote Engineering and Virtual Instrumentation" (REV), organized by the International Association of Online Engineering (IAOE) and Aristoteles University of Thessaloniki, Greece, March 1, 2023 [22]. The paper authors carried out the workshop. In addition to sharing insights on the topic, the activities aimed to collect comprehensive data on the international attendees and their ideas about University 4.0 following a systematic approach.

1. Pre-test Survey: an online form was used [23] to conduct a questionnaire with one open question: "What does the term University 4.0 mean to you?"

The survey was conducted at the beginning of the workshop to gather unbiased and authentic ideas that are not influenced by any speaker's presentations.

2. Post-Test: moderated group discussions using an Education 4.0 design dashboard with activities and thinking strategies by Ciolacu et al. [24]. The following criteria were mandatory: working in small groups (3–5 members), diverse empowered teams working together, and considering timekeeping. The moderators acted as coaches.

The Fig. 2 illustrates the assignment steps, activities, and thinking strategies:

- Introduce the activity: Build a small team (3–5 participants) and introduce each other. Participants wrote their name on a sticky note. Personal thinking (2 min.).
- Brainwriting: Define keywords. Each member writes thoughts related to the given topic. Define a problem. Creative thinking (5 min.).
- Exchange experience: Conduct conversations and know the topic from different perspectives. Discuss with the team your ideas. Creative thinking (15 min.).
- Invent: Develop new ideas. How could we do things differently? Use findings to formulate a precise and user-oriented problem. Future thinking (3 min.). e. Reflect: What should you do next? What has become clear about the situation? Where do you see potential/challenges? Reflection on action. (4 min. + 1 min. Plenum presentation).

Unlike the survey, this activity took place after a theoretical presentation by the speakers, showing the models systematized in the literature. In this case, the workshop attendees elaborated their model with some previous input on the topic.

Both corpora of data were processed through thematic analysis (TA) [24], a method for capturing patterns ("themes") around core concepts or ideas across qualitative datasets. The procedure consists of an open, iterative process named "coding," in which an initial code might be split into two or more different codes, renamed, or combined with others. The aim is to provide a coherent and compelling interpretation of the phenomenon grounded in the data.

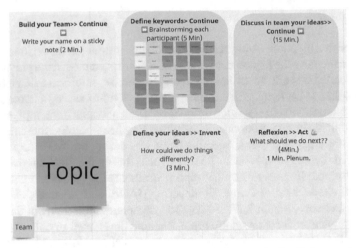

**Fig. 2.** Education 4.0 design dash with design thinking methods, Ciolacu et al. [25].

## 5  Results and Discussion

The results are presented as obtained by each data collection technique.

1) The audience responded to the Google survey (15 Min):

A total of 14 workshop attendees answered the survey. Since it consisted of only one question, 14 responses of different lengths were collected. In each one, themes were identified; thus, each answer was divided into as many themes as it contained (in this case, between one and four). They made 43 excerpts representing 24 themes grounded on the exact words the attendees wrote. In a second step, and as part of the iterative process, the themes were identified with a code and merged among the closest ones, reducing the initial 24 into 14 (as shown in Table 2).

Most responses (8) suggested "digitalization" as the foundation of University 4.0. In line with that theme, other excerpts highlight the role of technology, such as AI and AR and VR tools (4), online labs (3), remote-virtual-eLearning (3), hybridity (2), and better interconnectedness (2). All these can be merged into the code "technology," raising 22.

The second code as regards the number of excerpts is "links beyond university," with seven excerpts: 5 mentioning 'labor market, society and industry and' 2 'other x.4.0 areas'. The following code can be called "Time," which compromises the focus on the future (4 excerpts) and a reaction to the past (1). "New learnings" is the following code, with four excerpts alluding to the same features of learning: 'self-managed,' 'analytical,' 'flexible,' and 'Student-centered.' "New contents" complete the list with the following characteristics: 'interconnected' and '24/7 accessible contents through cloud platform' (3 excerpts). One last code summarizes "improved attitudes" (2 excerpts) comprising 'tending to better human beings' and a better -sustainable- world.'

2) The audience participated in moderated team discussion followed by written productions with collaboration (30 Min.). Engineering educators and researchers (n = 17, 78% male, and 22% female) worked in small groups with the Education 4.0 design

**Table 2.** Step 2 of Data Analysis: Maximizing Similarities

| Excerpts Grouped by Themes | Codes of Respondents | Number of Excerpts by Themes |
|---|---|---|
| **Code**: Technology | | 22 |
| Digitalization | (#1) (#2) (#4) (#7) (#10) (#11) (#12) | 8 |
| The significant role of technology such as AI, AR, and VR tools | (#9) (#12) (#14) | 4 |
| Online labs | (#4) (#11) (#14) | 3 |
| Remote-virtual-eLearning | (#1) (#11) | 3 |
| Hybridity | (#4) (#7) | 2 |
| Better interconnectedness | (#2) (#14) | 2 |
| **Code**: Links beyond the university | | 7 |
| It is integrated into the labor market, society, and industry | (#3) (#5) (#6) (#9) | 5 |
| In line with other x.4.0 areas | (#12) (#14) | 2 |
| **Code**: Time (future and past) | | 5 |
| Next generation of universities / future skills | (#3) (#5) | 2 |
| New pedagogical methods and up-to-date research | (#6) (#9) | 2 |
| Challenging old learning concepts | (#3) | 1 |
| **Code**: New learnings | | 4 |
| Self-managed, analytical, "flexible, Student-centered learning | (#13) (#13) | 4 |
| **Code**: New contents | | 3 |
| Interconnected, 24/7 accessible content through cloud platform | (#13) (#14) | 3 |
| **Code**: Improved attitudes | | 2 |
| Tending to better human beings and a better world (sustainable) | (#8) | 2 |
| **TOTAL OF EXCERPTS** | | 45 |

dash based on design thinking methods. They used Miro's virtual collaboration tool to archive their tasks (Table 3).

**Table 3.** Number of participants and the predefined topics.

| Team | Topics Education 4.0 Design Dash | Number of Participants |
|------|----------------------------------|------------------------|
| 1 | Propose models of University 4.0 | 5 |
| 2 | Characterize a potential model of University 4.0 | 6 |
| 3 | Compare and align Industry 4.0 learning objective models and academic learning objective models | no participants |
| 4 | Identify areas for continuous innovation | 6 |

The three moderated teams presented their findings on technology, links between industry, university, and society, new learning and teaching approaches, new skills and knowledge, and organizational infrastructure (Table 4).

A triangulation of data collected by the two techniques shows recurrent ideas on University 4.0 and provides more solid foundations about the perspectives of Engineering Education professionals. While this shows a broad set of data, several core themes have emerged that underpin the notion of the future of higher education and University 4.0. Foundation on digitalization and technology - the most emphasized aspect and one driving factor is the digitalization of education. This includes integrating current and emerging technologies such as AI, AR, VR, online laboratories, and technology-enabled hybrid learning environments. Technology will play an essential role in the move towards an integrated educational framework.

Universities and beyond – connections between universities and beyond universities with industry are becoming more critical. Technology also impacts the operation of universities and allows more efficient use of resources. Focus of education - Beyond changes relating to operations and technology, there is also a significant shift in the approach to education, away from teacher-centered education to student-centered education. This includes an emphasis on the ownership of learning, content co-creation, personalized and adaptive learning strategies, and collaboration. This also includes a focus on more altruistic drivers, such as contributing to a sustainable world and incorporating ethical, environmental, and societal considerations into the curriculum.

**Table 4.** Findings of the three moderated teams.

| Codes | Team 1 | Team 2 | Team 4 |
|---|---|---|---|
| **Technology** | Teaching development and technologies for 4.0 Motivating teachers to adopt new technologies | AI-based coach Emerging technologies | Acceptance of new technologies and methods |
| **Industry, University, Society** | Collaborate with the industry to deliver content | New private universities will rise like Code 42 Inter-institutional collaboration | Cross-university collaboration |
| **Skills and Knowledge** | Student-focused education Enhance digital student experiences New open-mind approaches | Personal development The Personal Footprint for Competencies Personal and subject-oriented competencies (instead of certificates/grades) How fast and how much they learned Address extended abstract understanding | New Mindset Mindset shift Individual learning outcomes Fostering interdisciplinary exchange |
| **Learning and Teaching Methods** | Using LMS not only for course management but also for instructions Flexible switch between in-person and virtual instructions | Scenario-based learning Hybrid and blended learning approaches | Content dictates methods Digital exams Paperless teaching |
| **Content Delivery** | Share content and course materials between institutions Put students first Students as co-creators of the educational process Give students responsibilities and guidance | Inventing flexible laboratories or learning platforms Address the 17$^{th}$ SDG of UN Flexible, usable personal Cross Labs Personal AI-based learning platforms | More flexibility with courses from other universities New evaluation forms Living collaborations in all studies |

*(continued)*

**Table 4.** (*continued*)

| Codes | Team 1 | Team 2 | Team 4 |
|---|---|---|---|
| **Organization Infrastructure** | Invest in human aspects | Pervasive of University 4.0 Model - the tendency to spread through everything | More budget and openness Paperless administration Non and digital workspaces |

## 6    Conclusions

Numerous calls for rethinking the design of the future university have been made in the literature. This university has been dubbed Fourth-Generation, Cyborg, Ecological, Entrepreneurial, Distributed, Singularity, Decentralized, Zoom, and the most general University 4.0. This paper seeks to initiate a discussion on the shape and type of the new generation of universities by collecting current ideas of them as either a straightforward extension of the existing traditional model or a transformation into a new paradigm shift. To put it another way, the question is whether the university will completely change in the future, both in terms of its contribution to society's advancement and its outward appearance or if it will continue to change gradually over time as it has in the past. To imagine the future university as described, this study aimed to achieve two main objectives. First, to explore and discuss various models related to University 4.0 and the 4th Industrial Revolution. An overview of University 4.0 will surely spark the discussion about what lies ahead for future generations. Second, to analyze the perceptions of Engineering Educators about the term 'University 4.0'. The data elicited from a non-probabilistic international sample of professionals in the field of Engineering Education show an emphasis on individual features, among which 'digitalization' (both in academic and administrative tasks) predominates. This tendency supports the idea of a continuation with partial changes (named University 4.0) rather than a profound alteration like higher education (named University 4.1) from their perspective. However, the reduced size of the sample is a limitation of this study, which could be enriched with further implementations in different contexts. By expanding its findings, future research can contribute to build upon a more comprehensive understanding of this important topic. Besides, this is an ongoing process and changing reality, so future studies should continue monitoring University actors' perspectives and concrete facts in higher education institutions worldwide.

## References

1. Borrageiro, K., Mennega, N.: Essential Skills Needed in the Fourth Industrial Revolution (4IR): A Systematic Literature Review, IST-Africa Conference, Tshwane, South Africa, pp. 1–13 (2023). https://doi.org/10.23919/IST-Africa60249.2023.10187815
2. Vijayalekshmi, S., Twetwa-Dube, S., Vinoth-Kumar, D., Gumbo S.: The Role of Higher Education Institutions in Enabling the Fourth Industrial Revolution: A Bibliometric Analysis,

2023 IEEE AFRICON, Nairobi, Kenya, pp. 1–3 (2023). https://doi.org/10.1109/AFRICON55 910.2023.10293692

3. Boyd, L., Lu, Y., Souppez, J.-B.R.G.: Pedagogy 4.0: Employability Skills and Computer-Aided Design (CAD) Education for Industry 4.0. In: 28th Int. Conference on Automation and Computing (ICAC), pp. 01–06. Birmingham, UK (2023). https://ieeexplore.ieee.org/document/10275222

4. Hussin, A.A.: Education 4.0 made simple: Ideas for teaching. Int. J. Edu. Lite. Stud. 6(3) (2018). https://doi.org/10.7575/aiac.ijels.v.6n.3p.92

5. Gregg, A., Park, J., Fenton, C., Lang, D., Handley, M.: Exploring the "Why" of Micro-Credentials and Digital Badges: Engineering Students' Motivations for and Perceived Utility of Learning Outside of Class. In: IEEE Frontiers in Education Conference (FIE), pp. 1–7. Uppsala, Sweden (2022). https://ieeexplore.ieee.org/document/9962376

6. Beloev, H., et al.: A vision of the university of the future. In: Proceedings of the 21st International Conference on Computer Systems and Technologies (CompSysTech '20), pp. 307–312. Association for Computing Machinery, New York, NY, USA (2020). https://doi.org/10.1145/3407982.3408027

7. Garretsen, H., van de Goor, I., van de Mheen, D.: Dutch experiences in new partnerships between science and practice in health promotion: toward a fourth-generation university. Health Promotion International 38(4), daab194 (2023). https://doi.org/10.1093/heapro/daab194

8. Cyborg University Website (2023). https://cyborguniversity.hcommons.org/

9. Barnett, R.: The Ecological University: A Feasible Utopia. Routledge (2017)

10. Boruck Klein, S., Mafra Pereira, F.C.: Entrepreneurial University: Conceptions and Evolution of Theoretical Models. Revista Pensamento Contemporâneo em Administração 14(4) (2020). https://doi.org/10.12712/rpca.v14i4.43186

11. Heller, R.F.: The Distributed University for Sustainable Higher Education. Springer (2022). https://doi.org/10.1007/978-981-16-6506-6

12. Diwan, P.: Is Education 4.0 an imperative for success of 4th Industrial Revolution? (2023). https://www.su.org/

13. Decentralized University Website (2023) https://decentralizeduniversity.org/

14. The Guardian: 'Zoom University': Is college worth the cost without the in-person experience? (2020). https://www.theguardian.com/world/2020/oct/06/zoom-university-college-cost-students-in-person-experience

15. García-García, F.J., Moctezuma-Ramírez, E.E., Yurén T.: Learning to Learn in Universities 4.0. Human Obsolescence and Short-Term Change. Theory of Education 33(1), 221–241 (2020). https://doi.org/10.14201/teri.23548

16. Harkins, A.M.: Leapfrog principles and practices: Core components of education 3.0 and 4.0. Futures Research Quarterly 24(1), 19–31 (2008)

17. Himmetoglu, B., Aydug, D., Bayrak, C.: Education 4.0: defining the teacher, the student, and the school manager aspects of the revolution. Turkish Online J. Dist. Edu. 21 (2020) https://doi.org/10.17718/tojde.770896

18. Diwan, P.: Is Education 4.0 an imperative for the success of the 4th Industrial Revolution? (2023). https://pdiwan.medium.com/is-education-4-0-an-imperative-for-success-of-4th-industrial-revolution-50c31451e8a4

19. Klaus, S.: The Fourth Industrial Revolution. Portfolio Penguin, London, England (2017)

20. Cervantes, M.: Higher Education Institutions in the Knowledge Triangle, Foresight and STI Governance. National Research University Higher School of Economics 11(2), 27–42 (2017). https://ideas.repec.org/a/hig/fsight/v11y2017i2p27-42.html

21. Etzkowitz, H., Zhou, C.: The Triple Helix: University–Industry–Government Innovation and Entrepreneurship, 2nd ed.. Routledge (2017). https://doi.org/10.4324/9781315620183

22. Ciolacu, M.I., et al.: Developing future skills in engineering education for industry 5.0: enabling technologies in the age of digital transformation and green transition. Open Science in Engineering, pp. 1019–1031. Springer
23. Pozzo, M.I., Borgobello, A., Pierella, M.P.: Using questionnaires in research on university. REIRE **2** (12), 1–16 (2019). https://doi.org/10.1344/reire2019.12.227010
24. Braun, V., Clarke, V., Terry, G., Hayfield, N.: Thematic analysis. In: Liamputtong, P. (ed.) Handbook of Research Methods in Health and Social Sciences, pp. 843–860. Springer (2018)
25. Ciolacu, M.I., et al.: Virtual Collaboration with Agile Methods in Engineering Education 4.0 - Jump to Digital Innovation Units in the New Normal. In: IEEE 9th Electronics System-Integration Technology Conference (ESTC), pp. 282–291. Romania (2022). https://doi.org/10.1109/ESTC55720.2022.9939480

# The Role of Online and Traditional Laboratories in the Context of Modern Engineering Curricula

Alexander A. Kist[1](✉) [iD], Catherine Hills[1] [iD], and Ananda Maiti[2] [iD]

[1] University of Southern Queensland, Toowoomba, QLD 4350, Australia
kist@ieee.org
[2] University of Tasmania, Launceston, TAS 7248, Australia

**Abstract.** Teaching laboratories play an essential role in engineering and science education. The role of online and traditional laboratories in delivering a conventional engineering curriculum is well understood. Learning outcomes that students can achieve undertaking laboratories have been clearly articulated. Partly driven by changes and technology, demands by industry and changing expectations by society, contemporary engineering education is shifting from traditional teacher-focused content-driven paradigms to student-driven programs that focus on professional and twenty-first-century skills. Project-based learning, among other learning activities, replaces lectures and tutorials. This raises the question of what role online and traditional laboratories play in the context of modern engineering curricula. This study compares and contrasts frameworks that capture laboratory and project-based learning to answer this question. The analysis at this high level concludes that many of the learning outcomes that have been traditionally addressed by teaching laboratory sessions are now covered by project-based and other activities. Laboratory learning activities are only required for niche skills and applications. Further work is required to make more specific conclusion about individual learning objectives. This study is a starting point to prompt further investigations.

**Keywords:** Laboratories · learning outcomes · engineering curriculum · 21st century skills

## 1 Introduction

Traditionally, teaching laboratories have played an important role in the delivery of engineering and science curricula. Laboratory classes are an essential part of engineering programs covering topics such as hydraulics and electronics. The affordances laboratories can provide have been captured by the thirteen learning outcomes for engineering laboratories that have been identified by an ABET colloquy [1]. A similar set of objectives has also been identified for science laboratories [2].

Modern engineering education does not only focus on technical knowledge but also the integration of a broad range of skills and real-world applicability. The importance of these non-technical skills is consistently raised by employers and professional bodies. This is reflected in the current International Engineering Alliance (IEA) Graduate Attributes and Professional Competencies [3].

M. E. Auer et al. (Eds.): STE 2024, LNNS 1027, pp. 135–145, 2024.
https://doi.org/10.1007/978-3-031-61891-8_13

As global challenges become increasingly complex, engineers require awareness and a working knowledge of topics such as sustainability, complex ethics, societal and human impacts, but also emotional intelligence. This has been documented by reports such as the scoping study commissioned by the Australian Council of Engineering Deans (ACED) exploring "the knowledge, skills and attributes of professional engineers required to meet anticipated changes in the nature of engineering work in Australia in the year 2035" [4]. Collaborative and project-based learning, work integrated learning and industry partnerships are important building blocks in a modern engineering curriculum to better prepare students for rapidly changing requirements of a modern workforce.

The digital revolution has also had a significant impact on both on content and skills that students need to master as well as how education and universities operate. In the context of laboratory learning, online laboratories have evolved as a viable alternative to face-to-face labs. Diverse learning platforms and hybrid learning models provide flexible options and cater to a wider range of learning styles.

While the boundaries of traditional engineering disciplines blur, agile, well-rounded engineers are required to tackle emerging challenges. Many classical engineering programs continue to follow the standard recipe of lectures, tutorials, laboratories, honours thesis and placements. In this context, the aim of this study is to unpack the role of teaching laboratories in a modern engineering curriculum. As a starting point, this work focuses on contrasting stated learning outcomes of laboratories with project-based delivery methods.

The remainder of the paper is organised as follow. Section 2 introduces the underlying frameworks that underpin this work. Section 3 discusses the research methodology that has been employed and Sect. 4 presents the data analysis. Conclusions are provided in the final section.

## 2  Frameworks

This section discusses relevant literature and frameworks in the context of laboratory learning activities, project-based learning, the changing context of engineering education and university education in general.

### 2.1  Practical and Laboratory Learning

Face-to-face laboratories have been the traditional way to deliver practical leaning activities in engineering. Where students have studied in traditional distance education, residential schools have often been used where distance (now online) students attend on-campus classes to undertake hands-on laboratory classes. With the advent of online technologies, other models have been employed for practical learning. This includes take-home kits, such as pocket labs [5], remote laboratories [6] and other educational tools to allow students to undertake practical activities. In the same context, online simulations are also employed.

The learning outcomes that laboratories are expected to delivery are intuitively well understood by faculty and have been formally identified. The ABET list of 13 potential learning outcomes that can be achieved with educational engineering laboratories is often used as practical starting point for individual engineering educators or institutions to considering the educational objectives for practical activities. "The goal of the colloquy was to determine through consensus a taxonomy of laboratory learning objectives, which could be validated and disseminated throughout the educational community" [1]. A summary of the learning outcomes is shown in Table 1. The taxonomy provides a structure for engineering educators. The taxonomy is widely quoted and other have proposed extensions to the original thirteen items.

Looking broader, a set of generic aims for traditional science laboratory learning have also been proposed [5]. These can also assist to focus how laboratories can contribute to course learning outcomes. The scope of teaching laboratories can also be expanded to cover the conceptual space, rather than a physical space [7].

While the frameworks and potential objectives for laboratory or practical work in engineering or science curricula well defined, in practise they are often not explicitly addressed [8]. "While course goals are often specified, the literature shows a general dearth of well-written student learning objectives for laboratories" [8]. From an educational perspective this can be problematic as "the pedagogical effectiveness of any educational activity is judged by whether or not the intended learning outcomes are achieved" [9]. For the discussion in the context of this research this is less problematic as the work focuses on the learning outcomes as a standard for comparison. Whether the instructional objectives for laboratories are well formulated does not impact on the overall role these activities are meant to deliver are part of the curriculum.

**Table 1.** ABET Colloquy Learning Objectives for Laboratories [1]

| Objective | Details |
|---|---|
| Instrumentation | Apply appropriate sensors, instrumentation, and/or software tools to make measurements of physical quantities |
| Models | Identify the strengths and limitations of theoretical models as predictors of real-world behaviours. This may include evaluating whether a theory adequately describes a physical event and establishing or validating a relationship between measured data and underlying physical principles |
| Experiment | Devise an experimental approach, specify appropriate equipment and procedures, implement these procedures, and interpret the resulting data to characterize an engineering material, component, or system |
| Data Analysis | Demonstrate the ability to collect, analyse, and interpret data, and to form and support conclusions. Make order of magnitude judgments and know measurement unit systems and conversions |
| Design | Design, build, or assemble a part, product, or system, including using specific methodologies, equipment, or materials; meeting client requirements; developing system specifications from requirements; and testing and debugging a prototype, system, or process using appropriate tools to satisfy requirements |

(*continued*)

**Table 1.** (*continued*)

| Objective | Details |
| --- | --- |
| Learn from failure | Recognize unsuccessful outcomes due to faulty equipment, parts, code, construction, process, or design, and then re-engineer effective solutions |
| Creativity | Demonstrate appropriate levels of independent thought, creativity, and capability in real-world problem solving |
| Psychomotor | Demonstrate competence in selection, modification, and operation of appropriate engineering tools and resources |
| Safety | Recognize health, safety, and environmental issues related to technological pro-cesses and activities, and deal with them responsibly |
| Communication | Communicate effectively about laboratory work with a specific audience, both orally and in writing, at levels ranging from executive summaries to comprehensive technical reports |
| Teamwork | Work effectively in teams, including structure individual and joint accountability; assign roles, responsibilities, and tasks; monitor progress; meet deadlines; and integrate individual contributions into a final deliverable |
| Ethics in the Lab | Behave with highest ethical standards, including reporting information objectively and interacting with integrity |
| Sensory Awareness | Use the human senses to gather information and to make sound engineering judgments in formulating conclusions about real-world problems |

## 2.2   Engineering Education and Accreditation Requirements

Significant changes are occurring in the engineering profession and in engineering education relating to the emphasis of required graduate competencies. The latest release the International Engineering Alliance (IEA) Graduate Attributes and Professional Competencies [3] reflects requirements for advance in technologies, changes in engineering disciplines, changes in pedagogies, and human-focused values such as sustainable development, diversity, inclusion, and ethics. Engineers in the future consciously contribute to building a more sustainable and equitable world.

Examples of relevant areas of focus includes impacts of design, net carbon zero outcomes, seeking to achieve sustainable outcomes as represented by the 17 UN Sustainable Development Goals (UN-SDG), a greater emphasis on meeting cultural requirements in all engineering activities, understanding of economic decision-making and application to individual work contributions. The widely cited report by Ruth Graham [10] on the global start of the art in engineering education has highlighted key tends and innovations in the field. Trends include a focus on student-centred learning to address contemporary societal, environmental, and technological challenges, and examples where institutions delivering such curricula at scale.

These changes are also outlined by a scoping study commissioned by ACED [11] that identified that a step change in engineering education is required to deliver graduates needed and valued by the industry. The study emphasised a move from I-shaped graduates with a strong technical focus to T-shaped graduates with the technical competencies complimented with skills and expertise in systems approaches, privileging lifecycle and societal considerations, social license to operate, problem finding, collaboration, creativity, and digital intelligence.

## 2.3 Digital Revolution

Changes in digital technology have had significant impact on society, the way we work and play. Consequently, the digital revolution has also had a significant impact in education. This includes formal university education, but also informal educational such as online course on Udemy and gamified language learning apps like Duolingo, for example. With references to published studies, Haleem [12] discusses 34 significant technologies in education. The list includes virtual classrooms but also promoting distance learning. Practical examples include blended laboratory class delivery[13].

Looking at digital transformation from a student-centred perspective provides several lenses that directly link to student learning experiences and learning opportunities. These include enhancing access and flexibility, personalised learning paths, engaging learning materials, collaboration, digital intelligence and multifaced assessment. Digital technology can also have negative connotations. Challenges include the digital divide, support needs and training, as well as mental health and well-being.

The European Commission (EC) has published a Digital Education Action Plan (2021–2027) that proposes that digital education should facilitate more personalised, flexible, and student-centred teaching [14]. Roo et al. [15] draw on the work of Bonwell and Eison [16] on active learning and the EC's initiative to propose Active Digital Learning Pedagogy which is underpinned by four pedagogical drivers: student-centredness, formative feedback, constructive alignment and flexible infrastructure. While these are different lenses, they all point in the same direction. Similar points have also been raised in other jurisdictions.

## 2.4 Project-Based Learning

Project-Based Learning in higher education are curriculum structures that centres around student project work where students take control of the learning over an extended period of time to explore complex and real-world problems. It is widely seen as an approach that improves student learning and is a common feature of all modern engineering curricula. While an extensive body of knowledge on both project-based exists, student learning processes are not well understood. Studies have investigated affective, cognitive and behavioural outcomes [17] in the context of student outcomes and project-based learning in higher education. The data analysis in this paper will provide more references and discussions on project-based learning.

## 3  Methodology

While the delivery of modern engineering curricula differ widely, one consistent feature is project-based learning as it lends itself to practice-based engineering projects. Therefore, this study uses the affordances of project-based learning as a benchmark for the analysis in this paper. As there is no generic taxonomy of learning outcomes that project-based curriculum can deliver, a methodology was required that systematically collect documented learning outcomes of project-based learning in engineering education. Outcomes identified can then be analysed in the context of generic learning objectives laboratory learning activities. By comparing and contrasting the frameworks, it is possible to develop a picture of the relationships.

As the methodology literature research was undertaken to locate appropriate studies. A systematic, comprehensive literature review that aims to locate all published literature in a particular field which was not necessary to support the argument of the scoping study. Instead, here limited search approach was used. IEEE Xplore, Scopus and Google Scholar were used as databases with PBL, PjBL and the objective in Table 1 as search terms. For each objective separate research was undertaken. The search horizon was 20 years. In some instances, the search terms had to modified to limit the number of results. To limit the matches for *models* the search term *modelling* was employed, for *experiment*, *experiment activity* was used.

## 4  Data Analysis

In this section we present the data analysis. The ABET colloquy learning objectives for laboratories are used as focal points. These are checked against published examples of where PBL is used to address similar learning outcomes. Given the inherent focus of PBL on design, creativity, teamwork, and communications are not included in the analysis. The subheadings below correspond to the objectives in Table 1. It was challenging to locate meaningful examples for *ethics in the lab* and this will require further investigation.

### 4.1  Instrumentation

Instrumentation has been a key focus of several PBL courses. In many cases a course-long project was used as a central theme for the course, with classes and lab sessions often directly supporting different subsystems of the larger project. In other cases, smaller, independent projects were posed in each weekly lab session.

A project was used in a biomedical instrumentation course to provide context and motivation [18] with traditional laboratory sessions being replaced by laboratory sessions where students worked on their open-ended design projects. The instrumentation-related learning objective was deemed to be satisfactorily achieved using this approach. A similar approach was used to implement a PBL approach in a bioinstrumentation course [19] using weekly laboratory sessions to build towards a larger project, reporting a measurable increase in student knowledge and skills in the circuitry design and microprocessor areas.

A weekly "mini-project" was used to provide authentic applications in an instrument techniques course [20], with a PBL approach used to demonstrate achievement of the instrumentation related course learning objectives. Student results showed a significant improvement in attainment of the course outcomes by students when comparing the same course results before and after the implementation of the PBL approach. This study also noted that the PBL version of the course also contributes to twice as many program learning outcomes, bringing in additional elements such as experimental design, data analysis and communication amongst others.

It is noted that these examples are predominantly related to electronics and traditional instrumentation areas and still utilise traditional laboratory facilities, though in a less structured manner. The PBL approach consistently enabled achievement of the instrumentation outcomes and also a marked improvement in attainment of objectives.

## 4.2 Models

The creation and use of models is found across a diverse range of PBL implementations, with many focusing on either fundamental physics models [21] or the use of computer modelling such as finite element analysis [22] and testing more complex models such as computational biology.

PBL was used to develop student's ability to identify suitable constraints-based modelling techniques to apply to real life biology applications [23]. Along with suitable support material and classes in the modelling software, students attained the intended learning outcomes and also learned from each other. Students were also highly engaged with the course and the course evaluation was favourable. In a completely different field, 3D VR models were developed in a mining engineering course [24] where students successfully created 3D models of components, along with an understanding of when the development of such models was economically justified. For a standard engineering mathematics course, Wedelin and Adai showed that students could successfully develop problem solving skills and mathematical modelling skills within a carefully designed problem based learning environment [25]. The style of course was proposed as an interim step towards project based or problem-based learning. These examples show that modelling from diverse areas can be included in PBL courses to attain the intended learning outcomes.

## 4.3 Experiment and Data Analysis

For the purpose of the discussion below both dimensions are discussed together. Experimentation here refers to designing and conducting experiments, and it can be seen that this approach can be used from a high school level [26] in a PBL biotechnology curriculum which directed students to conduct experiments in order to obtain solutions to the problems given. During this process the students conducted the experiments, analyses the results and the data in order to find fundamental concepts. Notable outcomes included opportunities for elaborating creativity as well as data analysis.

Also in Junior High School, a comparison of two cohorts, one using PBL and the other traditional teaching approaches [27] showed a marked improvement and subsequent difference in outcomes for the two cohorts despite a similar starting point. The PBL cohort was shown to have achieved a higher level of cognitive learning and also demonstrated superior scientific skills (including data interpretations, using equipment and experimenting) at completion of the study.

At university level, Technology PBL has been used to replace dictated laboratory experimentation with project learning in an engineering course for a cohort. The students who undertook the TPBL approach achieved higher outcomes than the control group

[28]. Students in an analytical chemistry class reported improvement in many skills (lab work, analytical methods, data acquisition, reviewing results) through PBL activities [29].

### 4.4  Learn from Failure

PBL has been successfully used in a chemical engineering course [30] to move students from the mindset of "lecturer-designed failproof step-by-step instruction" to "student-designed approaches with possibilities of failures and opportunities to learn from failure" by presenting problems in a real world context and providing opportunities for students to undertake self-directed learning. It is noted by Eugenia Etkina in [31] that PBL gives students an increased opportunity to learn from failure.

### 4.5  Psychomotor and Sensory Awareness

Given that the psychomotor domain can be considered to include interpreting "sounds, temperatures, smells and visual cues" [32], it is justifiable to combine the two elements.

A literature review [33] concluded that PBL affects psychomotor skills, though so do other approaches such as inquiry and demonstration methods.

In high school students, a PBL approach was shown to increase both the cognitive and psychomotor achievements of students when they were taught Hooke's Law using PBL experimentation compared to a direct instruction and experiment model [34].

### 4.6  Safety

Lab safety remains a common theme in PBL and traditional laboratory usages, with many PBL courses referring to lab safety rules as a key part of the PBL process such as in chemical labs [35], in agricultural engineering applications [36] and also in multidisciplinary engineering projects such as "Drones for Good" where structures, rules and lab safety training are critical [37].

### 4.7  Ethics in the Lab

A study at an Irish University [38] showed that students reported a high exposure to ethical behavior in PBL lab activities, rating it as the most frequently encountered learning objective. Interestingly the authors noted this was rated much lower by the lab instructors, and also considered the presence of a positive bias among students who felt that ethical behavior might have a direct impact on their grades.

## 5  Observations and Future Work

The data analysis in the previous section shows that for each objective of the ABET colloquy serval examples are available that demonstrate that project-based learning can address these objectives. The analysis did not identify any obvious gaps in the list of objectives. The generic laboratory learning outcomes articulate in Table 1 that are typically expected form teaching laboratory also have value in informing the educational design of learning activities in the context of project-based learning and engineering. To cover all to learning activities need to be purposefully designed. As a next step this project will contrast the traditional laboratory objectives with specific PBL outcomes [39] and the framework of Nikolic et al. [32].

## 6 Conclusions

The study demonstrates that modern engineering curricula delivery methods cover many learning objectives that have traditionally been reserved for teaching laboratories. In the current disruptive environment this will provide a better understanding of what curriculum changes are required and what role teaching laboratories play in a modern engineering curriculum.

As engineering educators, we have experience with traditional delivery models and some innovative technologies, such as remote laboratories. However, while the need for change has been widely acknowledged there is far less clarity on how this can be achieved. By looking at how learning outcomes that have been traditionally delivered by laboratory activities can be delivered as part of the curriculum proper, this study has taken a first step towards addressing this challenge. This study is not indented to be an endpoint of the conversation, but a starting point and a trigger for further work.

## References

1. Feisal, L.D., Peterson, G.D.: A colloquy on learning objectives for engineering education laboratories. In: American Society for Engineering Education Annual Conference & Exposition, pp. 1–12. American Society for Engineering Education (2002)
2. Johnstone, A., Al-Shuaili, A.: Learning in the laboratory; some thoughts from the literature. Univ. Chem. Educ. 5(2), 42–51 (2001)
3. International Engineering Alliance: International Engineering Alliance, Graduate Attributes and Professional Competences version 4 (2021). [Online]. Available: https://www.ieagre ements.org/assets/Uploads/IEA-Graduate-Attributes-and-Professional-Competencies-2021. 1-Sept-2021.pdf
4. Crosthwaite, C.: Engineering Futures 2035 Engineering Education Programs, Priorities & Pedagogies. Australian Council of Engineering Deans, Report (2021)
5. Cvjetkovic, V.: Pocket Labs Supported IoT Teaching. Int. J. Eng. Pedag. (iJEP) 8(2), 32–48 (2018)
6. May, D., Alves, G., Kist, A.A., Zvacek, S.: Online laboratories in engineering education research and practice. In: Johri, A. (ed.) International Handbook of Engineering Education Research. Routledge (2023)
7. Kist, A.A., Maxwell, A.D., Gibbings, P.D.: Expanding the concept of remote access laboratories. In: Proceedings of the 2012 ASEE Annual Conference and Exposition (ASEE 2012). American Society for Engineering Education (ASEE) (2012)
8. Feisal, L.D., Rosa, A.J.: The role of the laboratory in undergraduate engineering education. J. Eng. Edu. 94(1), 121–130 (2005)
9. Arango, F., Chenghung, C., Esche, S.K., Chassapis, C.: A scenario for collaborative learning in virtual engineering laboratories. In: 2007 37th Annual Frontiers In Education Conference - Global Engineering: Knowledge Without Borders, Opportunities Without Passports, 10–13 Oct. 2007, pp. F3G-7-F3G-12 (2007). https://doi.org/10.1109/FIE.2007.4417818

10. Graham, R.: The global state of the art in engineering education, Massachusetts Institute of Technology (MIT) Report. Massachusetts, USA (2018)
11. Crosthwaite, C.: Engineering Futures 2035: A scoping study. Australian Council of Engineering Deans, Report (2019)
12. Haleem, A., Javaid, M., Qadri, M.A., Suman, R.: Understanding the role of digital technologies in education: A review. Sustainable Operations and Computers 3, 275–285 (2022). https://doi.org/10.1016/j.susoc.2022.05.004
13. Kist, A., Hills, C., Maiti, A., Byrne, T., Landers, R.: Contrasting Delivery Models for Authentic, Team-Based IoT Learning Activities, presented at the expat'23 – 6th Experiment@ International Conference. Évora, Portugal (2023)
14. European Commission: Digital Education Action Plan 2021–2027: Resetting education and training for the digital age (2020)
15. Røe, Y., Wojniusz, S., Bjerke, A.H.: The digital transformation of higher education teaching: four pedagogical prescriptions to move active learning pedagogy forward, (in English). Frontiers in Education, Perspective 6 (2022). https://doi.org/10.3389/feduc.2021.784701
16. Bonwell, A.C., Eison, J.A.: Active learning: Creating excitement in the classroom. 1991 ASHE-ERIC higher education reports. ERIC (1991)
17. Guo, P., Saab, N., Post, L.S., Admiraal, W.: A review of project-based learning in higher education: Student outcomes and measures. International Journal of Educational Research 102, 101586 (2020). https://doi.org/10.1016/j.ijer.2020.101586
18. Long, A.S., McKay, T.G.: A design project based approach to teaching undergraduate instrumentation. In: 2014 IEEE International Conference on Teaching, Assessment and Learning for Engineering (TALE), pp. 41–44 (2014). https://doi.org/10.1109/TALE.2014.7062557
19. Kyle, A.M., Jangraw, D.C., Bouchard, M.B., Downs, M.E.: Bioinstrumentation: a project-based engineering course. IEEE Trans. Educ. 59(1), 52–58 (2016). https://doi.org/10.1109/TE.2015.2445313
20. Srikanth, I., Pulavarthi, V.B., Metri, R.A., Bhattar, C.L.: The learning perspective for the course of instrumentation technique through project based learning. J. Eng. Edu. Transformat. 33(Special Issue) (2020)
21. Blandin, A.: Learning physics: a competency-based curriculum using modelling techniques and PBL approach. In: Oral presentation at the GIREP–ICPE–MPTL International Conference, Reims, 22-27 August 2010 (2010)
22. Zhuge, Y., Mills, J.E.: Teaching finite element modelling at the undergraduate level: a PBL approach. In: Proceedings of the 20th Annual Conference for the Australasian Association for Engineering Education (AaeE 2009) (2009)
23. Sauter, T., et al.: Project-based learning course on metabolic network modelling in computational systems biology. PLoS Comput. Biol. 18(1), e1009711 (2022). https://doi.org/10.1371/journal.pcbi.1009711
24. McAlpine, I., Stothard, P.: Course design and student responses to an online PBL course in 3D modelling for mining engineers. Australasian J. Edu. Technol. 21(3), 10/14 (2005). https://doi.org/10.14742/ajet.1324
25. Wedelin, A., Adawi, T.: Warming up for PBL: a course in mathematical modelling and problem solving for engineering students. Högre utbildning 5(1), 23–34 (2015)
26. Wicaksono, I., Budiarso, A.S.: Validity and practicality of the biotechnology series learning model to concept mastery and scientific creativity. Int. J. Instr. 13(3), 157–170 (2020)
27. Citradevi, C.P., Widiyatmoko, A., Khusniati, M.: The effectiveness of project based learning (pjbl) worksheet to improve science process skill for seven graders of junior high school in the topic of environmental pollution. Unnes Science Education Journal 6(3) (2017)
28. Waks, S., Sabag, N.: Technology project learning versus lab experimentation. J. Sci. Educ. Technol. 13, 333–342 (2004)

29. Matilainen, R., Nuora, P., Valto, P.: Student experiences of project-based learning in an ana-lytical chemistry laboratory course in higher education. Chemistry Teacher International **3**(3), 229–238 (2020)
30. Choon, C.O.B., Qingxing, N.X.: Developing self-directed problem solvers in the chemical & biomolecular engineering course. In: 2017 7th World Engineering Education Forum (WEEF), pp. 818–822. IEEE (2017)
31. Feder, T.: College-level project-based learning gains popularity. Phys. Today **70**(6), 28–31 (2017)
32. Nikolic, S., Suesse, T., Jovanovic, K., Stanisavljevic, Z.: Laboratory learning objectives measurement: relationships between student evaluation scores and perceived learning. IEEE Trans. Educ. **64**(2), 163–171 (2021). https://doi.org/10.1109/TE.2020.3022666
33. Riyanti, L.E., Kuntadi, C., Arrafat, B.S., Kurniawan, I.E.: Project-based learning, inquiry methods, demonstration methods, and psychomotor abilities: a review of the literature. Dinasti Int. J. Edu. Manage. Soc. Sci. **4**(5), 822–833 (2023)
34. Leli, S.W., Yosaphat, S.: The effectiveness of problem based learning model to increase physics cognitive and psychomotor achievements through laboratory activity in Hooke law lesson. Jurnal Pendidikan Fisika **5**(2), 83–90 (2016)
35. Liu, J.-P., et al.: A simple demonstration of deoxygenation of carbonyl groups for under-graduates in an organic chemistry laboratory class through project-based learning. Journal of Chemical Education **100**(9), 3540–3546 (2023). https://doi.org/10.1021/acs.jchemed.3c00059
36. Yildirim, S.G., Baur, S.W., LaBoube, R.A.: Fundamentals of framing construction in architectural engineering; a hands-on learning experience, 24–26 (2014)
37. Hoople, G., Choi-Fitzpatrick, A., Reddy, E.: Drones for good: interdisciplinary project based learning between engineering and peace studies. Int. J. Eng. Educ. **35**(5), 1378–1391 (2019)
38. O'Mahony, T., Hill, M., Duffy, A.: Undergraduate engineering laboratories: a study exploring laboratory objectives and student experiences at an irish university, pp. 69–81. Springer (2022)
39. Holgaard, I.E., Kolmos, A., Winther, M.: Designing progressive intended learning out-comes for PBL: a workshop format for curriculum redesign. In: 8th International Research Symposium on PBL, pp. 331–340. Aalborg Universitetsforlag (2020)

# Work-in-Progress: Technology-Driven Introductory Phase in Engineering Sciences for Sustainable Individual Support of Students' Academic Success

Brit-Maren Block[(✉)], Jannis Dethmann, and Benedikt Haus

Institute for Production Technology and Systems, Leuphana University Lüneburg, 21335 Lüneburg, Germany
`brit-maren.block@leuphana.de`

**Abstract.** The work-in-progress (WiP) paper comprises the conception, implementation, and evaluation project of an integrated introductory phase in engineering sciences with the aim of providing personalized support to students before commencing their studies and during their initial two semesters. The didactic innovation lies in designing a digitally supported comprehensive introductory phase that effectively accompanies the transition from school to university and significantly contributes to enhancing students' academic abilities during their initial semesters. As a specific contribution to the ongoing research discussion on learning with and about technologies in engineering education, this paper aims to provide theoretical and practical insights into the following aspects: how to design an enhanced introductory phase based on theoretical foundations, how to agilely address forthcoming developments and assist freshmen using technologies in engineering educational environments, and how to conduct evaluations for quality assurance and validation processes.

**Keywords:** Engineering Education of the Future · Virtual Environments · Learning Analytics · Undergraduate · Mobile and Micro Learning

## 1 Introduction

Growing digitalization and global transformation processes lead to new fields of work and modified framework conditions that the engineering sciences have to face. Graduates need to be prepared to identify and describe problems and to develop appropriate solutions in order to contribute to change processes related to the future in a digital world, e.g. [1–3]. The content-related examination of new technologies is just as essential as the use of modern instructional technologies in the learning process. Moreover, securing young talent in the technical field remains a major challenge. This applies both to attracting students and to reducing drop-out rates, which in some cases exceed 30 per cent, e.g. [4]. Strengthening the transition from school to university and providing students

© The Author(s), under exclusive license to Springer Nature Switzerland AG 2024
M. E. Auer et al. (Eds.): STE 2024, LNNS 1027, pp. 146–154, 2024.
https://doi.org/10.1007/978-3-031-61891-8_14

with more individualised support during the introductory phase of their studies are therefore essential [5,6]. The EER community emphasises the need for
evidence-based teaching in the engineering sciences [7], especially in the upcoming transformation processes [8,9]. The aim of the WiP-paper is the evidence-
based design of an initial study phase that supports students' learning success
in a technology-based and individualised way. This paper contributes threefold
to that discourse: (1) by outlining the research-based re-design of the technology
oriented initial phase of the engineering bachelor degree programme in Sects. 2
and 3; (2) by highlighting two case studies of a technology based learning approach in Sect. 4; and (3) by presenting first experiences and further steps to go
in Sects. 4 and 5.

## 2    Methodology and Research-Based Objectives

The research objectives in this paper are standing for both, to generate implications for teaching practice and also to obtain empirical data on students' learning
strategies in order to provide even more effective and individualised support. In
order to design and to study within the same research process, see [10], Design
Based Research (DBR) was chosen. In the *analysis* part of the DBR model [11],
the data from an initial survey was analysed in order to draw conclusions about
the students and their initial qualifications. The new concept was developed
in the *design* phase, see Sect. 3. After the current implementation, the *evaluation* and reflection is planned, see Sect. 4.3. The longitudinal study of first-year
engineering students was conducted between 2012 and 2023. In addition to sociodemographic data (age, gender), questionnaires were used to enquire about entry
qualifications, completed apprenticeship and self-assessed deficits in their previous education. The analysis of the data shows that more and more first-year
students are transferring to university directly after leaving school. In addition,
the proportion of first-year students with a general higher education entrance
qualification (known as "Abitur" in Germany) has risen from 60% to 92%. This
development comes at the expense of practical prior education, such as through
vocational training or an apprenticeship. The self-perception of general deficits in
knowledge or competences was high at an average of 71% over the entire survey
period. The most frequently cited deficits are a lack of mathematical knowledge,
practical and technical skills and, increasingly, computer skills as shown in Fig. 1.
    In summary, the following aspects were surveyed via the entrance-evaluation:

- strong heterogeneity of the students, in regard to their prior training in the
  fields of mathematics and electrical engineering,
- strong deficits within field related practical knowledge and computer skills.

Based on these findings and the general importance of the introductory phase
for successful study [4–6], the objectives of the study are defined. The aim is to
provide individualised support for students at the start of their studies and to
improve their study skills in the first semester. Section 3 presents the research led
concept of the technology-supported study entry phase and its implementation.

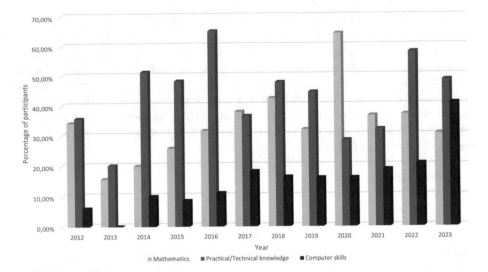

**Fig. 1.** Percentage of respondents who see knowledge deficits in mathematics, practical/technical knowledge and computer skills.

## 3    Design of the Technology Oriented Initial Phase

Figure 2 presents the three-part concept for the start of studies. A *hybrid preliminary course*, which is not the focus of this paper, is designed to strengthen the transition from school to university. It teaches the basics of maths and electrical engineering as well as introductions to current topics (energy, e-mobility, sustainability). The *holistic subject-related first semester* combines the subject-specific first semester modules (maths and electrical engineering) in terms of content and methodology, incorporating digital teaching and learning approaches. Through the *interlacing of electrical engineering and mathematics*, see Fig. 3, synergies can be created and mathematical training can be made more application-oriented. The Electrical Engineering 1 and Mathematics 1 courses each comprise 4 semester hours per week in the first semester; in the following summer semester, Electrical Engineering 2 and Mathematics 2 take place with the same concept. As shown in Fig. 3, the modules consist of lectures, exercises and tutorials. Practical laboratory sessions are integrated into the Electrical Engineering module, in which students carry out experiments in small groups.

The integration of learning scenarios with Matlab and Matlab Grader in both modules is anchored in the concept as *Digital Support and Analytics*. Matlab Grader is an online-based interactive environment for creating

**Fig. 2.** Design of the technology oriented initial phase.

tasks with automatic evaluation, allowing learning at an individual pace and depth for all students. For further details, please visit https://de.mathworks.com/products/matlab-grader.html. On the one hand, this strengthens practice-orientated teaching and enables the cumulative acquisition of skills in a tool environment as well as the systematic development of digital skills relevant to the subject. Another aim is to personalise the teaching-learning process as a contribution to improving self-learning, enabling individual support and research-based study progression. The collection and processing of learner-related data enables a better response to individual learning circumstances and heterogeneity in the first semester and should reduce the current drop-out rate. With data-driven adaptive interventions (e.g. early identification of "risk persons", recognising and actively working on possible deficits, assistance with exercises/tutorials, iteration loops), urgently needed innovations in the area of individualised student support are being developed and tested.

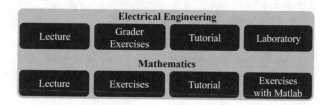

**Fig. 3.** Structural representation of the components of the Electrical Engineering and Mathematics modules.

## 4    Exemplary Concept Modules

### 4.1    Matlab Grader

Prior to the integration of Matlab Grader, the exercises were made available to students online in the form of a PDF after the lecture. Tutorials were held weekly to answer questions and provide assistance. By integrating Matlab Grader, the exercises are provided in a browser-based learning environment. Figure 4 shows an example of a task; schematically, all tasks are structured identically. Depending on the type of task, images can be added for explanation. Students can work flexibly in terms of time and receive feedback immediately after submitting a task. In addition, it is possible to give step-by-step instructions in the event of an incorrect solution. In Fig. 4, for example, only one of the three variables asked for contains the correct target value; a hint box appears for each of the incorrect variables. The number of possible submissions per task can be freely selected; we work with a maximum of two submissions for further investigations. The thematically related tasks are made available after the lecture and can be completed by students throughout the whole semester. In addition to the implementation of Matlab Grader, the tutorial will continue to be offered in presence.

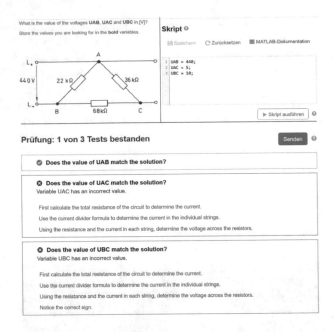

**Fig. 4.** Interface of a sample task in Matlab Grader.

## 4.2   Embedded Systems I/O: Interrupts and Polling

As a second implementation of the concept described in the prequel, a new portable low-cost experimentation platform was designed and is presented here as an example. An application that combines the fields of electronics, computer science, and mathematics is the field of embedded systems engineering. Inspired by [12], where a microcontroller development board is combined with an H-bridge driver shield in order to control the speed of a DC motor. The setup is extended so it can be used as a gentle introduction to the field of embedded systems engineering for students from the second or later semesters. The main idea is to create a *portable* kit that students can also use from home, with no more additional hardware than a PC/laptop, that does not cost too much. This way, students can get hands-on experience with real hardware even in case of lock-downs during pandemics, which is not possible using virtual labs [13]. To supply the required voltage, an inexpensive step-up/boost converter *MT3608* can be connected to a USB port of a PC. While the load unit can be any DC motor, a sensible choice is a 3-wire fan e.g. for PC cooling due to their omnipresence and safe, simple operation. By using a fan with integrated tachometer open collector circuit, the concept of event-based programming using interrupts can be taught, with the specific goal of estimating the rotational speed of the fan. The two base concepts of tachometer signal interpretation, namely time measurement and edge counting, can be combined by using a Kalman filter [14]. It combines model knowledge with measurement information to obtain denoised speed signals with minimal delay compared to traditional filters. Furthermore, Kalman filter theory is an excellent way to implement *sensor fusion* approaches because measurements for different state variables can be exploited at the same time. The software is translated to C code, cross-compiled and then flashed into the Nucleo. Additionally, it is possible to run the software stand-alone in order to monitor signals and tune parameters in real-time while the system is running. In Fig. 5, the whole Simulink software is shown, combining the two base paradigms of embedded systems engineering [15], which are: time-based programming, where the instructions are executed sequentially in loops with a fixed time period, and event-based programming, where external signals will interrupt the execution of the loop and execute the (obligatorily fast) interrupt service routine (ISR). Through the time measurement, it is possible to compute the period of the mechanical rotation, using a factor of 2 (which is a design choice by fan manufacturers and signifies two tachometer signal periods per mechanical rotation). This can be used to calculate the rotational speed (in rad/s), by dividing $2\pi$ through the measured time. On the other hand, for high speeds, it makes more sense to prioritize edge counting for the determination of the rotational speed. For this, the counter signal (each revolution will increment the counter twice), which is an integer number, is multiplied with $\pi$ to obtain a signal for the rotor angle (in 180° steps). This very coarse angle measurement is fed into the Kalman filter together with the angular velocity measurement (sensor fusion) in order to obtain a speed signal that is smooth enough for digital PID feedback control. The portable embedded systems experimentation kit was tested in multiple iterations of a 5-hour-workshop about speed control of DC motors targeted towards second-semester students. The content of the lecture part of the workshop (1.5 to 2 h) is based on [12] and includes a quick introduction to electrical motors, switched inverter (motor driver) technology, pulse-width modulation (PWM),

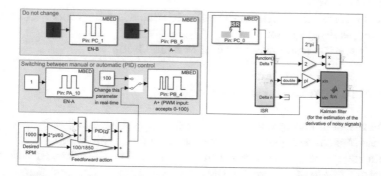

**Fig. 5.** Example Simulink program that is executed in classical time-based loops (*polling*) and also includes asynchronous, event-based elements (*interrupts*) in order to obtain a speed estimate from the tacho signal of the fan.

proportional-integral-derivative (PID) control, modelling of dynamical systems using differential equations and their time-discretization, and other related topics that arise in the highly interdisciplinary field of embedded systems engineering. After the lecture part, the students were provided with all relevant documentation and data sheets about the hardware, driver installation, etc. After that, a list of exercises was posed that instructs and guides the students through the process of designing the *control system* that in its conclusive form is depicted in Fig. 5. While the students were instructed to start working on their own for the first few preparatory exercises, they were later encouraged to work in groups or ask their peers for help in case of technical holdups.

### 4.3   Implementation and Further Research Focus

The two examples were implemented in the Electrical Engineering 1 and 2 modules. The workshop presented in Sect. 4.2 was implemented in the summer semester 2023, the integration of Matlab Grader takes place in the current winter semester. Initial implementation findings on implementation difficulties, usage behaviour and difficulty levels of the tasks are available and can be discussed at STE 2024. The first advantages of using Matlab Grader have already resulted from the personalised submission (during the submission of a task, the time and the submitted solution are recorded and assigned to the corresponding student). The data can be exported at any time and is available to the teaching staff, providing feedback to the lecture (e.g. repetition of difficult topics). The focus of the upcoming evaluation is on the use of Grader; the following research questions will be analysed using the data collected:

- Does the use of Matlab Grader have an influence on the grade average in the exam?
- Are there differences in grades between participants and non-participants in Matlab Grader?
- Is there a correlation between the point in time when the tasks were completed and the resulting grade?
- What is the students' perception of Matlab Grader?

## 5   Conclusion and Further Steps

Based on research findings, an innovative concept of technology-driven introductory phase in engineering sciences for sustainable individual support of students' academic success was derived and implemented. With the reorganisation of the introductory phase of the course, the objective of recording and supporting the individual learning path of the students was stringently pursued. Matlab/Grader is used as an integrative IT tool. Implementation experiences from the current semester provide valuable empirical data on students' individual learning strategies and their support needs and will be incorporated into a re-design in the following semester. First experiences, lessons learned and recommendations will be shared within the EER-community at the STE 2024 conference. Both the developed and implemented new concept as well as the empirical findings of the evaluation will provide an important contribution to the necessary change in the introductory phase of studies in the engineering sciences.

**Acknowledgement.** The research work is kindly supported by the Ministry of Science and Culture of Lower Saxony (Germany). The authors would also like to thank the participating students for their co-operation and feedback. The unknown reviewers are thanked in advance for their comments in the early stages of this work.

# References

1. Auer, M.E., Kim, K.S. (eds.): Engineering Education for a Smart Society: World Engineering Education Forum & Global Engineering Deans Council 2016, Advances in Intelligent Systems and Computing, vol. 627. Springer International Publishing and Imprint and Springer, Cham (2018)
2. Harteis, C. (ed.): The Impact of Digitalization in the Workplace: An Educational View, Professional and Practice-Based Learning, vol. 21. Springer, Wiesbaden (2018)
3. Uskov, V., Howlett, R.J., Jain, L.C. (eds.): Smart Education and e-Learning 2017. Smart Innovation, Systems and Technologies. Springer International Publishing, Cham, 1st edition 2018 edn. (2018)
4. Heublein, U.: Zwischen Studienerwartungen und Studienwirklichkeit: Ursachen des Studienabbruchs, beruflicher Verbleib der Studienabbrecherinnen und Studienabbrecher und Entwicklung der Studienabbruchquote an deutschen Hochschulen, Forum Hochschule, vol. 2017, 1. DZHW, Deutsches Zentrum für Hochschul- und Wissenschaftsforschung, Hannover (2017)
5. Heublein, U., Hutzsch, C., Schmelzer, R.: Die entwicklung der studienabbruchquoten in deutschland
6. Klöpping, S., et al.: Studienabbruch in den ingenieurwissenschaften: Empirische analyse und best practices zum studienerfolg (2017). https://www.acatech.de/wp-content/uploads/2018/03/acatech_STUDIE_Studienabbruch_Web.pdf
7. Borrego, M., Henderson, C.: Increasing the use of evidence-based teaching in stem higher education: a comparison of eight change strategies. J. Eng. Educ. **103**(2), 220–252 (2014)
8. The future of engineering education in a post-pandemic world, vol. Keynotes, pp. 1-22. Universitat Politècnica de Catalunya (2022). http://hdl.handle.net/2117/374092
9. Auer, M.E., Pester, A., May, D.: Learning with Technologies and Technologies in Learning, vol. 456. Springer International Publishing, Cham (2022)
10. McKenney, S., Reeves, T.C.: Educational design research. In: Spector, J.M., Merrill, M.D., Elen, J., Bishop, M.J. (eds.) Handbook of Research on Educational Communications and Technology, pp. 131–140. Springer, New York (2014). https://doi.org/10.1007/978-1-4614-3185-5_11
11. McKenney, S.E., Reeves, T.C.: Conducting Educational Design Research, 2nd edn. Routledge, New York (2019)
12. Block, B.M., Haus, B.: New ways in engineering education for a sustainable and smart future. In: 2020 IEEE Frontiers in Education Conference (FIE), pp. 1–9. IEEE, Uppsala, Sweden (2020)

13. Block, B.M., Haus, B., Von Geyso, T., Steenken, A.: Remote labs in electrical engineering to address heterogeneous student competencies in undergraduate engineering education. In: 2023 IEEE Global Engineering Education Conference (EDUCON), pp. 1–7. IEEE, Kuwait, Kuwait (2023)
14. Wan, E.A., Nelson, A.T.: Kalman Filtering and Neural Networks. John Wiley & Sons Ltd., Hoboken (2001)
15. Morris, N.M.: Interrupts and polling. In: Microprocessor and Microcomputer Technology, pp. 173–186. Macmillan Education UK, London (1981). https://doi.org/10.1007/978-1-349-16651-0_9

# Will My Job Be Automated? Fathoming Current and Persisting Impediments for Automation

Bastian Prell[✉], Simon Wilbers, Norman Günther, and Jörg Reiff-Stephan

University of Applied Sciences Wildau, Hochschulring 1, 15745 Wildau, Germany
bastian.prell@th-wildau.de

**Abstract.** The fear of mass unemployment due to automation has been looming since the first use of automation. At least the last two decades were characterized by the emergence of technological innovation that brought the possibilities of automation to most professions. More recently, AI promises to broaden the use of automation across many industries. Contrary to popular scientific predictions, employment has not yet declined as feared. Therefore, technology has not replaced the inherent human aspects of work. This raises the question of the reasons why it has not succeeded in doing so, and for general limitations of automation. This paper enriches the stream of research in this field by proposing a bottom-up approach. Eleven exemplary professions were chosen, representing roughly 60% of the US employees. Fourteen categories were developed to pay tribute to reasons of untapped automation potential. To overcome those obstacles, the authors propose the use of a checklist, conflating shop floor reality and strategic visions for automation. Hence, this work proposes a tool that is consistent with the paradigm shift towards a more human centered approach, rooted in the updated human, technology, organization model. This advances strategical considerations of the curriculum for automation education as well as insights of socio-economic factors that ought to be considered in automation projects.

**Keywords:** Automation · Acceptance · Socio-Economic · Work · AI

## 1 Introduction

In 1926, the Russian economist Nikolai D. Kondratiev proposed the concept of Kondratiev waves as an attempt to understand long-term fluctuations in the economy [1]. The Austrian economist Joseph Alois Schumpeter identified, at the beginning of every long-term economic upturn, a revolutionary technology that causes "creative destruction" in the economy [2]. Kondratiev waves have never truly gained traction in scientific discourse, partly because the pairing of technology and timing is vague at best and depends heavily on the location. England as the starting point of the industrial revolution, with Germany following with its industrialization decades later. This challenge to the Kondratiev model is currently more relevant than ever with the ongoing industrialization of China, India and parts of Africa. In addition to the regional uncertainty, there is the technology uncertainty. This gets vaguer for more recent times or even the

© The Author(s), under exclusive license to Springer Nature Switzerland AG 2024
M. E. Auer et al. (Eds.): STE 2024, LNNS 1027, pp. 155–166, 2024.
https://doi.org/10.1007/978-3-031-61891-8_15

current developments, as retrospective analysis seems far easier. Even though both concepts have proven useful for past transformations that usually started in the producing sector and revolutionized the overall way people lived, worked and consumed or finally structured their society. The mechanization, accessibility through railways, the electrification, automation and computerization can be stated as both past Kondratiev cycles and innovation that disrupted existing markets and influenced societies by destroying established ways of production and creating new ones.

Furthermore, for the last two decades, technological innovation has emerged that could be put to action in several areas. But for some reason, technology has not yet conquered many of these areas. Projecting into future application for automation, it may even raise the question of general limitations of automation. The authors suspect sociological aspects to play a major role. Therefore, this research aims at working out the socio-economic interaction with new automation technology by starting from the perspective of users or decision makers within companies, discussing what may indicate for using automation respectively against it. Actual usage of technology and therefore the overall advancement towards can be seen as continuous feedback loop iterations rather than the linear understanding of progress. The education of future automation technicians ought to reflect those insights, which motivates this study.

## 2  State of the Art

From the dawn of industrialization, automation and later digitization have evoked the fear of numerous jobs becoming obsolete resulting in mass unemployment. A working paper published by [3] of Oxford University around 2013 continued this narrative by concluding that 47% of jobs in the United States of America could be lost because of computerization [3]. Defined computerization as the application of inexpensive, widely available and potent computing power which enables the rise and widespread application of artificial intelligence. A 2015 study by Economic Research at Ing- DiBa transferred these findings to the situation in Germany and, primarily due to the higher degree of industrialization, concluded that 59% of jobs in Germany in their current form may become a victim of advances in automation and digitization [4]. In sharp contrast to the findings of [3] in the almost ten years since, the unemployment rate in Germany has not risen but slightly declined as data provided by [5] suggests, as it can be seen in Fig. 1. Nonetheless, an increased movement of employment from certain branches especially towards the service sector can be observed, usually accompanied with higher wage levels, that could indicate for professions that may require more qualification, generally speaking, are more complex [6].

As the bespoken paper by [3] has drawn a lot of interest since its first availability to the public in 2013, a wealth of studies has picked up the phenomenon of increasing computerization and its effects on employment. A shortcoming of this study was identified as follows: The authors used an occupation-based approach to measure automation potential. Therefore, for any job that provided automation potential, the authors concluded, that it would be eliminated.

Naturally, this assumption has been challenged by numerous studies. E.g, by analyzing the potential for substitution of human workers by computerization on a profession level and on a requirement as well as task level, or any combinations of those [7], overcame these shortcomings.

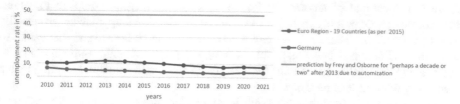

**Fig. 1.** Declining unemployment rate, data by Eurostat [5]

[7] state: "Frey and Osborne probably overestimate the technical automation potential of occupations because the results are based on expert assessments, which typically lead to an overestimated technical potential. Furthermore, when determining the technical potential, social, legal and ethical barriers to the introduction of new technologies are not considered. The technical automation potential is therefore likely to be lower. Above all, however, the results only relate to the technical automation potential. This must not be equated with possible employment effects, since machines can change job characteristics without replacing the human worker. Roughly 15% of all employees (subject to social security contributions) have a high potential for substitution. However, the study points out that existing professions and job profiles are also subject to increasing adaptation to digitization, so that a complete disappearance seems unlikely."

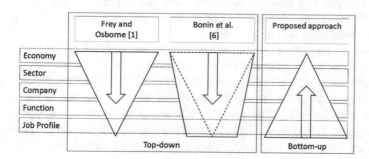

**Fig. 2.** Top-down vs. Bottom-up

Following this reasoning, [8] argues that work will change but not disappear as suggested by [3]. [9] even claim, that automation, might not greatly affect the overall employment. And similarly, to the different paces at which industrialization took place, a range of adoption and acceptance rates can be observed even between companies of the same branch, as it is lined out by [10]. Therefore, according to [11] it's not yet definite, who will benefit from these developments.

# 3   Method

As shown, earlier approaches have mostly relied on top-down analysis, assessing the application potential of automation for countries by evaluating complete branches. Instead, this paper proposes a bottom-up approach. Top-down approaches are usually based on structural data, such as employment rates. The authors argue that a micro-economic validation is beneficial. This is why, the proposed bottom-up approach, as depicted in Fig. 2, relies on methods for identifying automation potential inspired by the practically relevant lean methods.

Following currently observable trends it can furthermore be concluded that technical feasibility can only be seen as a necessity. Integration into organizations and foremost human acceptance ought to be considered as well. Human nature and insufficient integration may both prevent the actual application. As an example, the lavatory attendant can be stated, as their actual task can be seen rather in reminding their customers of the efforts it takes to keep up the respective facilities than the actual up-keeping.

*Thus, the main hypothesis of this paper constitutes that the organizational integration as well as the interplay with human counterparts is restraining the application of automation to a larger extent than actual technical issues.*

This is supported by prior findings from specific industrial context, that the established trinity of *technology, organization and human* is being reconsidered. The usual understanding is constructed so that the task defines the technology, which therefore dictates what organization and human workforce are needed. This might be turned around in the light of labor shortages, instead of human workers structuring their work according to the primacy of the machine, workers' needs become increasingly significant. This topic was addressed by a multi-method approach, that is depicted in Fig. 3. In a first step eleven professions were chosen from a broad spectrum to assess automation applicability technically but also socio-economically.

The initial assessment was structured into 14 categories (see Table 2). Eleven of those were designed by prior interviews with experts in the field of lean management and production consulting. Therefore, the starting point of this study, contrary to previous research, was the shop-floor relevance, instead of automation experts. The remaining 3 categories were deducted from the O*NET-framework which was also referred to by [3]. Reflecting the bottom-up theme, only the three categories focusing on the individual worker were included, while the remaining 3 categories were discarded as they once more had a rather structural focus.

The Occupational Information Network (O*NET), is a database of occupational characteristics issued and continuously updated by the United States Department of Labor. It is derived from survey responses of, representative samples of workers [12]. O*NET items have generally good correlations with US wages, which will be of relevance for the validation, to be discussed later. The three classifiers, that were mentioned earlier, directly chosen from the O*NET Content Model are: Worker Characteristics, Worker Requirements, Experience Requirements.

To assess 11 professions for each of these 14 categories comparison of couples - death match wise - were carried out separately by two researchers. Therefore, the initial assessments resulted in two rankings that were quite similar but nonetheless combined by summation. To assess the explanatory value of the findings, the exemplary professions

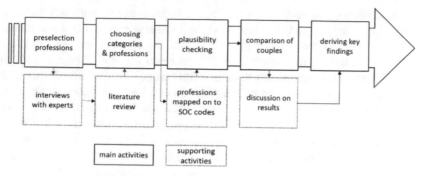

**Fig. 3.** Flow Chart Representing the Approach

were mapped on to standard occupational classification (SOC). This allows to deduce the number of people working in this field, as displayed in Table 1 As the work by [13] such data does not transfer entirely to e.g., Germany, but can yield a trend for industrialized regions. The eleven exemplary professions will be discussed in the remainder of this section. Three of them in more detail presented while all were analyzed in the same manner. It will be structured, so that an introductory statement will summarize how many SOC-professions were included and to what workforce volume this transfer. This will be followed by two statements of which the first lines out, why automation might be applicable at this profession, while the latter states why it is not.

**Barkeeper** - Aggregation of 9 occupations: Bars, restaurants, nightclubs, hotels, coffee shops, and other hospitality establishments serve customers drinks. Of which the majority are standardized products. This holds especially true for carbonated soft drink in the United States where, based on volume, the marked is split entirely between the standard products of three companies (e.g., Keurig Dr Pepper, PepsiCo and Coca-Cola) [14]. Machines could mix and dispense these standard products easily. Even when considering that Barkeepers serve cocktails that have more ingredients, these processes as well could be performed by programmed machines. Bartender typically work long hours, especially during the night, sometimes mostly waiting for customers. Night shifts have shown to have negative health consequences. Human labor could be used more profitably elsewhere. Customers rarely come to hospitality establishments only for the food or drinks. They mostly value the social contact and the atmosphere of the establishment. A good barkeeper or hostess can be an integral part of this experience. Generating the appearance of customization when a cocktail or drink is prepared specially to the wishes of a demanding customer. Even though, a machine could prepare the drink by measuring the ingredients even more precisely, an automated coffee maker has not reached the credibility of a seasoned barista. Relatively low wages in the hospitality industry have since put minimal incentives to automation. Additionally, a human worker when present helps to deter unwanted customer behavior.

**Driver** - Aggregation of 19 occupations: Traffic rules regulate how all parties in traffic should behave. Machines are best in following rules and therefore should replace the human driver. 94% of all road accidents are caused by human error [15]. Removing the cause for 94% of all accidents would result in great saving potential for in insurance premiums. Occurrences of unknown system states similar to those preventing the automation of control room crews apparently are too common on the open road to be neglected. Artificial intelligence does not yet seem capable of coping with them. In particular, when an autonomous vehicle should work under all imaginable weather conditions and all conceivable traffic situations. Additionally, autonomous vehicles and human drivers will have to exist alongside each other, where the latter not always follows traffic rules reliably. Some driving situations require human to human communication, e.g., four vehicles from four directions appearing simultaneously at a four way stop. This and other legal questions need to be considered before autonomous vehicles take to the roads.

**Lavatory Attendant** - Aggregation of 4 occupations: Receptionists as well as lavatory attendants are on standby, waiting for work for most of their workday. The actual value adding portion of their workday is comparable short. By traveling between multiple facilities, this portion could easily be increased, even if transfer times would counter this effect slightly. Money collection via turnstile or other means of access control requires some initial investment but will break even eventually. Everyone has the need to relieve themselves, which results in a large market allowing the recovery of the development cost of self-cleaning toilets or washrooms. Their actual task can be seen rather in reminding their customers of the efforts it takes to keep up the respective facilities than the actual up-keeping. Similar to security guards, the presence of lavatory attendants deters unwanted human behavior. In cleaning, unknown system states will occur that may prove challenging to automate.

**Security Guard** - Aggregation of 8 occupations: Alarm systems or systems to trigger alarms in case of non-compliance have been historically among the first examples of automation. The function of a security guard, checking whether a building or property is in order and to detect looming damage potential, is by definition extremely repetitive which would be the standard application for automation. Once the standard has been defined, the job of a security guard is to check whether the standard is held up or could be endangered in the future. Sensors checking for damages (e.g. water leakage or glass breakages), sophisticated CCTV cameras hooked up to computer vision algorithms should potentially be capable of accomplishing the task of a security guard. It could prove to be hard to identify all possible damage scenarios beforehand that a system would have to check for. Especially since most damage scenarios will rarely occur, possibly never. Designing and configuring an intelligent alarm system that can deal with yet unknown system states would be extremely costly and maybe not possible. Security guards, deter potential delinquent behavior by mere presence. This only holds true for a human or near human (animal) adversary. A flood or fire will not be deterred by a security guard. Many crimes of opportunity are not well planned, and the lack of visible security measures may even encourage, especially spontaneous crimes.

**Support** - Aggregation of 2 occupations: Call center agents provide support according to a script. Script loyalty, sticking to the predetermined phrases and information by itself is a metric, call center agents are measured against. Once this script is established and tested, chat or voice bots can do the work of a call center agent. The three-tier system has already been established, so that roughly 80% of inquiries are dealt with automatically, leaving the 20 most complex percentd to the human. The boundaries of what automation is capable of for feasible set up effort will surely be shifting. The bulk of the requests a support line answers are in no way special; this could be substituted by chat bots respective voice bots. What if the situation changes and the script is not updated? Or the requests presented to the customer service agent represents an unknown system state? Automation may be capable of giving us a self-adapting script, but do we want this? Will human executives trust AI enough to endow it with the responsibility of changing its script? Otherwise, the topics handled on support lines, may get more complex parallel to technological advancement, so that the most complex inquiries may always be a step ahead of the automation of its answering.

Also analyzed in a likewise manner were the following occupations **Cashier, Control Room Crew, Insurance Agent, Pharmacist, Post Clerk**, which couldn't be discussed in depth in this paper.

# 4 Results

This assessment resulted in a matrix with values for each exemplary profession and each category. Whether such qualitative assessments are robust could be validated by a category that was assessed but could also be underpinned with quantitative real-world data. The wage level can serve as such a validation variable. From the assessment a ranking can be generated in ascending order, while another rank can be formed from the hourly mean wage data depicted in Table 1. Both ranks were compared, and only minor differences detected. The top and bottom quartile had been estimated consistently to the data, which therefore holds true for the middle range, except one miss-estimation that build up, so that an offset was detected throughout this middle range. Nonetheless, this validation test was passed.

The remaining categories were grouped into those that were positively and those that were negatively connotated with automation possibilities. All categories were either positively or negatively connotated, as neutral categories were not included by method design. Therefore, the implications were inverted for the negative connotated categories, so that a higher ranking would be correctly counted as less suitability for automation application. The results were than displayed as a heat map in Fig. 4. Green coloring represents high chances and advantages for automation, while red indicates obstacles, risk and resistance. While yellow represents the continuum in between.

Furthermore, this supports the findings of [7] that automation will occur first at task or requirements level and not at job level. This is characterized by replacing or shifting the need from some tasks to others of a job rather than replacing the entire job.

The presented method could be useful to assess the current state by reviewing the automation potential from different angles. It is worth mentioning that all categories were assumed to be static. Over a long - or maybe even not so long - period, these categories and their impact will shift in their connotation, weigh or characteristics. Legal implications, wage levels, worker shortages, traditions, trust level, capital intensity are prone to change over time. A task that used to be not opportune for automation under shifting socio-economic conditions may become disposed for it.

**Table 1.** Data for Example Professions

| Profession | Employment | Hourly Mean Wage [$] | Share of workforce |
|---|---|---|---|
| Barkeeper | 1.609.830 | 14,67 | 1% |
| Cashier | 6.680.190 | 12,37 | 5% |
| Control Room Crew | 648.370 | 37,65 | 0% |
| Driver | 31.913.310 | 22,84 | 22% |
| Insurance Agent | 5.678.870 | 32,05 | 4% |
| Lavatory Attendant | 4.426.020 | 13,95 | 3% |
| Pharmacist | 769.680 | 31,11 | 1% |
| Post Clerk | 1.131.150 | 23,27 | 1% |
| Security Guard | 7.900.300 | 21,67 | 5% |
| Support | 2.892.520 | 17,43 | 2% |
| System Caterer | 23.494.940 | 15,3 | 16% |
| Sum | 87.145.180 | 23,73 | 59% |

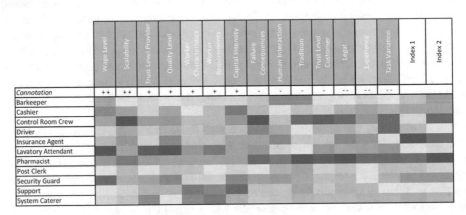

**Fig. 4.** Heatmap of Automation Feasibility

The proposed method could be regarded as a way to construct an understanding of the shop-floor, respectively application reality of automation. As deduced earlier, assessing the status quo of how much automation could expand into existing job profiles is rather complex. Future developments therefore are even harder to predict. In environments with high uncertainty the use of scenarios is advised. Nonetheless, the overall trend of whether a category signals positive or negative connotation with automation by itself is already a factoring. This equal weighting was considered and accumulated with respective positive or negative sign. Categories showed the same number for positively and negatively connotated categories. Also, the amplitudes were designed comparable. Thus, the expected value for symmetrical distributions was zero. Such generated values were defined as Index 1. For a second scenario the authors discussed the categories and could agree that some factors seem to be more relevant than others. Such categories were weighted twice as much for a second Index. Therefore, the connotation was expressed by the symbols + and −−in Fig. 4. Those that had twice the impact for Index 2 were marked with the same sign but twice, so that − and ++ can be found in the *Connotation* row.

For all professions the chosen scenario did not change the assessment expressed by the Indexes. Those that were positive for Index 1 stayed positive for the second Index. Thus, it can be concluded that also in between the two chosen scenarios a steady assessment is observed. This brings forward evidence for the robustness of the Index assessment, as estimating a certain weight for each factor may be quite difficult. From the Index review it becomes obvious that insurance agent as well as pharmacist are professions least prone to automation, while security guard and system caterer seem to provide least obstacles towards being automated. Low mean levels for wages, e.g., for security guard or lavatory attendant might offer less motivation for automation. Also, tradition and legal obstacles are relevant when automating the pharmacist job profile. An exception amongst the job profiles considered, is the driver. For all other considered jobs, mainly socio-economic reasons keep a job from being automated, not technical constrains. In contrast, for the driver, it is mainly the technical reason of very high variation in the task of driving and the relatively high consequences of failure that keeps the human driver from being replaced.

The 14 categories turned out to be highly significant for whether a job or task can be automated. This has become apparent not only through the validation step, but also through the detailed practical work on the 11 example professions. Due to the high relevance for automation of the 14 categories, a practical tool for decision-makers could be created. This Multi-Aspect Checklist for Automation as displayed in Table 2 is a practical tool that was produced by this research. An executive intending to automate a job or subtask should ask all 14 checklist questions derived from the 14 categories before the project start. If he must answer with "No" to one of the questions, this is an indicator that the selected task cannot be successfully automated. Or at least the question answered with "No" requires further investigation. In this case, we strongly suggest specifying the task marked for automation in greater detail, possibly decreasing the scope of the project respectively, or automating another task first.

**Table 2.** Multi-Aspect Checklist for Automation Projects

| | | | |
|---|---|---|---|
| Wage level | + + | Highly conducive to automation | Is the wage level for the task in question relatively high? |
| Scalability & Task Volume | + + | | The task is being performed relatively often? |
| Trust level required from provider | + | conducive to automation | Is the trust level required from provider of the service or product relatively high? |
| Quality Demands on the product or service | + | | Is the quality demands for the product or service relatively high? |
| Worker characteristics (O°Net Model) | + | | Are the worker characteristics relative to other tasks relatively specific? |
| Worker requirements (O°Net Model) | + | | Are the worker requirements relative to other tasks relatively specific? |
| Capital intensity of the operation | + | | Is the task, relative to others, capital intensive? |
| Consequences of failure | − | detrimental to automation | Are consequences of failure low? |
| Amount of human interaction | − | | Is the amount of human to human interaction low? |
| Tradition of the trade | − | | Is the amount of tradition of the task low? |
| Trust level required from customer | − | | Is the trust level required from customer low? |
| Legal Implications | − − | highly detrimental to automation | Are legal implications low? |
| Experience Requirements [O*NET® Content Model] | − − | | Are experience requirements low for the worker currently performing the task? |
| Variation in the Tasks | − − | | Is the variation in the tasks low? |

## 5   Conclusion

Primarily, this paper presents an addition to the discussion initiated by [3] by proposing a bottom-up approach, starting from job profiles rather than from structural data. This is a thought experiment, on what jobs to automate first in a company employing a variety of exemplary jobs. The representativeness of the exemplary jobs was proven by mapping them to SOC data. With about 59% of US employees working in the eleven exemplary jobs considered.

The assessment of suitability for automation was conducted by a variety of categories that were constructed by interviews with consulting experts for factory and value stream design. This was followed by a thorough plausibility and robustness check of the model. With this basis, two researchers pairwise compared the exemplary professions, always determining a profession who was more suitable or more prone to each category. This produced a ranking of the eleven professions by these categories.

None of the exemplary professions were "green" across all categories. Representing, that none of the professions was completely automatable. This supports argument by [13] to review tasks rather than entire professions. None of the exemplary professions exhibited "red" across all categories, indicating a total impracticability for automation. One notable exception among the exemplary professions it the driver. Where all categories are either green or red except of two technical categories. Hinting at the fact that mainly technical hurdles like the relatively high consequences of failure and the high task variation keep drivers from being automated. It should be noted that the researchers who prepared this model are not free from prejudice and misperception of professions and the people who practice them. For example, it is possible that the researchers were biased towards the scientifically educated pharmacist in comparison to the lavatory attendant, which might require substantially less training.

Lavatory Attendant and Pharmacist seem to be antithetical, in the proposed model. Category wise, they form an almost perfect pair of opposites. Since neither is due for automation, this leads to the question whether classic categories like white color vs. blue-collar workers are still the right dimension of thinking about automation. It is also worth mentioning that this study was designed prior to ChatGPT publications, which only support that latter statement and could break ground for further exploration. Possible scenario analysis could take into account trends such as changes of capital costs, wages and other labor costs as well as resource shortages or even supply chain disruptions. A practically useful result of this research is the multi-aspect checklist for automation projects. If a costly automation project is considered for implementation, managers should make sure to ask the questions on the checklist. Questions that must be answered with no represent a high risk of failure for the planned automation project. It warrants future research on the experience of those applying this checklist.

**Acknowledgments.** This publication was made possible through the funding of the PhD program Innovation and Career Center - Integrated Engineering by the state of Brandenburg's Ministry of Research, Education and Culture (Germany) and the *pro_digital* European Digital Innovation Hub (EDIH) which has received a co-funding from the European Union's DIGITAL EUROPE Programme research and innovation programme grant agreement No 101083754.

# References

1. Kondratieff, N.D., Stolper, W.F.: The long waves in economic life. Rev. Econ. Stat. **17**, 105 (1935). https://doi.org/10.2307/1928486
2. Schumpeter, J.A.: Kapitalismus, Sozialismus und Demokratie (Uni-Taschenbücher S) - ZVAB - Joseph A. Schumpeter: 3825201724. UTB, Stuttgart

3. Frey, C.B., Osborne, M.A.: The future of employment: How susceptible are jobs to computerisation? Technol Forecast Soc Change **114**, 254–280 (2017). https://doi.org/10.1016/j.tec hfore.2016.08.019
4. Brzeski, C., Burk, I.: Die Roboter kommen. IngDiBa Research Report, Frankfurt (2015)
5. Eurostat: Unemployment Data (2022)
6. Wolter, M.I., et al.: Industrie 4.0 und die Folgen für Arbeitsmarkt und Wirtschaft, IAB-Forschungsbericht 8/2015, Nürnberg (2015)
7. Dengler, K., Britta, M.: Folgen der Digitalisierung für die Arbeitswelt. IAB-Forschungsbericht 11/2015, hdl.handle.net/10419/146097, Nürnberg (2015)
8. Timpf, S.: Beschäftigungswirkungen der Digitalisierung und kein Ende der Arbeit? Dossier "Digitalisierung" Teil 4 für die Kommission "Arbeit der Zukunft" (2017)
9. Zika, G., et al.: Arbeitsmarkteffekte der Digitalisierung bis 2035: Regionale Branchenstruktur spielt eine wichtige Rolle. IAB-Kurzbericht 9/2018, hdl.handle.net/10419/185844, Nürnberg (2018)
10. Stettes, O.: Keine Angst vor Robotern. IW-Report 11/2018, hdl.handle.net/10419/176771, Köln (2018)
11. Lehmer, F., Matthes, B.: Auswirkungen der Digitalisierung auf die Beschäftigungsentwicklung in Deutschland. IAB-Bericht5/20017, doku.iab.de/aktuell/2017/aktueller_bericht_1705.pdf, Nürnberg (2017)
12. Boese, R., Lewis, P., Frugoli, P., Litwin, K.: Summary of O*NET 4.0 Content Model and Database at O*NET Resource Center., onetcenter.org/dl_files/summary_only.pdf, Raleigh, North Carolina (2001)
13. Bonin, H., Gregory, T., Zierahn, U.: Übertragung der Studie von Frey/Osborne (2013) auf Deutschland, ZEW Kurzexpertise Nr. 57, ftp.zew.de/pub/zew-docs/gutachten/Kurzexpertise_BMAS_ZEW2015.pdf, Mannheim (2015)
14. Statista: Leading carbonated soft drink (CSD) companies in the United States in 2020, based on volume share. In: Beverage Digest (2021)
15. Nielsen Marktreport: Nielsen Consumers Deutschland: Verbraucher - Handel - Werbung. The Nielsen Company (2018)

# Industry 4.0 and Education

# Implementation of an Industry 4.0 Module in the International Educational Project ERASMUS+ WORK4CE

Peter Arras[1] (ID), Anzhelika Parkhomenko[2,3](✉) (ID), and Illia Parkhomenko[2,3] (ID)

[1] KU Leuven, Campus De Nayer J.P. De Nayerlaan 5, 2860 Sint Katelijne Waver, Belgium
[2] University of Applied Sciences and Arts, Otto-Hahn Str. 23, 44227 Dortmund, Germany
anzhelika.parkhomenko@fh-dortmund.de
[3] National University "Zaporizhzhia Polytechnic", Zhukovskogo Str. 64, Zaporizhzhia 69063, Ukraine

**Abstract.** The ERASMUS+ WORK4CE (Cross-domain competences for healthy and safe work in the 21st century) is a European funded educational project, which aims at bringing patterns, methods and competences for the projectized and digitalized future of Industry and society. The Industry 4.0 educational module was developed as part of this project with the purpose of giving students introductory knowledge on the key new technologies that are the core of Industry 4.0 so that they can understand the technological challenges of digital transformation. The paper presents the results of developing a framework for implementing the case method for a multidisciplinary project when studying Industry 4.0 module.

**Keywords:** Industry 4.0 · Case Method · Wire Bending Machine · Multidisciplinary Project · Virtual and Physical Prototypes · Virtual and Augmented Reality · Human Machine Interface

## 1 Motivation and Purpose

Industry 4.0 is a way of fusion of advances in Artificial intelligence, Cyber-Physical Systems, Robotics, the Internet of Things, Additive manufacturing, Cloud Computing, Big Data, Augmented Reality and many other technologies [1, 2].

Connected technologies (networks, IoT, robots), modern technologies in design and production (digital twins, virtual and augmented reality, additive manufacturing) and new software solutions (machine learning, artificial intelligence, etc.) are omnipresent in industrial digital transformation projects. However, many enterprises and companies embarking on the path of digital transformation face numerous organizational, technological and operational challenges [3].

Therefore, an important educational method is an integrated approach in teaching technology, reflecting the real-world challenges in digital transformation [4, 5].

In the Work4Ce project a 10-credit teaching module is developed for Industry 4.0. It consists in an obligatory core course (Introduction to Industry 4.0 and pathways to Industry 4.0 project management) and several elective technical courses (Digital twins, Internet of Things (IoT), Additive manufacturing, Quality of industrial systems/predictive maintenance/condition monitoring, Data analysis for industry 4.0/Use of Big Data, Building information management (BIM)). Students can choose a set of electives on a more in-depth exploration of the key technologies of Industry 4.0. This makes it possible to adapt the educational materials to the changing needs of the market and the individual educational plan of each student [6].

Showing and experiencing the digital transformation process is a complex task in the education of (engineering) students as it has many facets. To cope with these challenges methods of case studies/problem-based learning can be used to give a round-up and as an example of implementation.

As known, a case is a description of a specific problem and is intended to teach students to analyse various types of information, summarize it, formulate a problem and develop options for solving it, usually as a team, in a short time [7].

The authors of the work [8] described the general principles of designing and conducting a case study with all necessary levels and stages and proposed the framework that can help researchers in conducting research in easy way.

The paper [9] provides an in-depth overview of the challenges researchers face when applying case studies and suggests solutions to these challenges from a practical perspective.

The authors of the paper [10] point out that the selection and application of case studies is a complex task and involves great responsibility that case study developers bear.

The authors of [11] showed that the era of the Industry 4.0 clearly demonstrates the close connection between cyber-physical and social systems, and presented a multidisciplinary case study of a socio-technical cyber-physical system.

The paper [12] describes the benefits of integrating multidisciplinary and team learning, where students from different faculties can work on real projects, develop working prototypes, present scientific and practical results, and demonstrate learning outcomes.

The article [13] discusses in detail the key benefits of multidisciplinary collaboration using a BIM-based shared building model approach, which arouses students' interest in learning, help improve communication skills and find solutions to problems.

An example of a multidisciplinary case study related to IoT technologies was presented in [14]. The work shows that in the process of implementing a case study, students master the principles of operation of traffic control systems, the features of organizing traffic light control, and develop hardware and software for a prototype of traffic light control system.

Although there are certain methods for developing a case study, nevertheless, teachers do not always decide to use this format due to certain difficulties in its implementation in practice. Therefore, the task of improving the technology of working with cases in the educational process is relevant.

The purpose of the work is the development and practical implementation of a case framework for a multidisciplinary project in the Industry 4.0 module.

# 2 Methodology

Recommendations for the digital transformation of an educational course [15] can be successfully used when developing a case framework. The proposed stages of preparing a case study for the educational process are as follows.

1. Analysis of course parameters. It involves identifying the target group, context and available resources for conducting the case study.
2. Determination of the objectives of the case study. It involves identifying the case study's contribution to the course in order to stimulate the learning process and add value to the course being studied.
3. Determination of case study activities. It involves the selection of types of students' activities (Acquisition, Inquiry, Practice, Discussion, Production, Collaboration) and their sorting by content and order of execution to achieve case study objectives. Also, to support each type of student activity, it is necessary to choose didactic formats (forms of work and associated tools) that can be used by teachers (for example, Concept map, Presentation, One-sentence summary, Facilitating research questions, Simulations, Discussion board, Group work, Group discussion, Peer instruction).
4. Determination of content and assessment methods. It provides for the definition of intermediate and final results, on the basis of which teachers will provide feedback to students and will be able to evaluate their achievements at all stages of the implementation of the case study.
5. Creating a case study design. It involves the development of a general plan of action for the implementation of case study by students and assessment by teachers, as well as the establishment of correspondence between learning activities and assessment.

The developed case study should be multidisciplinary, attractive and recognizable to students, understandable for implementation and offer opportunities for the integration of various Industry 4.0 technologies.

The case study "Research and development of virtual and physical prototypes of a Wire Bending Machine (WBM)" is an example of multidisciplinary project in which students have to combine different knowledge and skills over the complete design and prototyping phases of a real asset.

The case uses a step-by-step introduction from domain research, problem statement, creation of a virtual prototype forward to a physical prototype of WBM.

The case implements an interdisciplinary approach between mechanical (design and manufacturing, optimisation), automation/electronics (instrumentation, feedback) and software engineering (control algorithms, data-analysis), and also suggests opportunities for future integration of additional technologies.

The main goal of case study is to plan, design and implement a prototype of WBM, to evaluate different options and technologies for implementation, and to manage the project in Agile sprints. The management can be done in Atlassian Confluence (all development documents and artefacts) and Atlassian Jira. These management documents and artefacts will be reviewed during the assignment presentations.

Proposed stages to implement the case study:

1. Analysis of the current state and problems in the area under study, research of exist-ing prototypes of WBM. Didactic formats are facilitating research questions, peer instruction, group discussion, one-sentence summary.
2. Analysis of the requirements to WBM prototype, development of the concept of a prototype and business plan for its creation. Didactic formats are concept map, group discussion, group work, presentation.
3. 1st sprint. Implementation of WBM virtual prototype using virtual and aug-mented reality technologies, engineering calculations, programming and simulations. Didactic formats are group work, group discussion, simulations, presentation.
4. 2nd sprint. Implementation of a physical prototype of WBM prototype using additive manufacturing technologies. Didactic formats are a group work, presentation.
5. Testing and optimization of the prototype. Planning for future work opportunities. Didactic formats are group work, group discussion, presentation.

## 3   Actual and Anticipated Outcomes

Wire bending technologies are widely used in the aerospace industry, automotive, shipbuilding, mechanical engineering and other important areas [16].

Existing technological systems for wire bending are designed for bending wire of different diameters made from a wide range of alloys. In the general case, the WBM with computer numerical control (CNC) includes an accumulation mechanism, an alignment mechanism, a feeding mechanism, and a wire bending mechanism.

The WBM prototype is small machine to bend metal wires in 2D and 3D shapes. It consists of electromechanical and control subsystems (see Fig. 1). Thus, such object allows students joined studying of mechanical, electrical and software engineering related to manufacturing systems and Industry 4.0 processes.

The electromechanical subsystem consists of several assembled mechanical parts such as frame, shaft, gears, three stepper motors and one servomotor. Main components of the control subsystem are Arduino Nano V 3 board and CNC Shield V 4 that can be used as a drive expansion board for CNC machines.

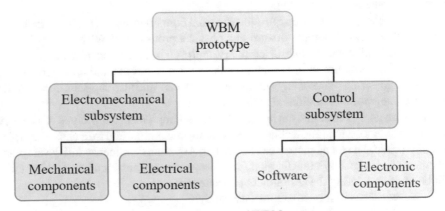

**Fig. 1.** General structure of WBM prototype.

Several examples of such physical prototypes for studying the main principles of wire bending technological process and ways to control it are presented in [17–20].

The difference between prototype and professional equipment is huge. Prototypes have parameters and characteristics that do not meet real industrial requirements, but are useful for finding ways to modernize technological systems and their control algorithms. Therefore, the topic of the case study is relevant.

Virtual and physical prototypes are often used both in the field of design and in the field of technological preparation of production. Students perform this case study using virtual and augmented reality technologies in PTC Creo environment (see Fig. 2 a, b). After that they use 3D printing technology (see Fig. 2, c), assembly and test WBM physical prototype (see Fig. 3).

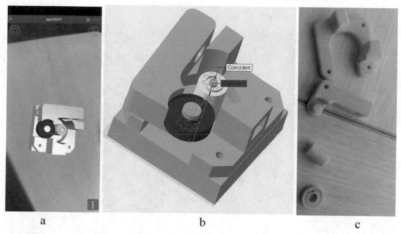

a                                 b                                 c

**Fig. 2.** Application of augmented (a), virtual (b) reality and 3D printing (c) technologies for development of feeding mechanism for WBM.

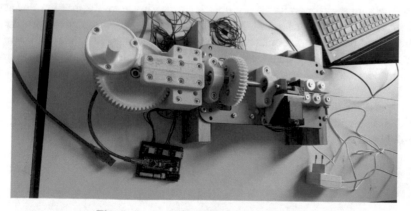

**Fig. 3.** Testing of WBM physical prototype.

In the future, changing all of the components to metal analogues would surely make the whole prototype more professional, reliable and higher quality.

The goal of the programming process is to create a control code for three stepper motors of the machine that provides the possibility to bend different geometrical shapes: the feeder motor feeds the wire further in the machine, the bender motor bends the wire in 2D shapes, the Z-axis motor allows creating 3D shapes.

Computer numerical control is carried out by user interaction with the prototype controller through a graphical interface to ensure the selection of the geometric shape and its parameters. The software for the WBM prototype control can be developed in the Arduino IDE and human machine interface – in the Processing environment (see Fig. 4). For electronic circuit simulation online system Tinkercad can be recommended [21].

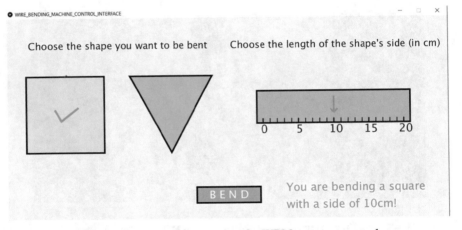

**Fig. 4.** Human machine interface for WBM prototype control.

The methodology for evaluating the concept, assessing learning progress, and achieving the case study objectives is presented in Table 1. It shows what students are expected to do in different stages, what outcomes they reach and how to show intermediate results.

So, during this project the peculiarities of functioning and interaction of electromechanical, electronic and software components of complete production system were studied.

Thus, the challenge of making a comprehensive case which is compact to fit in the available timing of students' curriculum while still containing several important technologies enabling Industry 4.0 module has been realized.

**Table 1.** Methodology of case study progress assessing

| Stage | Intermediate results | Outcomes |
|---|---|---|
| 1 | Conceptual design, setting the different functions for the design, look for similar machines, solutions | Show schematic solution |
| 2 | Selecting the hardware (steppers, controller, gears) and design calculations (gear ratios, forces) | Design parameters |
| 3 | Virtual model (embodiment design) in CAD (PTC CREO) system with design dimensions and kinematic possibilities of the design | Virtual model |
| 4 | Connecting the Arduino and testing the motions | Functional check of the system |
| 5 | Software development for controlling the steppers and motions | Concept of the software solution, functional check of software |
| 6 | Design of HMI and final delivery of the WBM | Optimizing for easy programming by user, usability |
| 7 | Report and presentation | Explaining solution to non-experts, overview of lessons learned |

# 4 Conclusion

Multidisciplinary projects allow students to combine the study of mechanical, electrical and software engineering in relation to Industry 4.0 production systems and processes.
As a result of this case study, students will:

– Discover the interrelations between technologies and existing industrial installations/machines/plants/production processes and understand how projects can span and connect a whole range of different applications for digital transformation;
– Get knowledge on the key emerging technologies, which are at the core of Industry 4.0, and understand what the technological challenges are for a digital transformation towards Industry 4.0.
– Be able to propose reasonable steps that can be taken to implement digital transformation in an Industry 4.0 and understand how to overcome the challenges of digital transformation projects.

**Acknowledgements.** This work is partly carried out with the support of Erasmus+ international educational project WORK4CE "Cross-domain competences for healthy and safe work in the 21st century" (619034-EPP-1-2020-1-UA-EPPKA2-CBHE-JP), as well as DAAD projects EU-ViMUk and Uk4DigiTrans.

# References

1. Gomez, C., Guardia, A., Mantari, J.L., Coronado, A., Reddy, J.N.: A contemporary approach to the MSE paradigm powered by artificial intelligence from a review focused on polymer matrix composites. Mech. Adv. Mater. Struct. **29**(21), 1–21 (2021)
2. Elnadi, M., Abdallah, Y.O.: Industry 4.0: critical investigations and synthesis of key findings. Manag. Rev. Q **74**(2), 711–744 (2023). https://doi.org/10.1007/s11301-022-00314-4
3. Suleiman, Z., Shaikholla, S., Dikhanbayeva, D., Shehab, E., Turkyilmaz, A.: Industry 4.0: clustering of concepts and characteristics. Cogent Eng. (2022). https://doi.org/10.1080/233 11916.2022.2034264
4. Arras, P., Tabunshchyk, G.: Implementation of digital twins for Industry 4.0 in the engineering study program. In: Tonkonogyi, V., Ivanov, V., Trojanowska, J., Oborskyi, G., Pavlenko, I. (eds.) InterPartner 2021. LNME, pp. 677–686. Springer, Cham (2022). https://doi.org/10.1007/978-3-030-91327-4_65
5. Tabunshchyk, G., Arras, P., Wolff, C.: Development of Cross-Domain Competences for Work 4.0. In: Auer, M.E., Rüütmann, T. (eds.) ICL 2020. AISC, vol. 1328, pp. 205–211. Springer, Cham (2021). https://doi.org/10.1007/978-3-030-68198-2_18
6. Tabunshchyk, G., Parkhomenko, A., Subbotin, S., Karpenko, A., Yurchak, O., Trotsenko, E.: Work-in-Progress: framework for academia-industry partnership in Ukraine. In: Auer, M.E., Pachatz, W., Rüütmann, T. (eds.) Learning in the age of digital and green transition. ICL 2022. LNNS, vol. 634, pp. 938–945. Springer, Cham (2023). https://doi.org/10.1007/978-3-031-26190-9_96
7. Cherry, K.: What is a case study? An in-depth study of one person, group, or event. https://www.verywellmind.com/how-to-write-a-psychology-case-study-2795722. Last accessed 1 Mar 2024
8. Ebneyamini, S., Moghadam, M.R.S.: Toward developing a framework for conducting case study research. Int. J. Qual. Methods **17**(1), 160940691881795 (2018). https://doi.org/10.1177/1609406918817954
9. AlBalushi, Z.T.: Challenges of a case study. In: Baron, A., McNeal, K. (eds.), Case study methodology in higher education, pp. 323–343. IGI Global (2019)
10. Glette, M.K., Wiig, S.: The headaches of case study research: a discussion of emerging challenges and possible ways out of the pain. The Qual. Report **27**(5), 1377–1393 (2022)
11. Martinez-Canas, R., Rubio, R., Mondéjar-Jiménez, J., Ruiz-Palomino, P.: Multidisciplinary case study on higher education: an innovative experience in the business management degree. Contemp. Issues Educ. Res. **5**(5), 349–354 (2012)
12. Prabhu, V., Lim, P., Wee, K. S., Gardner, H.: A case study of integrated multidisciplinary project-based learning in polytechnic education. In: The 14th International CDIO Conference, pp.1–11. Kanazawa, Japan (2018)
13. Udeaja, C., Aziz, Z.: A case study of fostering multidisciplinary in-built environment using BIM. In: 31st Annual ARCOM Conference, pp. 701–710. Lincoln, UK (2015)
14. Hunko, E., Gladkova, O., Parkhomenko, A.: Investigation and development of traffic light control system prototype for serious game, In: 2020 IEEE European technology and engineering management summit (E-TEMS), pp. 1–6. IEEE, Dortmund, Germany (2020)
15. KU Leuven Leaning Lab: www.kuleuven.be/english/education/leuvenlearninglab/leuven-learning-lab. Last accessed 3 Jan 2024
16. Tang, W., Zhu, H., Zhu, M., Li, Q., Zhang, J.: Research on key technology of wire-bending and equipment development. Manag. Sci. Eng. **10**(1), 14–20 (2016)
17. Yadav, S., Karpe, K., Shinde, G., Arkas, H., Somvanshi, Y.: Arduino 3D wire bending machine. Int. J. Adv. Sci. Res. Eng. Trends **5**(6), 70–74 (2020)

18. Petiwala, N.S., Shinde, R.S., Ateeque, S.M., Patil, S.S., Singhania, A.: 3D wire bending machine. Int. Multidisc. J. **6**(3), 1–4 (2021)
19. Arduino 3D wire bending machine. https://howtomechatronics.com/projects/arduino-3d-wire-bending-machine/. Last accessed 3 Jan 2024
20. Diy Arduino based 2D wire bending machine. https://electricdiylab.com/diy-arduino-based-2d-wire-bending-machine/. Last accessed 3 Jan 2024
21. Autodesk Tinkercad. https://www.tinkercad.com/. Last accessed 3 Jan 2024

# Educational Advancement of an Industrial Automation Course: Combining Simulation with Physical Experience

Amélie Chevalier[(✉)] [ID]

University of Antwerp, Antwerp, Belgium
amelie.chevalier@uantwerpen.be

**Abstract.** This paper presents an education innovation in industrial automation by combining classical programmable logic controller (PLC) simulation exercises with physical setups. Five new physical setups are introduced in a third year bachelor course. The new teaching method is evaluated using a student feedback survey with closed-ended background questions and questions using a Likert scale. The variety in background knowledge of the students poses an extra challenge in this course. This is shown in the results as a lack of knowledge experienced by the transfer students coming from a professional bachelor program. The overall results show a keen interest in combining the new setups with the classical PLC sessions.

**Keywords:** industrial automation · education · PLC programming · Setups · Student feedback

## 1 Introduction

Industrial automation is rapidly evolving within the framework of industry 4.0 and plays a fundamental role with respect to productivity, quality and quantity of products [1,2]. This has given rise to changes in educational techniques and expectations from both teachers and students. In order to provide more challenging technologies to the students, reformations of curricula and teaching methods are needed. Industrial automation courses require a balanced combination of theory and practice as the students need to link the theoretical concepts to the real applications. Moreover, these applications originate from different sectors of industry [3,4].

In [5], it was shown that programmable logic control (PLC) programming is among the most desired skills in industry. A fundamental part in industrial automation is the PLC programming with a particular focus on sequential control [2]. There is a need for hands-on experience with PLCs and input/output devices in an industrial context. In [6], the main advantages of physical setups are identified as an increasing interest in the subject and the directly visible results.

M. E. Auer et al. (Eds.): STE 2024, LNNS 1027, pp. 178–185, 2024.
https://doi.org/10.1007/978-3-031-61891-8_17

A teaching strategy which combines both simulation laboratories with physical laboratories, specifically for PLC programming within an industrial automation course is proposed. This strategy should be able to cover different areas of application ranging from food industry and manufacturing to packaging and many others. An extra challenge is the rising number of students registered in engineering classes and the limited number of available contact hours. The goal is to formulate a teaching strategy for a growing group of students where the students have more practical experience.

In this work, an industrial automation course given in the third year of the bachelor program in electromechanical engineering is transformed. Beforehand, the course consisted of a theoretical part and a practical part on a Siemens S7-1500 PLC using TIA Portal. The students had only access to TIA Portal during the lab sessions under supervision of the teacher. All concepts were simulated using the Input-Output simulators connected to the PLC.

The new teaching strategy provides the students with access to the TIA Portal environment outside the class room using student licenses. This allows the students to prepare the concepts of PLC programming at home and simulate their program using the PLC SIM feature. The students also get access to the Beckhoff Twincat 3 software to program PLCs and to the PLC next Engineer software from Phoenix Contact. All three platforms are available in the lab for the students. In this manner, the exercises to get familiar with the concepts of PLC programming can be done at home with the opportunity to ask questions during one live session in the lab. Making use of work at home, clears up the schedule in the lab for the introduction of five physical industrial setups ranging in all areas of industrial applications. The students will be divided in five groups, each having a different setup. During a synthesis assignment at the end of the course, each group will educate their peers on the obtained PLC solution and the pitfalls they have encountered.

Using questionnaires with a Likert Scale, student feedback is collected to determine the advantages of the new teaching strategy and the limitations.

This paper is structured as follows: Sect. 2 presents the newly introduced setups, Sect. 3 presents the educational value with the application in lectures and the student feedback. Section 4 discusses the presented results from Sect. 3 and the last section presents a conclusion.

## 2 Materials and Methods

### 2.1 Setups

In this section, each of the setups used in the education innovation project, is shortly described.

**Mixing Tanks**
The plant is the IPC-201C from SMC and represents the production and mixing of liquids (Fig. 1). It has three tanks: two at the side which store the raw material (liquid) and another in the middle where the mixing takes place. It also includes

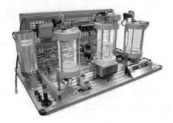

**Fig. 1.** Mixing tanks setup [7].

**Fig. 2.** Bottle filler setup [8].

an auxiliary tank module which can provide the raw materials. The station has six capacitive level sensors, two for each tank (minimum and maximum level). The station has a DC pump which pumps the water around. Fluid solenoid valves are used to control the in and outflow in the system. In the two outer tanks, the pressure can be changed to force the fluid out of the tanks and into the middle tank. At the top of the middle tank there is a stirrer actuated by a 24 V DC motor that homogenizes the liquid. The level in the middle tank can be measured with an analog input.

The following sequence needs to be programmed in the setup:

– Fill the outer tanks with the raw material provided in the auxiliary tank.
– Bring 1 L from product A from the left tank in the middle tank.
– Add 0,5 L from product B from the right tank in the middle tank.
– Mix the solution during 10 s and empty the middle tank after mixing.

**Bottle Filler**
The plant is the SMC IPC-202 bottle filler (Fig. 2). The process consists of different stations and actions to get the final result of a closed jar which is filled with fluid.

The following sequence needs to be programmed in the setup:

– Jar falls out of the warehouse onto the turning table.
– Turning table turns to the next station which is the water filling station.
– Tap of water is opened during a time period of 2.5 s in automatic mode and until you press the stop button in manual mode.
– The filled jar turns towards the next station which is the warehouse of caps.
– A cap is dropped onto the jar.
– The turning table turns towards the next station which is where the cap is pressed onto the jar.
– The cap is pressed to close the jar by a puncher.
– The turning table turns to the last station.
– A gripper arm takes the filled jar from the turning table.

**Palletizer**
The plant is the SMC IPC-203 palletizer (Fig. 3). The plant has two linear stepper motors to control the axes with the corresponding drivers. These motors have

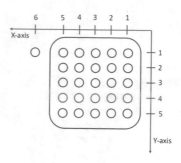

**Fig. 3.** Palletizer setup [9].

**Fig. 4.** Pallet matrix.

**Fig. 5.** Level tanks setup [10].

**Fig. 6.** Pick and place setup [11].

homing sensors to indicate the zero position. A vacuum pump is used to pick the filled jar up. The working principle of the positioning system can be explained as follows. The pallet is a matrix of 5 by 5 shown in Fig. 4. This schematic representation shows how the system names the coordinates. Location (1, 1) is located at the bottom right. The place where the jars are placed by the filling installation is at coordinates (6, 1). The coordinates in X and Y are sent to the drives in the form of 3 bits in each direction forming a binary number. The palletizing action needs to be programmed in this setup.

### Level Tanks

The plant is the Festo MPS PA Compact Workstation (Fig. 5). For this setup, the task is to implement 2 control loops: level and temperature. The actuators used in these control loops are: the centrifugal pump (P101) and the heating element (E104). The analogue sensors used in the setup are: an ultrasonic level sensor and PT-100 sensor. These sensors are used to measure, respectively, level and temperature. An on-off control strategy is asked to be designed for the level in the upper tank and the temperature in the lower tank with a bandwidth of 0.4 l on the level control and a bandwidth of 4 °C on the temperature control.

### Pick and Place

In the pick and place setup (Fig. 6), a metal piece will perform the trajectory shown in the Fig. 7. Cylinder B, moves the piece to the pickup zone (with limit switches $b1$ and $b0$). After picking up the piece using cylinder F and limit switches $f1$ and $f0$, the horizontal slides goes to position G4 where it waits for the input of the user. The user can select cylinder C, D or E (option 1, 2 or 3). The

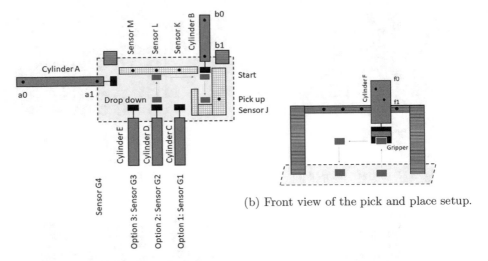

(b) Front view of the pick and place setup.

(a) Top view of the pick and place setup.

**Fig. 7.** Schematic representation of the pickt and place setup.

horizontal slider moves piece to this selected zone. The cylinder C, D or E moves the piece in front of cylinder A and cylinder A moves the piece back to the start position (with limit switches $a1$ and $a0$). The horizontal slider uses a check valve. This sequence needs to be programmed in the setup.

## 3   Educational Value

The educational values of the education innovation project in an industrial automation course is addressed. First, the applications of the setups in the course is discussed. Second, the student feedback is presented.

### 3.1   Application in Lectures

The third bachelor course in electromechanics is taught to a group of 85 students. The course theory is addressed in 8 h of traditional ex cathedra lectures. Afterwards, in four practical sessions, students get experience with sequential PLC problems, combinatoric PLC problems and PLC programming using standards such as the IEC-61131 (International Electrotechnical Commission) and the PackML standard of OMAC (Organization for Machine Automation and Control). During these practical sessions, the students get acquainted with PLC programming languages such as structured text (ST), ladder (LD), sequential function charts (SFC) and function block diagram (FBD). The tasks can be performed by connecting to a real PLC or using PLC simulation programs.

As part of the education innovation, an extra four hour session is introduced where the students need to perform the task mentioned in Sect. 2.1 on one of the setups. The students can perform this task in groups of 4 students to stimulate collaboration. The students are given a PLC program with the Input-Output list of the physical variable and are given the assignment to prepare the setup session beforehand at home. Introducing the physical setups in this course, gives the student the opportunity to work with industrial sensors and actuators which was in previous years not possible.

## 3.2   Student Feedback

At the end of the setup session, the students filled in a feedback survey to evaluate the use of the setups in this industrial automation course. A total of 63 students participated in the student feedback. Different backgrounds are present in the student group. Regular students are students following the line of industrial engineering since finishing high school. Transfer students are students who already have a professional bachelor's diploma. By taking a transfer year with fixed courses, they are allowed to enter the master's program in industrial engineering. These types of students have different experiences with PLC programming (different languages, different platforms and different hardware). The survey consists of a set of closed-ended questions to poll the background of the students and a series of questions using a Likert Scale, i.e. scoring between 1–5, with '5' indicating very satisfied and '1' indicating very dissatisfied. These questions are grouped in four topics: interest in the setups, prerequirements to work with the setups, technical documentation provided with the setups and quality of education.

The results on the background survey are shown in Table 1. Table 1 shows the diversity in student background in this class. There are students who know multiple PLC programming platforms compared to students who have never worked with a PLC before. The results of the student feedback survey using the Likert Scale are shown in Table 2 with the mean value and standard deviation over all students in the column 'all'. The mean and standard deviation results over all the regular students and all the transfer students are also reported in the columns 'regular' and 'transfer' in order to investigate if the background of the student influences the feedback outcomes.

**Table 1.** Student feedback survey: Background

| Study program | | Experience with PLC | |
|---|---|---|---|
| Regular | 68.25% | Beckhoff | 50.79% |
| Transfer | 31.75% | Siemens | 23.81% |
| | | Siemens & Beckhoff | 22.22% |
| | | No experience | 3.17% |

**Table 2.** Student feedback survey using the Likert Scale: mean (M) and standard deviation (SD). How satisfied are you with ...

| Topic | Question | M ± SD all | M ± SD regular | M ± SD transfer |
|---|---|---|---|---|
| Interest Setups | Q1A: ... the combination of classical practica and setups? | 4.19 ± 0.89 | 4.10 ± 0.86 | 4.40 ± 0.92 |
| | Q1B: ... using industrial setups to gain insight in automation techniques? | 4.56 ± 0.69 | 4.48 ± 0.70 | 4.75 ± 0.62 |
| Prerequirements | Q2A: ... the necessary knowledge to work with the setups? | 3.84 ± 0.86 | 3.88 ± 0.85 | 3.75 ± 0.89 |
| | Q2B: ... the time needed to prepare the setup task at home? | 4.24 ± 0.78 | 4.38 ± 0.72 | 3.95 ± 0.80 |
| | Q2C: ... the time allocated for the setup lab? | 4.20 ± 0.74 | 4.34 ± 0.52 | 3.90 ± 0.99 |
| Technical documentation | Q3A: ... the setup explanation document? | 3.69 ± 0.89 | 3.57 ± 0.85 | 3.95 ± 0.92 |
| | Q3B: ... the setup task? | 4.10 ± 0.84 | 3.98 ± 0.89 | 4.37 ± 0.66 |
| | Q3C: ... the explanation video on installing the PLC programming software? | 4.44 ± 0.66 | 4.40 ± 0.66 | 4.53 ± 0.66 |
| | Q3D: ... the explanation video on simulating in the PLC software? | 4.34 ± 0.67 | 4.31 ± 0.64 | 4.40 ± 0.73 |
| Quality of education | Q4A: ... the difficulty of the setups? | 3.81 ± 0.69 | 3.79 ± 0.71 | 3.85 ± 0.65 |
| | Q4B: ... the limitations of the setups? | 3.66 ± 0.84 | 3.71 ± 0.80 | 3.55 ± 0.92 |
| | Q4C: ... the variety of the setups? | 4.15 ± 0.86 | 4.23 ± 0.71 | 4.00 ± 1.10 |
| | Q4D: ... preparing the setups at home? | 3.95 ± 0.74 | 3.81 ± 0.79 | 3.95 ± 0.74 |

## 4    Discussion

The results for the entire group of students indicate both strong point as weak points in the education innovation project. Points of improvement are shown in the lowest mean values in Table 2: the setup explanation document ($3.69 \pm 0.89$) and the limitations of the setups ($3.66 \pm 0.84$). The strong points are indicated with the highest mean values in Table 2: using the industrial setups to gain insight ($4.56 \pm 0.69$) and the explanation video on the installment of the PLC programming software ($4.44 \pm 0.66$).

If however, the background of the students is taken into account, an extra point of improvement is indicated in the group of transfer students: the necessary knowledge to work with the setups ($3.75 \pm 0.89$). This lower mean value indicates that the transfer students lack some knowledge which the regular students do not.

## 5    Conclusion

This paper presented an education innovation in industrial automation courses by introducing new practical setups along side the more traditional PLC labs. The new teaching method is tested and scored using a student feedback survey. The conclusion of the survey is that there is keen interest to provide practical experience in the training of PLC programming. It is hypothesized that the new teaching strategy will increase the learning effect and result in an increase in registered students in the master in Automation Engineering. Future work will investigate this hypothesis.

# References

1. Rojas, A.M., Barbieri, G.: A low-cost and scaled automation system for education in industrial automation. In: 2019 24th IEEE International Conference on Emerging Technologies and Factory Automation (ETFA), Zaragoza, Spain, pp. 439–444 (2019). https://doi.org/10.1109/ETFA.2019.8869535
2. Vazquez Gonzalez, J.L., Barrios Aviles, J., Rosado Muñoz, A., Alejos Palomares, R.: An industrial automation course: common infrastructure for physical, virtual and remote laboratories for PLC programming. Int. J. Online Biomed. Eng. (iJOE) 14(08), 4–19 (2018). https://doi.org/10.3991/ijoe.v14i08.8758
3. Bayrak, G., Cebeci, M.: An automation platform designs for electrical and electronics students: an application study. Procedia. Soc. Behav. Sci. 47, 950–955 (2012). https://doi.org/10.1016/j.sbspro.2012.06.762
4. Özerdem, O.C.: Design of two experimental setups for programmable logic controller (PLC) laboratory. Int. J. Electr. Eng. Educ. 53(4), 331–340 (2016). https://doi.org/10.1177/0020720916630325
5. Hsieh, S.: Design, construction, and evaluation of portable programmable logic controller (PLC) kit for industrial automation and control education. Int. J. Eng. Educ. 39(4), 823–835 (2023)
6. Naef, O.: Real laboratory, virtual laboratory or remote laboratory: what is the most efficient way? IJOE Int. J. Online Eng. 2(3), 1–7 (2006)
7. SMC Training IPC201c. https://www.smctraining.com/en/webpage/indexpage/427
8. SMC Training IPC202. https://www.smctraining.com/en/webpage/indexpage/428
9. SMC Training IPC203. https://www.smctraining.com/en/webpage/indexpage/429
10. Festo Didactic. https://www2.festo.com/MPS_PA_CW
11. Lagae, B., Jacobs, V.: Festo pick and place - redesign. Bachelor thesis, University of Antwerp (2023)

# Enhancing Industrial Automation and Control Systems Cybersecurity Using Endpoint Detection and Response Tools

Filip Katulić[(✉)] [iD], Stjepan Groš[iD], Damir Sumina[iD], and Igor Erceg[iD]

Faculty of Electrical Engineering and Computing, University of Zagreb,
10000 Zagreb, Republic of Croatia
`filip.katulic@fer.hr`

**Abstract.** In today's rapidly evolving cybersecurity landscape of industrial automation and control systems (IACSs), threat detection and prevention are more critical than ever. Compared to information technology (IT), the cybersecurity maturity of IACS is still low. Therefore, one way to increase the cybersecurity of IACS is to adopt and implement proven IT protection methods. Endpoint detection and response (EDR) and extended detection and response (XDR) tools have emerged as the means of protecting IT systems. Their exceptional threat detection, analysis, and response capabilities make them indispensable components for protecting critical infrastructures. The paper addresses the potential integration of EDR and XDR tools into IACS environments to strengthen their cybersecurity and protect them from new and existing threats. To demonstrate the feasibility of such a solution, we have implemented a chosen EDR tool on an industrial programmable logic controller (PLC). We have also analyzed the impact of such a solution on real-time PLC performance, which shows that such tools could be implemented in real-life IACSs.

**Keywords:** Industrial automation and control systems (IACS) · Operational technology (OT) · Endpoint detection and response (EDR) · Extended detection and response (XDR) · Programmable logic controllers (PLC)

## 1 Introduction

Industrial automation and control systems (IACSs), such as programmable logic controllers (PLCs), supervisory control and data acquisition (SCADA), and process control systems (PCSs) enable us to control and supervise a wide variety of processes safely and reliably. Some examples of processes where IACSs, often referred to as operational technology (OT) systems, are commonly found are the energy and water supply sectors, two of the fifteen critical infrastructure (CI) sectors as defined in the NIS2 directive [1]. Although IACSs are responsible for controlling such critical systems, the maturity of industrial cybersecurity is low compared to information technology (IT) systems [2]. One of the main reasons for this is that IACSs often rely on legacy equipment that was developed without cybersecurity in mind. For this reason, connecting legacy OT systems

© The Author(s), under exclusive license to Springer Nature Switzerland AG 2024
M. E. Auer et al. (Eds.): STE 2024, LNNS 1027, pp. 186–197, 2024.
https://doi.org/10.1007/978-3-031-61891-8_18

to more modern IT systems automatically increases the risk of cyberattacks [2, 3]. The attackers usually target both the insecure communication protocols, such as Modbus-over-TCP (Modbus/TCP) [3], and the endpoint devices, e.g., PLCs [4]. An example of such an incident is the Stuxnet attack, in which an unknown attacker used the Stuxnet malware to disrupt the Iranian nuclear program even though the targeted PLC was not directly connected to the Internet [4]. One way to address the cybersecurity risks intro-duced is to adapt IT-proven cybersecurity measures to legacy IACS devices, which is often not possible due to limited processing power, proprietary software and hardware, and most importantly the risk of compromising the real-time performance of the devices [3]. The other way to protect such systems is by changing the critical devices in IACS (e.g. PLC) with those that support cybersecurity features if the cybersecurity risk analysis shows that such an investment is justified. One of the IT-proven cybersecurity tools that can be used to protect IACS endpoints are endpoint detection and response (EDR) and extended detection response (XDR), tools that monitor and analyze the data collected from the device and to detect suspicious activity.

This study analyzes a novel IACS cybersecurity enhancement by implementing an EDR/XDR tool on PLC endpoints within the industrial network. In contrast to pre-vious studies that use tools to detect PLC program integrity violations, we propose a solution that additionally analyzes the underlying PLC operating system (OS) to pro-tect IACS from complex threats. The potential impact on the chosen PLCnext AXC F 2152 PLC real-time performance was tested on a custom experimental water tower cyber-physical system. It was shown that the implemented EDR tool enables improved protection without compromising real-time execution.

This paper is organized as follows. In Section 2, the related work is presented. Section 3 gives an overview of the used PLC platform, while the Section 4 analyzes the implemented water tower cyber-physical system. Within Section 5, the implementation and the impact of the EDR tool was analyzed. Finally, Section 6 presents the conclusions and future work.

## 2   Related Work

PLCs have been shown to be vulnerable to various cyber-attacks by researchers and real-world threat actors [2–10]. In [5], researchers categorized attacks on PLC-based systems into fifteen subcategories against availability, integrity, and confidentiality. In addition, they conducted a vulnerability analysis in which they concluded that most PLC vulnera-bilities lie in the firmware, memory, and IEC 61131-3 programs. It should be noted that if the PLC is compromised, the attacker has already used one or more vulnerabilities and methods to gain initial access to the IACS and evade potential cybersecurity protection mechanisms [6]. From this point of view, the endpoint protection of PLC-based systems should be considered as the *last line of defense*, as stated by the researchers in [7].

In [8], the authors developed an IACS-specific malware called *Harvey* that targets the PLC firmware, primarily through the firmware update mechanism or manual SD card firmware deployment. Once the malware infects the PLC, it then modifies the inputs and outputs of the PLC to maximize damage to the organization. The authors also emphasize that the malware can be detected if the firmware integrity can be verified.

Another common approach to defending PLC-based systems is the utilization of intrusion detection systems (IDSs), tools that monitor and analyze data (network traffic, host logs) to detect unusual or suspicious activity [9]. Depending on the data source, a distinction between network-based and host-based IDSs can be made. Since network-based IDSs only analyze network traffic, they cannot detect vulnerabilities in PLCs and, therefore, fall outside the scope of this work.

Host-based IDSs, on the other hand, are tools specifically designed to analyze suspicious activity on devices and enable the detection and mitigation of different cybersecurity threats. Several research articles [7, 10–13] analyze the implementation of host-based IDSs in PLC-based systems. Most researchers analyze the internal real-time memory and input/output states of the PLC and compare them with the expected states. The given principle is based on the fact that most industrial processes can be modelled as state-based machines, which means that any combination of inputs, outputs and internal memory can be predicted depending on the current state. In [10], the authors propose a security monitoring tool called *Snapshotter* that logs the physical input and output signals of the PLC. After the data is captured and analyzed, an additional system compares it to the expected states which are generated by a PLC simulator. One of the drawbacks of the work is that the authors implemented the *Snapshotter* agent on a non-industrial Raspberry Pi platform. Still, they also point out the possibility of implementing such solutions on modern PLCs with more computing power.

Furthermore, in [7], the researchers proposed a host-based IDS that checks the integrity of the PLC program and the validity of control program variables depending on the state of the industrial process. The advancement of their work when compared to the state-of-the-art is that they used an advanced Siemens S7–1518 PLC which allows users to create C/C++ programs that run in an embedded Linux OS. The authors have developed a program called *Goosewolf* that monitors PLC status from the embedded Linux OS, enabling real-time detection of IT and OT denial of service attacks and PLC control logic modifications. Although they point out that the two PLC's runtimes (C/C++ and PLC control logic) are "somewhat dependent", the authors did not analyze the situation in which the threat actors target the embedded Linux OS.

Finally, in [11], the authors have presented a tool named *APTSHIELD*, an EDR tool specifically designed for Linux hosts that enables real-time threat detection and mitigation, while the authors in [12] tested different EDR tools on how they protect systems against complex advanced persistent threats (APTs). Although their results suggest that none of the EDR tools provide complete security, they indicate that such solutions can provide additional help when dealing with complex and advanced threat actors. Based on the given statement, we attempted to implement a selected EDR tool on an embedded Linux OS and find out whether this has impact on the real-time performance of the PLC.

## 3   Chosen PLC Platform

As part of this study, we selected the Phoenix Contact AXC F 2152 PLC as a suitable candidate for EDR implementation. The analyzed PLC is part of the PLCnext technology ecosystem, Phoenix Contact's product line of industrial automation devices and technology focused on Industry 4.0, which means more robust integration and interconnectivity of IT and OT environments [13]. Its enhanced cybersecurity (secure-by-design paradigm), guaranteed by an IEC 62443 standardization [14], and its modularity and openness, make the PLC a suitable candidate for such an analysis.

Figure 1. Shows the simplified representation of the PLCnext Control system architecture, while Table 1. Provides more information about the system properties. As can be seen, the basis of the multi-layer PLCnext Control architecture is the proprietary PLCnext Linux, a specially created Yocto-based Linux OS that has an OSADL PREEMPT_RT real-time patch applied [13]. The PREEMPT_RT real-time patch makes the Linux OS a multitasking real-time OS, allowing Linux to be used in IACSs where real-time operation is required [13]. In addition, the Linux OS comes with preinstalled software, such as *tcpdump* for analyzing network traffic, *htop* as task and process manager, and many others. The PLCnext Linux OS does not support any available package managers but requires users to add verified packages by using PLCnext Store, which only has a limited set of tools [13]. However, it is possible to install software by installing a pre-built binary directly or by using Open Container Initiative (OCI) containers [15]. More on OCI containers will be given later in the work.

The Middleware and external function extensions are located on top of the Linux OS. The Middleware layer acts as a software communication hub between user programs, internal function extensions, and vice versa. The Middleware layer also ensures data consistency when different programs and functions manipulate the same process values (PLC inputs and outputs, memory locations, etc.) [13].

As for the PLCnext components layer, it consists of various system and service firmware components that allow real-time execution, ensure support for different communication protocols, enable access to process data, and provide all other necessary functions for the safe and reliable operation of the PLC.

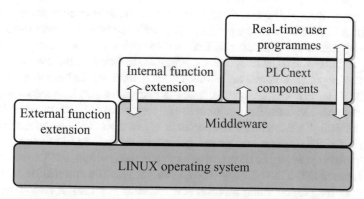

**Fig. 1.** Structure of the PLCnext Control system architecture.

As for internal and external functions extensions, they are usually used for non-real-time tasks, such as database access, cloud communication, etc. The internal functions can be written in C++, while the external functions support different programming languages, such as Python, Node-RED, etc. Finally, real-time user programs are used

**Table 1.** PLCnext AXC F 2152 PLC properties.

| Property | Information |
|---|---|
| Processor | ARM Cortex-A9 CPU 2 × 800 MHz |
| RAM memory | 512 MB DDR3 SDRAM |
| ROM memory | 512 MB |
| SD card support | Yes (up to 32 GB) |
| TPM platform | Yes |
| OS architecture | Yocto-based Linux OS, armv7l, 32-bit |

for industrial automation control and supervising tasks. They can be written in standard PLC programming languages as defined in IEC 61131-3 and in high-level programming languages, such as C++, C#, or MATLAB Simulink.

## 4   Water Tower Cyber-Physical System

To provide the experimental environment for PLC performance and cybersecurity evaluation, we have selected a water tower IACS. Water towers are structures used in water distribution systems, mainly for maintaining the constant pressure of water within the system. The reason behind choosing a water tower IACS is its simplicity, as most have only one or several control loops for maintaining the water level within the water tank, thus the water pressure within the distribution system. It should also be noted that water towers, being part of water supply systems, are part of the CI as defined in NIS2 directive [1].

A detailed process and control diagram (P&ID) of the developed water tower IACS is given in Fig. 2. As can be seen, the process starts with pumping the water from the water distribution system into the water tank. Before it enters the tank, water flows through the measuring station, where the temperature, pressure, and water flow are measured. The PIC 1 control loop measures and maintains the water level within the water tank. Before entering the water distribution system, water pressure, temperature, opacity, and flow rate are measured. The control program of the implemented IACS also has an integrated safety mechanism. For example, if the pressure within the water tower system rises to a critical level, the relief valves are opened, protecting the system from overpressure.

The complete cyber-physical system was implemented within the PLCnext controller by using standard IEC 61131-3 programming languages. The automation tool used for the creation of the programming solution was *PLCnext Engineer 2023.9*. One function block, called *Water tower process*, served as an industrial process simulation. The other function block called the *Control loop* was used to automate and control the industrial process.

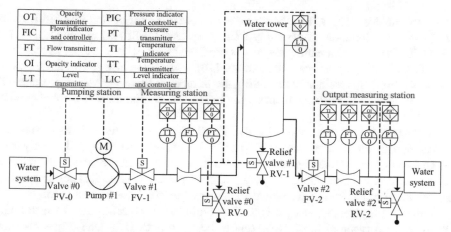

**Fig. 2.** P&ID diagram of the developed cyber-physical system model.

## 5 Implementation of the Chosen EDR Solution

There are several ways to access the command line interface (CLI) of PLCnext Linux OS. One way is to use Secure Socket Shell (SSH), a network protocol for remote authentication and communication. Suppose the device is accessible via a local network. In that case, the user can connect to a device by using an *admin* user, the network address of the PLCnext controller and a default password printed on the PLC housing. After accessing the CLI, the user can work with preinstalled tools (e.g. *nftables, tcpdump, htop, sqlite*) and install new ones in several ways.

PLCnext does not currently support package managers, apart from user initiatives that attempt to implement them. Implementing package managers would substantially simplify the installation process of programs, but unfortunately, our attempts to implement such solutions have failed. The reason behind the failure is that the online implementations were created for older PLC firmware versions, introducing different compatibility issues. As mentioned above, other ways of installing applications are through PLCnext Store and by directly installing pre-built binaries.

On the other hand, PLCnext AXC F 2152 comes with preinstalled *Podman* (from firmware version 2021.9), a tool for OCI containerization of applications and software. In computing, containerization is a virtualization technology that allows application isolation without hardware emulation/requirements. Basically, containers run their own OS, file systems, and network stacks, but share the kernel with the underlying OS [15]. More on containers and their advantages is given in [15]. It should also be noted that *Podman* is believed to be more secure than its alternative, *Docker*, for its support of rootless containers, meaning that *Podman* allows users to run and manage containers without administrator rights [15].

As for the PLCnext controller vulnerabilities, they are regularly updated and published on the Phoenix Contact Product Security Incident Response Team (PSIRT) website [16]. One of the most recent advisories that disclose multiple Linux component vulnerabilities dates back to February 2023, in which the AXC F 2152 PLC was one of the vulnerable devices [16]. Within the advisory, more than 60 different vulnerabilities were disclosed, with Common Vulnerability Scoring System (CVSS) 3.1 ranging from 3.7 to 9.8. CVSS 3.1 is a vulnerability severity score framework, where severity ranges from 0 to 10, later meaning critical vulnerability [16]. PSIRT has also released a firmware update that mitigates vulnerabilities. Still, the number of them indicates that Linux OS is quite vulnerable to different attacks, showing a need for a solution that could monitor PLC performance.

It should also be noted that PLCnext Runtime, a part of the OS responsible for managing real-time user programs (IEC 61131-3, C, C++) can be stopped directly from the CLI using the command `sudo /etc/init.d/plcnext stop`. If the attacker gains access to the Linux OS CLI by e.g., stealing user credentials, it is possible to stop the execution of PLC programs directly and bring the IACS to a critical state.

Based on the given, we analyzed possible solutions for protecting the PLCnext Linux OS, other than recommended cybersecurity best practices [16]. One of the potential solutions was EDR, a relatively new technology introduced in 2013 that collects data from endpoint devices and transmits it to a centralized server. The centralized server then analyzes the data to detect potential suspicious activity, alerting the user in real-time. The main difference between classic antivirus programs and EDRs is that EDRs do not rely exclusively on potentially malicious files in the system but also analyze and detect suspicious or anomalous activity on the device [17]. It should be noted that EDR tools are usually distributed, meaning that endpoint devices have agent programs that monitor, log information, and execute limited analysis, while the in-depth analysis is performed on a standalone EDR server [17].

Currently, there is a large number of EDR and XDR software on the market [18]. For this reason, it can be quite difficult for an average user to select the tool that suits their needs, as most of the information available online is written in the form of *Top X EDR tools available*, without almost any argumentation on why a particular EDR tool ended on the list. Our criteria for selecting an EDR tool were that it should be open-source, customizable, free to use, and have a large user community. One such tool is *Wazuh*, a security monitoring program which provides integrity monitoring, detection, analysis, and response to various threat scenarios [17]. In addition to EDR/XDR usage, Wazuh can also be used as a security information and event management (SIEM) solution, allowing user to gain real-time information about endpoint activity (e.g. security alerts and notifications, status, vulnerability scan results).

The Wazuh EDR tool consists of two main components – the Wazuh agent, a cross-platform tool that runs on the endpoints that the user wants to supervise, and the Wazuh server, a central component that analyzes the data received from the Wazuh agents [17]. The deployment of the Wazuh server is relatively simple, as the main requirement is that it must be deployed on a 64-bit Linux OS. For this purpose, the *Kali Linux 2023.3* OS was deployed using *VMware* virtualization tool on a general-purpose workstation. After setting up the virtual machine, the Wazuh server was deployed using the quick start guide in [17]. IP address of the deployed Wazuh server VM was 192.168.0.148/24.

As for the Wazuh agent, we analyzed several possible methods for deploying an agent on a PLCnext controller Linux OS. The Wazuh documentation describes how to deploy an agent using package managers, but PLCnext PLC does not support this deployment type [13]. The documentation in [17] lists Linux distributions and architectures for which pre-built package images for manual deployment exist. Unfortunately, no package image is explicitly created for Yocto-based OSs, making the installation on the PLCnext platform complicated. Other alternatives for deploying the Wazuh agent include installing from sources by compiling the Wazuh source code and copying the binaries to the Linux OS, which could also be difficult for an average OT engineer. Finally, deploying Wazuh agent inside a container is possible, allowing it to analyze the host for potential threats. This approach enables the user to encapsulate the Wazuh agent within a container, making the deployment process easier and prone to fewer errors. More on the disadvantages of such deployment will be given later in this section.

The first step in deploying Wazuh agent via an OCI container is to find the Wazuh agent image that supports the PLC Linux OS architecture. Since the Podman tool supports images created for Docker, we have used *Dockerhub*, a library for container images to find a suitable Wazuh agent image [19]. Fortunately, we have found one up to date Wazuh agent image which supports the required armv7 architecture, allowing the deployment of the Wazuh agent on the PLC host. To do so, the PLC needed Internet access, and after writing the command `podman pull coralogixrepo/wazuh-agent-docker:latest`, the Wazuh agent image was downloaded to the host. The latest available Dockerhub Wazuh agent image is 4.2.6, first released in March 2022.

After downloading the Wazuh agent image, the next step is to share data between the host and the Wazuh agent container. Data sharing between the host and the container is achieved through volume mapping (command $-v$), associating directories and files between the host and the container [15]. This way, the Wazuh agent container can scan and analyze the host logs for threats and send the information to the centralized server. In addition to volume mapping, environment variables (command $-e$) allow the configuration of options for the agent, such as the server (manager) IP address, the agent's name, and the network ports for the Wazuh agent connection.

To start the Wazuh agent container, we have used the following command:

```
podman run --network host \
-e WAZUH_MANAGER=192.168.0.148 \
-e WAZUH_AGENT_NAME=PLCnext_PLC \
-v /var/log/syslog:/var/log/syslog \
-v /var/log/auth.log:/var/log/auth.log \
coralogixrepo/wazuh-agent-docker:latest
```

which starts a Wazuh agent named *PLCnext_PLC* on a PLCnext controller and establishes a connection with the previously created Wazuh server. The $-v$ commands assign the host logs to the container, allowing the agent to access PLC's logs. The Wazuh server dashboard, as shown in Fig. 3, shows that the Wazuh agent has established a connection to the server. Although the agent is connected to the server, this implementation has several problems. First, the Wazuh agent OS is detected as Ubuntu 22.04.03 LTS because we did not transfer the information about the PLCnext Linux version to the Wazuh container.

Because of this, the Wazuh agent uses its own OS inside the container, which is in fact Ubuntu 22.04.03 LTS. To solve this problem, one could transfer the information about the running PLCnext OS through environment variables ($-e$), after which the

**Fig. 3.** Wazuh server dashboard.

information could be used when configuring Wazuh rules inside the container. This approach would require significant changes in the Wazuh agent container configuration. Similarly, Center for Internet Security (CIS) benchmark, a set of community-developed secure configuration recommendations [17], uses Debian/Linux 10 rules to determine whether the OS is configured to be as resilient against cyber-attacks as possible. As this work deals with the analysis of EDR tool impact on PLCs performance, it is out of scope to solve the given problems. However, later on we will provide possible solutions to the identified problems.

### 5.1  Impact of the EDR Tool on Real-Time Performance of the PLC

After implementing the Wazuh EDR tool on the PLCnext controller, the next step was to determine the memory and CPU utilization of the Wazuh agent container. To do this, we used the `podman stats -i= 1` command, which gives the running container's real-time CPU and memory impact on the host at one-second intervals. The real-time data was then logged and formatted by a custom `bash` script. It should be noted that the given command returns the resident set size (RSS) memory, which represents the physical memory used by a process. As for the CPU usage, the `podman stats` command returned some results that were above 100%. The reason behind this is that the command displays the CPU usage as a percentage of a single CPU core [20]. For this reason, we have divided the CPU percentage by a factor of 2 as our PLC has 2 CPU cores. More on the commands that are used is given in [20]. Figure 4 shows the performance impact of the Wazuh agent, while Table 2. Contains a detailed statistical analysis.

The impact on performance was analyzed for three different states: the *startup phase* (from T = 0 s to T = 25 s), in which the Wazuh agent and the server establish connection; the *intense working phase* (from T = 25 s to T = 109 s), in which the Wazuh agent analyzes the PLC state; and the *normal working phase* (from T = 109 s to T = 180 s),

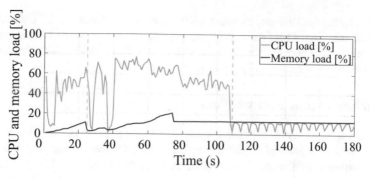

**Fig. 4.** Wazuh agent container CPU and memory usage.

**Table 2.** CPU and memory usage statistics of the Wazuh agent container.

| States | startup | intense | normal |
|---|---|---|---|
| Average CPU load [%] | 39,16 | 54,55 | 9,48 |
| Maximum CPU load [%] | 66,07 | 77,02 | 12,69 |
| CPU load deviation [%] | 18,46 | 18,02 | 3,24 |
| Average memory load [%] | 4,82 | 10,89 | 12,67 |
| Maximum memory load [%] | 11,24 | 21,04 | 12,76 |
| Memory load deviation [%] | 3,39 | 4,57 | 0,02 |

in which the Wazuh agent only communicates with the Wazuh server. The timing was determined by analyzing the log files (OSSEC.log) within the Wazuh agent container.

A high and volatile CPU load strain can be observed within the startup phase. This is due to the Wazuh agent and the server trying to authenticate themselves and establish a connection. After connecting, the intense working phase begins, in which the log files, file integrity, and root check scans are executed. The high volatility of the PLC's CPU and memory usage is due to the different start and end times of the checks. To make the phase more intensive, we have pre-loaded the analyzed /var/log/syslog file with a number of logs (2000), which were created by a rule that detects ICMP ping packets using the command sudo iptables -AINPUT -picmp --icmp-type echo-request -jLOG --log-prefix "Ping Packet:". It should be noted that none of the later scans (the Wazuh agent was configured to analyze system every hour) were not as intense as the first scan, regardless of the number of logs in log files). Later in the *working phase*, CPU and memory usage are almost constant, with the load values being quite low. This is because the agent and the server only maintain the connection while waiting for the next scan. During the execution and testing of the Wazuh agent container, the water tower PLC program was executed within the PLC runtime. We could not detect any latency or longer program runtimes due to the agent. There were also no warnings in the Linux logs (/opt/plcnext/logs) of the PLCnext controller about the impairment of the real-time performance. To further reduce the risk of impacting the

PLC's real-time performance, Podman allows the user to limit the CPU and memory usage of the container [20]. These hard constraints can guarantee that the implemented tools do not interfere with the standard operation of the PLC, making the containerization concept even more suitable for IACS environments.

## 6  Conclusion and Future Work

In this study, a novel IACS cybersecurity solution was presented, introducing an additional layer of security to PLCs utilizing the Linux OS. The core of the proposal involved the implementation of a Wazuh EDR tool on a chosen PLCnext AXC F 2152 PLC. Additionally, this work gave a methodology for implementing OCI containers, a relatively new technology in IACS environment. Containerization could be essential in IACS cybersecurity, as pre-created platform images do not need much configuration, lowering the deployment time. One of the crucial aspects of this study was the analysis of the impact on the PLC's real-time performance. To do so, a water tower cyber-physical system was implemented using IEC 61131-3 programming languages, providing the experimental setup for the performance analysis. The results demonstrated that implementing the chosen EDR tool did not compromise the real-time performance, even when subjected to a substantial working strain. This finding is pivotal as it theoretically allows the implementation of such tools in existing IACS equipment that supports the EDR tool, with a low risk of impacting the real-time execution. However, it is important to note that the proposed solution is not without challenges. One notable limitation is that the Wazuh agent image does not automatically detect the PLCnext Linux OS. This is because the PLCnext Linux OS is not one of the supported OSs by Wazuh, allowing only limited protection. However, this problem can be solved by creating custom configurations for vulnerability detection, CIS benchmarking, log collection, and other scans supported by Wazuh. An even better solution would be for the PLC vendors to create the specifications of their proprietary OSs, allowing further integration of such solutions in IACSs.

Future research will focus on refining and expanding the proposed cybersecurity solution. As the Wazuh supports file integrity monitoring, we will investigate possible protection of real-time user programs by continuously monitoring file integrity, allowing integrated protection of the PLC's OS and the automation process. This approach could prevent APTs from targeting IACS control programs, as they would need to attack multiple different systems successfully. Furthermore, the impact on various industrial PLCs that support Linux OS will be analyzed, such as the PLCnext AXC F 1152 PLC, which has a lower amount of resources compared to the PLC used in this study. This analysis is crucial for evaluating the scalability of the proposed solution. Lastly, an investigation will be conducted into the potential use of different tools that could improve IACS cybersecurity, with a particular focus on leveraging containerization technology.

## References

1. European Commission, Directive (EU) 2022/2555 of the European parliament and of the Council of 14 December 2022 on measures for a high common level of cybersecurity across the Union, amending Regulation (EU) No 910/2014 and Directive (EU) 2018/1972, and repealing Directive (EU) 2016/1148 (NIS 2 Directive), Off. J. Eur. Union, vol. L 333, pp. 80–152 (2022)

2. Pliatsios, D., Sarigiannidis, P., Lagkas, T., Sarigiannidis, A.G.: A survey on SCADA systems: secure protocols incidents threats and tactics. IEEE Commun. Surveys Tuts. **22**(3), 1942–1976 (2020)
3. Katulić, F., Sumina, D., Groš, S., Erceg, I.: Protecting modbus/TCP-based industrial automation and control systems using message authentication codes. IEEE Access **11**, 47007–47023 (2023)
4. Langner, R.: Stuxnet: dissecting a cyberwarfare weapon. IEEE Secur. Privacy Mag. **9**(3), 49–51 (2011)
5. Wang, Z., Zhang, Y., Chen, Y., Liu, H., Wang, B., Wang, C.: A survey on programmable logic controller vulnerabilities attacks, detections, and forensics. Processes **11**(3), 918 (2023)
6. Jadidi, Z., Lu, Y.: A threat hunting framework for industrial control systems. IEEE Access **9**, 164118–164130 (2021)
7. Allison, D., McLaughlin, K., Smith, P.: Goosewolf: an embedded intrusion detection system for advanced programmable logic controllers. Digit. Threats: Res. Pract. **4**(4), 1–19 (2023)
8. Garcia, L., Brasser, F., Cintuglu, M.H., Sadeghi, A.-R, Mohammed, O.A., Zonouz, S.A.: Hey, My Malware Knows Physics! Attacking PLCs with Physical Model Aware Rootkit. Netw. Distr. Syst. Sec. (NDSS), pp. 1–15 (2017)
9. Katulić, F., Sumina, D., Erceg, I., Groš, S.: Enhancing Modbus/TCP-based industrial automation and control systems cybersecurity using a misuse-based intrusion detection system. In: Proceedings of the International Symposium Power Electronics Electrical Drives Automation Motion (SPEEDAM), pp. 964–969 (2022)
10. Jin, C., Valizadeh, S., van Dijk, Snapshotter, M.: Lightweight intrusion detection and prevention system for industrial control systems. In: 2018 IEEE Industrial Cyber-Physical System (ICPS) (2018)
11. Zhu, T., et al.: APTSHIELD: a stable, efficient and real-time APT detection system for linux hosts. IEEE Trans Depend. Sec. Comp **20**(6), 5247–5264 (2023)
12. Karantzas, G., Patsakis, C.: An empirical assessment of endpoint detection and response systems against advanced persistent threats attack vectors. J. Cybersecur. Priv. **1**(3), 387–421 (2021). https://doi.org/10.3390/jcp1030021
13. Phoenix Contact, PLCnext Technology Training Documentation, PLC200 (2022)
14. Phoenix Contact. https://www.phoenixcontact.com/en-pc/products/controller-axc-f-2152-2404267. Last accessed 14 Nov 2023
15. The Linux Foundation. https://opencontainers.org. Last accessed 14 Nov 2023
16. Phoenix Contact, https://www.phoenixcontact.com/en-pc/psirt. Last accessed 14 Nov 2023
17. Wazuh Inc: https://documentation.wazuh.com/current/index.html. Last accessed 14 Nov 2023
18. G2: https://www.g2.com/categories/endpoint-detection-response-edr. Last accessed 14 Nov 2023
19. Coralogix Ltd: https://hub.docker.com/r/coralogixrepo/wazuh-agent-docker. Last accessed 14 Nov 2023
20. Podman: https://github.com/containers/podman. Last accessed 14 Nov 2023

# Digital Twins and Model-Based Design for New Vehicle Engineering

Raghuveer Rajesh Dani[1]🆔, Benjamin Geiger[2]🆔, Galyna Tabunshchyk[1(✉)]🆔,
Carsten Wolf[1], and Friedbert Pautzke[2]🆔

[1] Fachhochschule Dortmund, Otto-Hahn-Str. 23, 44227 Dortmund, Germany
{galyna.tabunshchyk,carsten.wolff}@fh-dortmund.de
[2] Hochschule Bochum, Am Hochschulcampus 1, 44801 Bochum, Germany
{benjamin.geiger,friedbert.pautzke}@hs-bochum.de

**Abstract.** Software-defined vehicles have been the main trend in the automotive industry for the last decades. Digital twins (DT), which are used for the entire design-execute-change-decommissioning lifecycle of real assets and in real-time, are a powerful instrument that could be used both in R&D and engineering education. Model-Based System Engineering (MBSE) approach shows its effectiveness in a variety of application domains, one of which is the development of the Electric Vehicle. In this work authors present the modular based digital twin for the powertrain.

**Keywords:** Model based design · digital twin · virtual vehicle · design and simulation

## 1 Introduction

The continuous evolution in connected technologies and the availability of big data in Industry 4.0 has speeded up the innovation in machine building and process control.

As a result is implementation of so-called digital twins (DT) in various domains, which are used for the entire design-execute-change-decommissioning lifecycle of real assets and in real-time.

Model-Based System Engineering Digital Twin (MBSE DT) approach considers DT as a container of models, of an actual physical asset in operation. For EV (Electric Vehicle) platforms it is essential to have a modular representation of the assets as it is a market requirement established by OEM [1].

## 2 Literature Review

In recent, many frameworks and solutions have been made available for the development and usage of the Digital twins for EV and its components.

© The Author(s), under exclusive license to Springer Nature Switzerland AG 2024
M. E. Auer et al. (Eds.): STE 2024, LNNS 1027, pp. 198–205, 2024.
https://doi.org/10.1007/978-3-031-61891-8_19

In [2], the DT of the permanent magnet synchronous motor (PMSM) is proposed. Real-time connections between the testbench and Matlab/Simulink were used.

The automatic generation of DT is considered by [4] from the previously collected datasets.

The solution for the simulation of the Advanced Driver-Assistance Systems (ADAS), Advisory Speed Assistance (ASA), is presented by [5]. However, the models of vehicle dynamics and the drivetrain are much more complex than they could be modeled by the game engine used by the authors.

A literature search in IEEE Xplore on the developing of DT for Battery Management Systems yields a set of publications covering a wide range of topics, for example: a framework with a DT for EV for predicting SoH (state of health) and estimating SoC (state of charge) is proposed by [5]. The proposed framework was trained and validated on open-source battery data collected by NASA.

DT for improving the performance of Lithium-ion batteries is suggested by [5], where authors proposed an enhanced Q-learning EMS (Energy Management System) using a DT model for an EV with both a battery and an ultracapacitor [6].

In this work, there is suggested a general framework based on MBSE DT approach, which considers DT as a container of models, of an actual physical asset in operation. Previously, the concept of the modular vehicle platform [1] and the validation and verification toolchain [8] were developed, which transform the purely virtual testing of the electric drive into the X-in-the-loop.

The database of component models was also developed as it has a significant impact on accuracy [9–14].

In the suggested framework for the DT development, a multidimensional approach based on the Mathworks and DSpace toolchain is used [8].

## 3    Research Methodology

The modular approach, which is implemented for each level of the DT architecture, forms the core. At the device level, modularity is provided on the one hand by the selection from a variety of components (as different components of different manufacturers, as different types of components), but on the other hand also by the selection of different vehicle topologies (e.g. whether 1, 2 or 4 engines). In addition, different detailed models are used for the components.

Since an automotive vehicle is a combination of a vast range of subsystems and components, designing everything from scratch is a cumbersome task. Moreover, automotive architecture comprises of subsystems designed and developed by engineers with different backgrounds such as Mechanical, Electrical, Electronics, and Software. Model-Based Software development (MBSD) improves the whole process of designing, simulating, and analyzing automotive software.

Simulation is achieved with the usage of the MathSoft toolchain [15]. The Virtual Vehicle Composer (VVC) is considered as a good starting point for setting up a Virtual Vehicle simulation environment from the design requirements in minutes, and further mapping it with the actual physical models [16]. However, since it's a packaged application, researchers don't have full control over the generated models. It is complicated to make detailed changes in the virtual vehicle (model) generation process and must comply with a predefined set of engineering parameters.

In MBSD there are hundreds of subsystems put together for simulation of the functionality of a whole system. These subsystems are further divided into software functionalities and sub-modules. For example, vehicle software has sub-modules representing each ECU (Electronic Control Unit) like VCU (Vehicle Control Unit), BMS (Battery Management System) ECU, Drivetrain ECU, Powertrain ECU etc. Ultimately, every part of the vehicle is controlled by an ECU. That's why there were developed OMAx modular MBSE DT approach, which should enable vehicle manufacturers in small series to test different combinations of components at minimum costs or to estimate the variety of external effects on the vehicle performance. There are several methods to ensure the modularity, interchangeability, and interoperability of components. Our approach is to provide a technique, which able to switch between a variety of models of actual vehicle components, within the subsystem.

The same subsystem can be used for similar methods that basically behave in the same way but have different properties (e.g. motor power) and only the parameters need to be adapted. This is implemented by means of scripts and parameter sets stored. However, if different types are involved (e.g. PMSM and ASM motors), different topologies (1, 2 or 4 motors) or different data bases (more complex models can be used for increased accuracy), different subsystems must also be used. To implement this automatically, special variant subsystems must be used.

## 4    Results and Discussion

### 4.1    Hardware Setup

For the physical level of the OMAx DT, an original modular platform is used. This platform is based on the commercially available Tropos Able. Existing drivetrain components were removed to integrate new ones more accessible on the loading platform and therefore quicker to replace with other components (Fig. 1).

1. High voltage Battery   2. Vehicle Control Unit (VCU)   3. Distribution box (high voltage)
4. Motor Control Unit (MCU)   5. Electrical Machine   6. On-Board Charger   7. DC/DC-Converter
8. Distribution box (low voltage)

**Fig. 1.** Physical modular vehicle setup

Components, which are used for the validation are shown on Fig. 1:

1. **High-voltage battery:** Energy supply
   Mapped parameters: Voltage, Current, Isolation measurement, controls (contactor incl. feedback, current limits)
2. **DC/DC-Converter:** Converts High voltage to 12 V for board electronics supply
   Mapped parameters: Controls (Start/Stop incl. Feedback, Input and Output voltage)
3. **Electrical machine:** Converts electrical energy in mechanical energy (motor) or vice versa (generator)
   Mapped parameters: Speed, Torque, Current, Voltage, Temperature
4. **MCU:** Controls motor and provides specific supply
   Mapped parameters: Input and output voltage and current, controls (start/stop, torque or speed control, limitations)
5. **VCU:** Central control unit for controlling drivetrain/vehicle
   Mapped parameters: Controls (OBC, MCU, DC/DC-Converter, Battery)
6. **On-Board-Charger:** Converts supply voltages to charge battery
   Mapped parameters: Input and Output Voltage and Current, Controls
7. **Distribution boxes (high voltage/low voltage):** Distribution for different components and component setups

## 4.2   General Approach

The common scenarios for the DTs functionality are all evenly structured in terms of input, process, and output. The interpretation of the component level architecture is shown in Fig. 2.

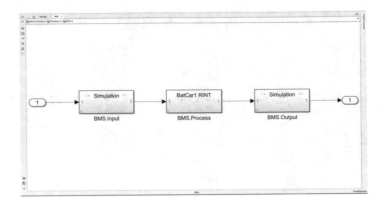

**Fig. 2.** Overview component architecture (example: battery)

To support MBSE DT approach there were developed different subsystems and the possibility of swapping between them, as well as there was developed functionality for evaluating the whole system performance. Moreover, a breakthrough was achieved by being able to control this swapping programmatically through a script in MATLAB thereby enabling us to perform the swapping without even touching the Simulink model. Also, with our method, the input output ports of the subsystem are automatically routed to the external systems and other components.

The input and output variant subsystem mainly depend on the intended use (pure simulation or x-in-the-loop). In case of X-in-the-Loop application, the inputs are taken from physical components (via CAN, I/O, etc.). In case of simulation, the values from the remaining simulation are just passed through.

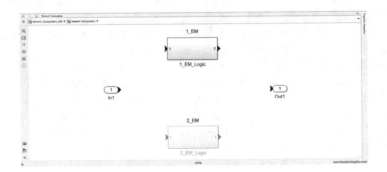

**Fig. 3.** Example of subsystem modelling

Let's further consider some possibilities on the data flow between components and subsystems.

As shown in the Fig. 3, our approach utilizes the power of variant subsystem modelling in Simulink. For example, if the above shown variant subsystem is representing the model for an electric machine (Motor) in an electric drivetrain. There could be some possible optional subsystems such as 1EM (1 motor) or 2EM (2 motors) drivetrains. Therefore, these two combinations could be modeled and placed inside the variant subsystem blocks. However, at a time, one and only one of all optional subsystems will be enabled and shall be taking part in the simulation. Also, the I/O ports in this example are not physically connected but are virtually mapped by the environment. The subsystems can be swapped noninvasively and programmatically by a MATLAB script, so it is possible to run the model for the whole system to act as 1EM drivetrain or 2EM drivetrain.

Depending on the available data basis, differently detailed models are used. For example for the battery, a RINT model can be used if only a small amount of data is available, but a Thevenin model can also be used if more detailed information is available Fig. 4.

In addition (not shown here), different components (e.g. ASM and PMSM for the motors) or different functionalities (e.g. insulation measurement integrated into the battery) can be used.

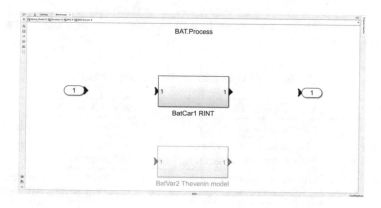

**Fig. 4.** Process subsystem (example: battery)

For the signals also, Simulink buses could be used. So each input and output signal contains different signals subdivided in sub-buses for a better overview:

– CAN as usual automotive standard for communication
– I/O for analog or digital inputs and outputs to control the component or subcomponents.
– Physical Inputs and Outputs correspond to the physical connections between the components, like voltage and current from the battery supplying MCU.

## 5  Conclusions

In this work, solutions for the MBSE DT approach for the new EV technology were considered. The physical vehicle or powertrain is constructed in such a way that components can be easily replaced. This also allows various setups to be tested in practice and thus validates the DT. It enables a complete simulation of the electrical drive train as well as hardware testing of individual components with hardware-in-the-loop.

One limitation is the electric motor. The mechanical connection to the drive axle is not implemented in the setup, as this is very complex for each motor individually.

For the different components, corresponding parameter sets are created, from which it is then possible to freely select which ones are used for the simulations. The models (whether complex or detailed, depending on the components used) are then used automatically. The same applies to the desired different topologies. Depending on which one has been selected, corresponding models are automatically used.

The models for the different topologies were created and modularity was implemented, which allows subsystems to be interchangeable with minimum time costs. Such approach completely aligned with the DT taxonomy and allows to implement all types of analysis for the vehicle including data acquisition, data manipulation, preventive maintenance, health evaluation.

In the Table 1, there is a comparison of the VVC app and our OMAx modularity approach.

**Table 1.** VVC vs OMAx Modularity Solution

| Parameter | VVC Application | OMAx Solution |
|---|---|---|
| Format | Software Package (app) | Design approach/method |
| Flexibility | Not possible to configure individual subsystems | Possibility to configure individual subsystems |
| Programmability | Cannot control the configuration programmatically | Can control the configuration through a MATLAB script |

With this approach, the researcher can simply swap between those subsystems and evaluate the whole system performance.

Developed DT could be used both for industry as at the design as at the validation stage and for the education as the source of information for manipulation.

**Acknowledgment.** This work was funded by the project OMAx-Vehicle (Funding-IDs 13FH0E63I, 13FH0E61IA, 13FH0E62IA) of the German Federal Ministry for Education and Research.

## References

1. Fruhner, D., et al.: OMEx-DriveTrain: an open, modular experimental platform for the electric powertrain. In: IDAACS 2021, pp. 307–312 (2021). https://doi.org/10.1109/IDAACS53288.2021.9660902

2. Ibrahim, M., et al.: EV-permanent magnet synchronous motor control strategy evaluation based on digital twin concept. IEEE (2023). https://doi.org/10.1109/CPE-POWERENG58103.2023.10227410

3. Garcia, F.A., et al.: Automatic generation of digital-twins in advanced manufacturing: a feasibility study. In: ICCAD (2023). https://doi.org/10.1109/ICCAD57653.2023.10152364

4. Wang, Z., et al.: Driver behavior modeling using game engine and real vehicle: a learning-based approach. IEEE Trans. Intell. Veh. **5**(4), 738–749 (2020). https://doi.org/10.1109/TIV.2020.2991948

5. Sahoo, A., et al.: Improved electric vehicle digital twin performance incorporating detailed lithium-ion battery model. In: IEEE CONECCT, Bangalore, India (2022). https://doi.org/10.1109/CONECCT55679.2022.9865710

6. Eaty, N., et al.: Electric vehicle battery management using digital twin, pp. 1–5 (2022). https://doi.org/10.1109/COINS54846.2022.9854955

7. Ye, Y., et al.: Reinforcement learning-based energy management system enhancement using digital twin for electric vehicles (2022). https://doi.org/10.1109/VPPC55846.2022.10003411

8. Tabunshchyk, G., Fruhner, D., Mutyala, S.B.C.: An open modular approach for the design and verification of the electric vehicles. In: Auer, M.E., Langmann, R., Tsiatsos, T. (eds.) REV 2023. LNNS, vol. 763, pp. 561–568. Springer, Cham (2024). https://doi.org/10.1007/978-3-031-42467-0_51

9. Miri, I., et al.: Electric vehicle energy consumption modelling and estimation–a case study. Int. J. Energy Res. **45**, 501–520 (2020). https://doi.org/10.1002/er.5700

10. Scholz, T., et al.: Algorithm for parameter identification of lithium-ion batteries. In: SIMVEC - Simulation und Erprobung in der Fahrzeugentwicklung, Baden-Baden, Deutschland (2018). https://doi.org/10.51202/9783181023334-389

11. Hameyer, K., et al.: The art of modelling electrical machines. ICS Newslett. Int. Compumag Soc. **19**(2), 3–19 (2012)

12. Simeon, K.-T., et al.: A comprehensive overview of the impacting factors on a lithium-ion-battery's overall efficiency. Power Electron. Drives **7**, 9–28 (2022). https://doi.org/10.2478/pead-2022-0002

13. Nikdel, M., et al.: Various battery models for various simulation studies and applications. Renew. Sustain. Energy Rev. **32**(C), 477–485 (2014)

14. Cittanti, D., et al.: Modeling Li-ion batteries for automotive application: a trade-off between accuracy and complexity. Politecnico di Torino Porto Institutional Repository (2017)

15. Mathworks. Simulink. Homepage. https://www.mathworks.com/products/simulink.html. Accessed 10 Nov 2023

16. Virtual Vehicle Composer app. https://de.mathworks.com/help/vdynblks/ref/virtualvehiclecomposer-app.html. Accessed 10 Nov 2023

# Evaluation Design for Cross-Reality Laboratories

Konrad Boettcher[1]([✉])[iD], Claudius Terkowsky[1][iD], Marcel Schade[1][iD],
Nils Kaufhold[1][iD], Louis Kobras[3][iD], Doreen Kaiser[2][iD], Pierre Helbing[4][iD],
Franziska Herrmann[3], Tobias R. Ortelt[1][iD], Ines Aubel[2][iD], Jan Haase[3][iD],
Bernhard Meussen[3], Johannes Nau[4][iD], and Detlef Streitferdt[4][iD]

[1] TU Dortmund University, 44227 Dortmund, Germany
Konrad.Boettcher@tu-dortmund.de
[2] Technical University Bergakademie Freiberg, 09599 Freiberg, Germany
[3] NORDAKADEMIE gAG Hochschule der Wirtschaft, 25337 Elmshorn, Germany
[4] TU Ilmenau, 98693 Ilmenau, Germany

**Abstract.** The evaluation of learning objectives, teaching-learning activities and learning outcome monitoring is the fundamental basis for their further iterative improvement. This paper outlines the development of a mixed-method evaluation instrument designed for cross-reality laboratories. The instrument measures the dimensions of Constructive Alignment (a framework for creating high quality teaching and learning at university level), effectiveness, efficiency, and appeal. In addition, the instrument measures how the laboratory addresses learning outcomes relevant to Work 4.0. For this, 88 initial quantitative items were reduced to 40 through a process involving pre-face validity testing, pre-pilot testing, and pre-pre-testing. The pre-face validity test considered both expert and student feedback. A statistical analysis of the results of a pre-pre-test was carried out to remove further questions from the questionnaire. For this purpose, the Cronbach's alpha as well as the pairwise correlation of the pre-pre-test results were calculated. In the questionnaire's continuing development, the number of questions will be further reduced.

**Keywords:** Engineering Education · Evaluation Design · Cross-Reality Laboratories · Laboratory Instruction

## 1 Introduction

### 1.1 Principles of Education

Constructive Alignment [1] is based on constructivist learning theory. It provides an instructional framework for the design of teaching and learning processes. It aims for the optimal coordination of intended learning outcomes, teaching-learning activities, and learning outcome monitoring and thus also includes ideas from instructional design theory [2]. Crucial to the design of teaching and learning processes is the formulation of intended learning outcomes. These are written statements that, from the students'

M. E. Auer et al. (Eds.): STE 2024, LNNS 1027, pp. 206–216, 2024.
https://doi.org/10.1007/978-3-031-61891-8_20

perspective, indicate the intended learning gain they should achieve by participating in predefined teaching and learning activities and whose achievement is measured or monitored through related assessments [1].

CA was developed by John Biggs for the design of university teaching. It is employed worldwide, especially in the tertiary sector of education [3]. CA combines three functions of pedagogical planning and implementation [1]):

- transformative reflection as eliciting a starting state and designing a target state, which consists of further developing the starting state and reflecting on strengths and weaknesses;
- pedagogical design to implement the results of transformative reflection in a concrete course design;
- action and practice research for formative evaluation of the implemented course to be able to optimize the constructive alignment.

CA is an approach to Outcome-Based Teaching and Learning (OBTL). As such, the traditional approach of an educator imparting knowledge moves into the background with CA focusing on the learning outcomes that learners should achieve through a set of teaching-learning activities. CA thus serves to develop teaching-learning activities and teaching-learning assessments that are both aligned to the intended learning outcomes in ways that are usually not achieved in traditional lectures, tutorials and examinations [1].

Successful coordination requires constant reviewing and, related to this, an iterative improvement, even if it is only due to the changing reality of life for different age groups. One changing reality of life is the ongoing digital transformation. Therefore, novel skills for Learning and Working 4.0 – so-called future skills – are especially needed in the STEM sector. Learning objectives for these future skills can be better addressed in laboratories than in e.g., lectures. They are supposed to extend the fundamental laboratory objectives of Feisel and Rosa [4]. Within the CrossLab project [5], these additional intended learning outcomes presented in Table 1 originated from industry expectations on laboratory teaching [6] and educators within CrossLab [7]. Some of the Work 4.0 learning objectives overlap with the fundamental learning objectives of [4]. This issue is intended to be solved within the validation process of the evaluation scheme.

Laboratory learning objectives can be assigned to specific domains. The taxonomy of Bloom (Table 2) describes the cognitive domain [8], while [9] describes the affective and [10] the psychomotor one.

The revised taxonomy by [11] distinguishes between four dimensions of knowledge: factual, conceptual, procedural, and metacognitive knowledge and assigns each of these to 6 stages of a cognitive process: remember (1), understand (2), apply (3), analyze (4), evaluate (5), create (6).

In order to investigate the quality of an instructional design, [12, 13] and [14] have proposed three priorities to be able to evaluate the learning efficacy and instructional outcomes of the instructional design. These three priorities are:

- effectiveness – defined as a measure of student achievement and the amount of learning related to an instructional objective;

**Table 1.** Fundamental learning objectives of laboratories in undergraduate engineering education and work 4.0-related learning objectives addressed by the CrossLab project.

| Feisel and Rosa [4] | | Work 4.0 [6] | Work 4.0 [7] |
|---|---|---|---|
| Instrumentation | Psychomotor | Know industry environment | Develop personality |
| Models | Safety | Overview over larger context | Improve learning style |
| Experiment | Communication | Working mindset/soft skills | Critical thinking and acting sustainably |
| Data Analysis | Teamwork | | Think out of the box |
| Design | Ethics in the Laboratory | | Self-directed learning |
| Learning from Failure | Sensory awareness | | Work with cyber-physical systems |
| Creativity | | | Organize and manage data with new methods |

**Table 2.** Taxonomies of Bloom et al. (1956), Krathwohl et al. (1964) and Simpson (1966)

| Level | Cognitive domain [8] | Affective domain [9] | Psychomotor domain [10] |
|---|---|---|---|
| 1 | Knowledge | Receiving | Perception |
| 2 | Comprehension | Responding | Set |
| 3 | Application | Valuing | Guided response |
| 4 | Analysis | Organizing | Mechanism |
| 5 | Synthesis | Characterizing | Complexovert response |
| 6 | Evaluation | | Adaption |
| 7 | | | Origination |

- efficiency – defined as a measure of time, money, and/or resources invested in developing and/or implementing a learning experience (LX);
- appeal or attractiveness – defined as a measure of how well a LX engages stakeholders and how much stakeholders like the experience. According to constructivist pedagogy, pleasure in learning is particularly associated with positive effects, which can have a positive influence on learning motivation and actual learning.

These three priorities form an "iron triangle" in which improving the achievement of one priority can lead to a reduction in the achievement of another. Consequently, different instructional designs and related instructional media are preferable for different priorities concerning the relative importance of effectiveness, efficiency, and attractiveness. This means that priorities are critical to good media selection and decisions [15, 16].

## 1.2  Cross-Reality Laboratories

The term "cross-reality lab" [17] describes all types of digitally enriched or online-supported teaching-learning labs, as they use different types of reality provided or created by technology. The term cross-reality thus stands for a connection between the physically existing reality and different types of technically created realities. In the context of the teaching-learning lab, the term can be used both as an umbrella term for all different types of online labs and as a term for combining different online and on-site labs into a coherent learning experience for students across different types of realities [18].

## 1.3  Evaluation of Laboratory Education

In a mixed-method approach, quantitative and qualitative questions are used in tandem to offer the possibility of more detailed or deeper insight into the results than either method on its own. For example, quantitative methods provide an insight into the strength or importance with which students view a question item on average, e.g., the difficulty of a task. Qualitative methods, on the other hand, provide additional information about what exactly caused the difficulties, which can hardly be achieved with purely quantitative methods.

Some prior evaluation schemes already exist. For example, [19] published the LLOM evaluation scheme, which is based on the Measuring the Learning Outcomes of Laboratory Work scheme of [20]. In LLOM, the 13 laboratory objectives from Feisel & Rosa [4] are mapped to the specific levels of the cognitive, affective, and psychomotor domains from Bloom's taxonomy. However, the proposed instrument shows several shortcomings as the questions mostly focus on lower levels of the domains[1]. Some mappings are somewhat doubtful. For example, *learn from failure* is mapped firmly to the affective domain level *characterizing; break down situations and respond accordingly* to *personal behavior*.

The fundamental learning objectives of [4] cannot be addressed at different levels because they are fixed. Also, different LLOM evaluation items were assigned to several learning objectives (which makes sense in their specific laboratory experiments), complicating the identification of the individual learning objectives. Learning objectives specific to Work 4.0 are not addressed and the dimensions effectiveness, efficiency, and appeal should be included. Some items do not meet the quality criteria according to [21], e.g., by linking several questions with "and". Furthermore, the development of the evaluation scheme is based on many students but lacks a broad variety of different laboratory types as well as addressed learning objectives.

## 1.4  Aim of this Paper

The aim of this paper is

- to create an evaluation design
- with respect to the guidelines of evaluation tailored to teaching with cross-reality laboratories, while

---

[1] 15 times level three or lower, 8 times levels 4 or higher, only once the level 6.

- addressing the fundamental laboratory learning objectives of [4] as well as extensions for Learning and Working 4.0.

## 2  Method

### 2.1  Evaluation Design Development

Questions and the categories to which they belong are required for the development of an evaluation form. Thus, qualitative and quantitative questions concerning CA, effectiveness, efficiency as well as appeal and Learning and Working 4.0 were collected.

In the first iteration, initial question items were gathered in group works by members of the CrossLab project. The first iteration of the evaluation questionnaire contained all items gathered this way.

The first draft was handed to students as well as members of the project who were not involved in the drafting of the questions for pre-face validity purposes. The project members were considered experts for the purpose of discussing the evaluation sheet since they were knowledgeable about the application context. Seven of those experts assessed all question items whether they were worded clearly, whether the wording fits the intention (i.e., it measures the underlying concept) and whether the item is relevant. The cutoff-point was set to 2/3. Thus, all items not reaching 2/3 of votes claiming the item to be valid, were discarded. Concurrently, six students assessed the same survey items as the experts in a pre-pilot test to ensure the target group would understand the survey items, with the same item retention criterion of at least 2/3 of all surveyed students confirming the item quality.

After that, 60 students assessed a robotics laboratory in a pre-pre-Test with a 5-point Likert-Scale. This aims at finding and discarding items without a distribution of answers (always yes neutral or no; standard deviation $< 0.5$), as well as items that show a strong correlation with another (correlation $> 0.95$, $< -0,95$) to eliminate very similar items. Thus, the overall number of items may be reduced in the pre-pre-Test, increasing the assumed acceptance of students to participate in a long questionnaire.

Participation in a survey of this length will likely require explicit motivation, explaining briefly the stage of development. Furthermore, making clear that the students putting in the effort is important for the development and that the evaluation will be used, in return, to improve upcoming laboratories which yields a direct benefit for future students as well.

In the next iteration loop the experts of the advisory board of the CrossLab project assess the face validity of the condensed questionnaire. The questionnaire is then to be employed in a pre-Test spanning over the CrossLab universities encompassing a broad range of laboratories [22].

Within the process of validation, if a student takes part in several laboratories, it is important to only include one completed questionnaire for the student. This reduces the number of possible returns but ensures that a personal bias is not counted multiple times while validating. In terms of evaluating the laboratory itself, these "duplicate" voices can of course be considered.

Quantitative items utilize an 11-point Likert-scale. This justifies the assumption of continuous data while the difference between two levels is still distinguishable. The

Likert-scale includes a neutral element to not force students to pick a side if they are indifferent. Since the questionnaire is supposed to be used in a wide range of laboratories in different courses and progress of studies, *not applicable* is added as a 12th option if the item does not make sense in the explicit laboratory employing this evaluation. No items are mandatory, thus providing students the opportunity to skip items they are not willing to answer. Not applicable option is pre-selected within the questionnaire to identify skipped items.

The results from the pre-Test will be used to carry out a confirmatory factor analysis to assess whether the items actually fit into the categories they are placed in. This will require between 200 and 400 students. In addition, the selectivity and item difficulty will be examined in order to validate the questionnaire and make sure the evaluation assesses what it is supposed to assess. The range of the student sample who can be reached within the German project CrossLab and therefore be used to evaluate laboratory types or to validate the evaluation scheme is shown in Table 3 with at least 1200 students per year.

**Table 3.** Student sample for upcoming evaluations. EQF is the European Qualifications Framework, with level 6 being Bachelor's and level 7 being Master's level studies, respectively. The Study Program corresponds to the different fields of STEM the respective course addresses.

| Students/Year | Level EQF/Semester | Study Program | Institute |
|---|---|---|---|
| 10 | 7/7 | S | TU Bergakademie Freiberg |
| 18 | 6/5 | S | TU Bergakademie Freiberg |
| 10 | 7/5 | S | TU Bergakademie Freiberg |
| 10 | 7/2 | S | TU Bergakademie Freiberg |
| 50 | 6/3–5 | S, T, E, M | TU Bergakademie Freiberg |
| 25 | 7/1 | E | TU Bergakademie Freiberg |
| 150 | 6/3 – 6 | T, E | TU Dortmund |
| 50 | 6/2 | S, T, E | TU Dortmund |
| 20 | 7/2 | E | TU Dortmund |
| 50 | 6/2 | T, E | TU Dortmund |
| 60 | 6/2 | T, E | TU Ilmenau |
| 300 | 6/2 | S, T, E, M | TU Ilmenau |
| 40 | 6/5 | T, E | TU Ilmenau |
| 20 | 6/2–6 | S, T, E, M | TU Ilmenau |
| 16 | 6/5 | S, T, E | NORDAKADEMIE |
| 16 | 6/5 | S, T, E | NORDAKADEMIE |
| 128 | 6/7 | E | NORDAKADEMIE |
| 128 | 6/5 | E | NORDAKADEMIE |
| 128 | 6/3 | E | NORDAKADEMIE |

The number can be increased by further laboratory experiments that are being developed at the universities as part of the CrossLab project with other STEM departments. The EQF denotes the level of the European Qualifications Framework for the degree in which the experiment takes place. The subject areas are indicated by the individual letters of the STEM term. All branches of STEM can be covered as well as the levels of the EQF addressed in higher education. The laboratories are mostly on-site, augmented, virtual, remote, and ultra concurrent remote laboratories and contain mainly script-based, project-based, and scenario-based learning [22].

## 3　Results

### 3.1　Quantitative Items

The questionnaire developed contained a total of 88 initial quantitative items. The developed items are classed into Constructive Alignment (16), perceived effectiveness (8) perceived efficiency (15), perceived appeal (33) and the laboratory objectives (13) of [4]. In addition, one item each was developed regarding the learning objectives of Learning and Working 4.0. Example items for the individual categories are listed in Table 4.

**Table 4.** Quantitative questions

| Dimension | Item |
| --- | --- |
| Constructive Alignment | The learning objectives are clearly defined |
| Perceived Effectiveness | I was able to achieve the learning objectives [of the course] |
| Perceived Efficiency | [The learning materials/tasks] were clearly formulated |
| Appeal | The laboratory activities were interesting |
| Intended Learning Outcomes | The laboratory activity was aimed at developing my skills in creativity |

### 3.2　Pre-face Validity

For the pre-face validity, seven experts in (cross-reality) laboratory education were asked about the validity and comprehensibility of the individual items. Items that were classified as invalid/incomprehensible by at least 1/3 of the respondents were removed from the questionnaire. This process eliminated a total of 48 items. After the pre-face validation, eight items for Constructive Alignment, two items for perceived effectiveness, three for perceived efficiency and 15 for appeal remained. Questions concerning the perceived effectiveness in the learning objectives according to [4] could be reduced to eleven items. Of the three additionally defined learning objectives (see Table 1 and Sect. 1.1), only the item concerning personal growth remained in the questionnaire.

### 3.3  Pre-pilot Testing

For the pre-pilot testing, six students were asked to rate the comprehensibility of the initial items, similar to the experts in Sect. 3.2. It is striking that only two items were rated as incomprehensible by the students (with a 1/3 cut-off). These two items were already removed by the pre-face validity, leaving the number of remaining items unchanged.

### 3.4  Pre-pre-Testing

For the pre-pre-testing, the evaluation of a robotics laboratory with 60 participating students was analyzed. To test the reliability of the questionnaire, Cronbach's alpha is used separately for each dimension. According to [21], an analyzed dimension is considered reliable for basic research from an alpha greater than or equal to 0.7, however, it should be noted that this parameter is not entirely reliable [23]. The dimensions of perceived efficiency, appeal and CA fulfilled this criterion. However, perceived effectiveness falls short with $\alpha = 0.53$.

Furthermore, the result of the CA and effectiveness cannot be considered, as the learning objectives were not communicated to the students at the beginning of the course. The CA must be checked in pre-testing under the correct conditions.

In addition, the correlation between each item was used to search for items that can be considered not useful for the instrument. If the correlation between any two items is either greater than 0.95 or smaller than $-0,95$, both items can be considered too similar for both items to be useful. Our analysis showed that the correlation between any two items is within these limits. Thus, we were not able to remove any items here.

### 3.5  Qualitative Items

The qualitative items contained in the evaluation form are shown in Table 5. As with the quantitative items, every question can be assigned to one of the five dimensions, with the addition of the dimensions general and creativity. There are two reasons for adding the new dimensions: These two dimensions are intended to assess the students' general impression regarding the completed laboratory experiment, and creativity is needed for new suggestions to improve the laboratory experiment.

However, using all ten open questions seem to be too much. Therefore, we reduced the qualitative questions to four questions (see grey-colored items Table 5). One consideration in this was that students should not be asked things they probably do not know or that might be unclear (e. g. what is Work 4.0?, What is Constructive Alignment?).

**Table 5.** Qualitative questions.

| Dimension | Item |
| --- | --- |
| Effectiveness | How could the experiment be designed to learn more? |
| Appeal | How could the experiment be more fun? |
| General | What was good? |
| General | What was poor? |
| General | Where did problems occur during the experiment? |
| Efficiency | Which parts of the entire experiment are inefficient (take a long time and offer little learning gain)? |
| Work 4.0 | How could the experiment be designed to learn more for the world of work 4.0? |
| CA | How can the alignment between the learning objectives, the experiment and the assessment be improved? |
| Creativity | If a University-fairy could grant you one wish, what would it be? |

## 4 Discussion

When evaluation questionnaires are developed at short notice or without enough input from experts, there is a danger that results from that instrument answer the underlying questions inaccurately, insufficiently, or not at all. In the worst case, such an instrument might even result in wrong answers due to an incorrect design. The use of already validated evaluation schemes avoids these problems. However, the existing evaluation patterns are often precisely tailored to specific use cases. The creation of a validated instrument is very time-consuming and involves many persons, such as experts for face validity as well as large numbers of students for pre-testing. In the case of general applicability for cross-reality labs, the pre-test must be performed on as many lab types and different experiments as possible to ensure generalizability.

In our first step, a questionnaire with 88 initial quantitative questions was developed, based on the five dimensions of effectiveness, efficiency, appeal, Constructive Alignment, and Learning and Working 4.0. The first step of validation, as described in Sect. 3, enabled the questionnaire to be reduced to 40 comprehensible items. It is important to note that the experts and students interviewed for pre-face validation were not involved in the development process of the initial questions. This ensures not only many different perspectives, but also that any bias of the initial question developers will not remain in the final instrument.

Since qualitative questions require a lot of time to answer, the number of dimensions asked is limited to effectiveness, appeal and two general questions, which should be enough to get the most important feedback from students.

## 4.1 Next Steps

The pre-face validity, pre-pilot testing and pre-pre-test have resulted in a template for the evaluation form. We are currently in the process of creating face validity and performing a pilot test involving all partner universities in CrossLab. In this step, laboratories at TU Bergakademie Freiberg, TU Dortmund University, TU Ilmenau, and NORDAKADEMIE will be evaluated: In Freiberg, laboratories from the Institute of Chemical Technology as well as from the Institute of Computer Sciences; in Dortmund, laboratories from the Departments of Biochemical and Chemical Engineering, Chemistry, Mechanical Engineering, Electrical Engineering, and Information Technology; in Ilmenau, laboratories from (Technical) Computer Science; in the NORDAKADEMIE, laboratories from the Department of Computer Science and Industrial Engineering. Once the pre-test has been completed with 200 to 400 students across all universities, a confirmative factor analysis will be carried out with the goal to determine whether the items actually fit into the categories. Furthermore, the most sensitive items will be selected, and the questionnaire will be validated. The final, validated evaluation form will be publicly available to university teachers. Educators using cross-reality laboratories are highly welcomed to use the instrument and join further research in this area.

**Acknowledgments.** The work presented in this paper is part of the research project "CrossLab-Flexibel kombinierbare Cross-Reality Labore in der Hochschullehre: zukunftsfähige Kompetenzentwicklung für ein Lernen und Arbeiten 4.0" funded by Stiftung Innovation in der Hochschullehre.

# References

1. Biggs, J.B., Tang, C., Kennedy, G.: Teaching for Quality Learning at University, 5th edn. McGraw-Hill/Open University Press, Maidenhead (2022)
2. Biggs, J.B.: Enhancing teaching through constructive alignment. High. Educ. **32**, 347–364 (1996). https://doi.org/10.1007/BF00138871
3. Biggs, J.B.: Constructive alignment in university teaching. HERDSA Rev. High. Educ. **1**, 5–22 (2014)
4. Feisel, L.D., Rosa, A.J.: The role of the laboratory in undergraduate engineering education. J. Eng. Educ. **94**, 121–130 (2005). https://doi.org/10.1002/j.2168-9830.2005.tb00833.x
5. Aubel, I., et al.: Adaptable digital labs – motivation and vision of the CrossLab project. In: 2022 IEEE German Education Conference (GeCon), pp. 1–6. IEEE, Berlin, Germany (2022)
6. Soll, M., Boettcher, K., Expected Learning Outcomes by Industry for Laboratories at Universities. In: 2022 IEEE German Education Conference (GeCon), pp. 1–6 (2022)
7. Boettcher, K., Aubel, I., Ortelt, T., Terkowsky, C., Soll, M.: Did you check it? Checklist for Redesigning a Laboratory Experiment in Engineering Education addressing Competencies of Learning and Working 4.0. In: Auer, M.E., Langmann, R., Tsiatsos, T. (eds.) Open Science in Engineering, pp. 601–609. Springer Nature Switzerland, Cham (2023)
8. Bloom, B.S. (ed.), Engelhart, M.D., Furst, E.J., Hill, W.H., Krathwohl, D.R.: Taxonomy of educational objectives: the classification of educational goals. Handbook 1: Cognitive Domain. David McKay, New York (1956)
9. Krathwohl, D.R., Bloom, B.S., Masia, B.B.: Taxonomy of educational objectives: The classification of educational goals. Handbook II: The Affective Domain. David McKay, New York (1964)

10. Simpson, B.J.: The classification of educational objectives: Psychomotor domain. Illinois J. Home Econ. **10**(4), 110–144 (1966)

11. Anderson, L.W., Krathwohl, D.R.: A Taxonomy for Learning, Teaching, and Assessing: A Revision of Bloom's Taxonomy of Educational Objectives: Complete Edition. Addison Wesley Longman, Inc. (2001)

12. Reigeluth, C.M., Honebein, P.C.: Will instructional methods and media ever live in unconfounded harmony? Generating useful media research via the instructional theory framework. Educ. Tech. Research Dev. (2023). https://doi.org/10.1007/s11423-023-10253-w

13. Jahnke, I.: Quality of digital learning experiences – effective, efficient, and appealing designs? Int. J. Inf. Learn. Technol. (2023). https://doi.org/10.1108/IJILT-05-2022-0105

14. Reigeluth, C.M. (ed.): Instructional Design Theories and Models. Routledge (1983). https://doi.org/10.4324/9780203824283

15. Honebein, P.C., Reigeluth, C.M.: The instructional theory framework appears lost. Isn't it time we find it again? Revista de Educación a Distancia **64**(20), (2020). https://revistas.um.es/red/article/view/405871/290451

16. Honebein, P.C., Honebein, C.H.: Effectiveness, efficiency, and appeal: pick any two? The influence of learning domains and learning outcomes on designer judgments of useful instructional methods. Education Tech. Research Dev. **63**(5), 745–765 (2015). https://doi.org/10.1007/s11423-015-9396-3

17. May, D.: Cross reality spaces in engineering education-online laboratories for supporting international student collaboration in merging realities. Int. J. Onl. Biomed. Eng. **16**(03), 4–26 (2020). https://doi.org/10.3991/ijoe.v16i03.12849

18. May, D., Terkowsky, C., Varney, V., Boehringer, D.: Between hands-on experiments and Cross Reality learning environments – contemporary educational approaches in instructional laboratories. Eur. J. Eng. Educ. **48**(5), 783–801 (2023). https://doi.org/10.1080/03043797.2023.2248819

19. Sasha, N., et al.: Laboratory learning objectives: ranking objectives across the cognitive, psychomotor and affective domains within engineering. Eur. J. Eng. Educ. (2023). https://doi.org/10.1080/03043797.2023.2248042

20. Salim, K.R., Rosmah, A., Hussain, N.H., Haron, H.N.: An Instrument for Measuring the Learning Outcomes of Laboratory Work. Paper presented at the International Engineering and Technology Education Conference, Ho Chi Minh City, Vietnam (2013)

21. Kember, D., Ginns, P.: Evaluating Teaching and Learning. Routledge, London (2012). https://doi.org/10.4324/9780203817575

22. Schade, M., Terkowsky, C., Boettcher, K., Ortelt, T.R.: Work in progress: cobbler, stick to your last! on providing engineers constructive alignment. In: Auer, M.E., Cukierman, U.R., Vidal, EVendrell, Caro, E.T. (eds.) Towards a Hybrid, Flexible and Socially Engaged Higher Education: Proceedings of the 26th International Conference on Interactive Collaborative Learning (ICL2023), Volume 3, pp. 284–292. Springer Nature Switzerland, Cham (2024). https://doi.org/10.1007/978-3-031-53022-7_29

23. Schmitt, M.: Uses and abuses of coefficient alpha. Psychol. Assess. **8**, 350–353 (1996)

# Introducing Aviation Technological College Students to Online Collaborative CAD Design

M. Greenholts⬤, I. Verner(✉) ⬤, and A. Polishuk⬤

Technion – Israel Institute of Technology, 3200003 Haifa, Israel
{mosgrini,ttrigor}@technion.ac.il, alepole@ed.technion.ac.il

**Abstract.** This paper argues for incorporating learning online collaborative design (LOCD) in technical colleges. It reports a case study of a LOCD workshop given to students majoring in aviation technology as part of the CAD course. In the implemented approach, the students, who already had basic CAD skills, were engaged in team projects in a setting where they performed the analysis tasks while sitting together, and performed the design assignment through online collaboration while sitting apart from each other and. The teams were divided into two groups: one used the desktop platform Solid Edge and the other the cloud-based platform Onshape. In the study, we identified several characteristics of learning online collaborative design and showed that the cloud-based platform was better suited to provide them. The students who used Onshape had less difficulty communicating with each other and managing their design projects. Our data analytics inquiry into the division of work and collaboration within Onshape teams indicated that during the design process, team members actively interacted with each other, alternating between the functions of executing design operations and monitoring their implementation. We call for further research towards the development of learning environments and instructional strategies for incorporating online collaborative design in college education.

**Keywords:** Technology Education · Collaborative Design · Online Learning

## 1 Introduction

The development of new digital technologies and concepts that revolutionize the industry leads to the dramatic transformation of engineering systems and their computer-aided design [1]. One of the most significant changes in CAD resulting from the progress of communication technologies and social networking is the creation of cloud-based design platforms and their rapidly growing popularity. Due to improvements in internet speed and network bandwidth, cloud-based CAD platforms such as Onshape are now available as Software-as-a-Service (SaaS) systems, which allow users to model and assemble 3D parts via a web browser [2].

Onshape enables access via any web browser on a computer or mobile device, requires no download or installation, and is free for academic use. Several designers can work simultaneously on the same CAD document, and the designs are automatically

M. E. Auer et al. (Eds.): STE 2024, LNNS 1027, pp. 217–225, 2024.
https://doi.org/10.1007/978-3-031-61891-8_21

updated and saved in the cloud [3]. Due to its qualities, Onshape is gaining popularity in introductory CAD courses [4]. The data analytics function of Onshape provides access and analysis of data related to users' design activities. Today, engineering designers usually work alongside partners and collaborators and cloud-based CAD systems have been growingly adopted in the modern industry [5]. Therefore, it is essential to extend CAD education and introduce students to online collaborative design using such systems.

Learning by design can be considered from individual and social perspectives. Design cognition research conventionally focused on individual cognitive processes in which designers performed projects while working on personal CAD stations [6]. When considering team design projects, design is based on collective intelligence when individuals collaborate in the production of solutions and their synthesis [7].

In collective design projects, the learners aim to construct knowledge of the learned subject and develop a usable product demonstrating their learning outcomes [8]. In such projects, one of the main responsibilities of the design leader is to organize and lead the collaborative work of team members through effective communication [9]. In guiding collective design projects, teachers seek ways to organize students in design teams to capitalize on the potential of each student and facilitate their collective work [10]. They need to follow up and evaluate the individual contributions of the students to the project and regulate the distribution of project work to avoid overloading [11].

There are difficulties in managing and guiding students who remotely work on collaborative design projects by using conventional desktop CAD platforms. It seems propitious to overcome these difficulties by developing a methodology for learning online collaborative design (LOCD) with cloud-based platforms, like Onshape [4].

This study proposes an approach to educating aviation technology students in online collaborative design and explores its implementation using the desktop and cloud-based design platforms Solid Edge and Onshape. We consider two student groups that participated in the CAD workshop and performed design projects, one with Onshape and the other with Solid Edge.

The research questions were:

1. What are the characteristics of LOCD by students who perform design projects using desktop or cloud-based CAD software platforms?
2. How did the students evaluate the contribution of the workshop to their learning and online communication during the CAD design process?

## 2   The Workshop

The workshop "Collaborative Design of Air Gliders" was conducted in the spring semester of 2022–2023 at the Israel Air Force (IAF) Technical College in the framework of the Unmanned Aerial Vehicles (UAV) course. The course was developed and delivered by the first author (Greenholts) as part of the Aviation Practical Engineers program. It addressed second-year students who have already learned the basics of aviation technology and CAD. The course covered a wide range of topics in the subject of aviation, with a focus on unmanned aircraft. It comprised lectures on UAV systems and the principles of digital design and making.

The workshop that was the practical part of the course involved students in the design and construction of air glider models. In the workshop, the students, through analytical analysis, determined the aeronautical parameters of the glider and designed the prototype ready for 3D printing.

The workshop was conducted for a group of eighteen practical engineering students majoring in aviation at the IAF technical college. In this 8-h workshop, the students carried out a design project. The project was conducted in a lab that had a drone testing area and was equipped with PCs. The project assignment was to design an air glider model while working in teams of two.

The teams were divided into two groups that performed the design assignment using two different CAD software platforms. The group of four teams (eight students) used Onshape, and another group of five teams (ten students) practiced with Solid Edge. It is worth noting that the students in both groups had basic knowledge of Solid Edge and Onshape from their high school CAD course. The outline of the 8-h workshop is presented in Table 1. The table specifies the topics and activities studied at the four workshop sessions.

**Table 1.** Workshop outline

| Workshop sessions | Topics and activities |
| --- | --- |
| 1. Introduction (4 h.) | Basics of UAV and flight physics, refreshing CAD skills |
| 2. Glider characteristics (2 h.) | Glider aerodynamic forces and COG |
| 3. Glider CAD design (1 h.) | Online collaborative design of a glider |
| 4. Final presentation (1 h.) | In-class presentations of the projects |

In the first session, the students were introduced to unmanned aircraft vehicles and learned the concepts of aerodynamic forces and center of gravity, and about the application of the Newton laws in aviation mechanics. The groups practiced CAD design using Solid Edge and Onshape correspondingly. In the second session, the students designed air glider models by making analysis and determination of characteristics. During this 2-h session, the students learned about the glider flying characteristics including aerodynamic forces and center of gravity.

Restrictions were placed on the necessary glider to concentrate on the design and learning objectives in the given period. The glider had to be made of PLA and weigh up to 30 g, length and width up to 15 cm, and height up to 3 cm. In both groups, the students determined the wing aerofoil profile of the designed glider and calculated its lift force, drag, and moment using the software tool JavaFoil. They saved the profile drawings of the wing generated by the software for further use in CAD design.

Then the students used the AERY Glider Design software to determine the characteristics of the glider aligned with the calculated characteristics of the wings. Figure 1A shows one of the student teams determining the characteristics of their glider model using the AERY Glider Design software.

The third session engaged the students in team design through distance collaboration and was organized in the following way. The students worked in two computer classes, and it was set up so that members of each team sat in different classes. Each team member carried out the glider design assignment while using his/her workstation and collaborating with the team peer who sat in another class. The peers collaborated via distance communication tools. In this session, the students were required to design and draw the model of the air glider. Figure 1 demonstrates a model of a glider designed by one of the teams using Onshape. In the last session, the students sat together in one classroom. Each team presented their progress in the assignment and answered the teacher's questions about their learning outcomes and their experience in the workshop.

**Fig. 1.** The designed glider.

## 3   The Study

Our exploratory case study followed up the students' learning practice focusing on their team collaboration in performing the workshop assignments. To gain insight into the learning process, we applied a mixed method and combined conventional data collection tools such as a post-workshop questionnaire, observations, worksheets, and project reports with the data analytics tools of Onshape.

The participatory observations during the design project focused on the distance collaboration of the team members with special attention to the use of different communication channels, quality of internet connection, collective access to project files and documents, division of roles and responsibilities, team discussions, and effectiveness of collaboration.

The CAD design activities were guided by two online worksheets. The sheets included knowledge questions related to the activities. Students' answers to these questions provided an additional source of research data. Using the first worksheet, the students documented their inquiry about existing methods of air glider design, determined the glider's wing profile using the JavaFoil and AERY Glider Design software, and answered knowledge questions in glider aerodynamics. The questions asked about the requirements, criteria, and constraints in glider design. The participants had to compare three alternative designs and select the most suitable design solution. Then they had to sketch the solution.

The students used the second worksheet in the third session of the workshop. The worksheet consisted of two sections. The first section guided them to make a CAD model of the air glider based on the sketch they made and the determined wing profile. In the second section, the students documented the specifications of their designs. We evaluated students' answers to the questions asked in the worksheets based on the following criteria: correctness of the answers, quality of glider design, and assignment completion.

At the end of the CAD design session, each student reported in a checklist the design operations done by her/him, by the teammate, and by them together. The student also indicated the number of times he/she remotely communicated with the teammate during different design operations. The internal consistency of the checklist, estimated using Cronbach's alpha, was $\alpha = 0.806$. In the analysis of remote CAD design operations made by the team members, we looked at the division of roles and responsibilities in the teams working with Solid Edge and Onshape.

The post-workshop questionnaire included three sections. The first section asked students to evaluate on the Likert scale the workshop's contribution to their learning of STEM concepts and applied skills related to the air glider project. In the second section, the students evaluated on the Likert scale their collaboration and communication in the project. In the third section, they answered open-ended questions about the pros and cons of the CAD system they used and about their satisfaction with the workshop.

From the project reports, we collected and analyzed data regarding the design processes and solutions developed by the teams.

Data analytics tools were used only for the Onshape teams, as Solid Edge lacks such tools. In Onshape, data on the design activities of all users are automatically recorded and can be obtained for analysis in the Excel table format indicating the time spent, the object addressed, and the specification of each action (creation of a part, viewing the model, etc.).

In Onshape, only one user can create the same design document at a time, while others can watch it from their computer screens. Therefore, when carrying out team design tasks, the students worked in turns, being alternatively in the roles of a maker and an observer. In the analysis of the table data, we looked at how the roles of each student changed during the design process. We characterized students' collaboration in design teams using two parameters, namely, the average time spent by each participant on the task and the number of times they switched from maker to observer.

The observations conducted by the course teacher focused on the episodes of students' online collaborative behaviors during the CAD design practice and supplied additional evidence for understanding the LOCD processes.

# 4 Findings

## 4.1 Characteristics of LOCD

The analysis of data collected in the workshop indicated the characteristics of learning online collaborative design some of which are presented below.

A. Shared design ideas and processes
   Seitamaa-Hakkarainen, Viilo, and Hakkarainen [12] emphasized that the development of one workable idea of the solution through the negotiation of team members is crucial for the success of a collaborative design project.

Our observations and project reports indicated that in all the teams the team members elaborated original design ideas and implemented them while working together. For example, one team designed a glider with rectangular and straight wings, and tail stabilizers inclined upward to provide horizontal stability to the glider. In the project report, the team described their collective work as follows:

> *"To determine the glider's dimensions that fit the predetermined constraint, we first calculated them analytically. Then, we tested the model using the AERY software, identified some glitches, and fixed them. After the repair, we repeated the test and after receiving a positive result used these dimensions in the model."*

B. The challenge of remote collaborative design
   Our observations of the design process indicated that the teams had difficulties in managing design collaborative processes and in meeting the project deadline. The project communication difficulties during the CAD design session were noted in the post-workshop questionnaire. 67% of the students from the Solid Edge group had such difficulties, and 33% reported that the difficulties were high.

The difficulties in the Onshape group were lower: they were noted by 63% of the students and 25% evaluated them as high. The difference between the groups in design communication can be seen also in students' reflections. From the Solid Edge group:

> *"During the project, we had difficulty reaching an understanding between us. We didn't have a camera and during the joint work a teammate didn't know what I was doing".*

From the Onshape group:

> *"The explanation through the screen is more difficult than a face-to-face conversation. It was not effective to show the computer screen through a phone."*

C. Automatic version control
   Version control is a critical feature of collaborative engineering design that allows all team members to work on the same up-to-date project documents [13]. In our workshop, the students saw first-hand that the success of collaborative design projects depends on version control. The difficulties in collaboration with their teammates and the problems caused by using outdated design versions were especially prominent

in the Solid Edge group. In the post-workshop debrief the students from this group noted:

*"My teammate and I worked on different design versions without coordination and saved them on our computers. We did not lose the design solutions but had difficulties in finding the best ones in the versions we made."*

From our observations, the Solid Edge teammates tried to share real-time changes made by each of them by visual communication via their smartphones, but this did not supply a sufficient version control solution. The above difficulties were rarely mentioned by the students who worked with Onshape, used its built-in automatic version control, and saw real-time changes on their computer screens.

D. Dynamic role changing

An important characteristic of online collaboration in the design of real and virtual systems is dynamic role-changing, indicating that the team members exchange their roles and responsibilities during the design process [14, 15]. In design education, such role-changing contributes to learning collaborative design by enhancing the exchange of knowledge and collaboration among the team members [16].

The Onshape data analytics tools enabled us to analyze the roles of each of the team members in the collective design process. This analysis revealed two active roles of students' participation in the design process: operator and observer. As the operator, the student used the commands: Import, Insert, and Create to manage files and documents, and performed CAD drawings using Add, Undo, Update, and Edit functions. As the observer, the student examined and verified the drawing actions done by the operator using the Tab and Show functions.

During the design project, the team members changed between the operator and observer roles, as presented in Table 2.

**Table 2.** Operating and Observing activities per team.

| Team and member numbers | Role changes | Average time (min.) | |
|---|---|---|---|
| | | Operator | Observer |
| Team 1 Member 1 | 24 | 3.3 | 1.7 |
| Team 1 Member 2 | 24 | 1.7 | 3.3 |
| Team 2 Member 1 | 17 | 1.9 | 4.8 |
| Team 2 Member 2 | 17 | 4.8 | 1.9 |

The first column of the table lists the members of the two teams. The second column displays the number of role changes. The last two columns present the average time in minutes spent by each participant as operator and Observer. Table 2 indicates the high intensity of role changes in both teams. This indication is in line with supported by the evidence of the checklist evaluations given by the students of the Onshape group. From the table, it also follows that the times in the operator and observer roles were unequally divided between the team members.

### 4.2 Workshop Contribution

In the post-workshop questionnaire, the students evaluated the workshop contribution while addressing learning and application of STEM subjects and experience in using the CAD system. Most students in both groups noted the contribution of the project to learning the aviation subject. The evaluations given by the students of the OnShape group (86%) were higher than those of the SolidEdge group (60%). 59% noted that the workshop provided them with opportunities to apply mathematics, physics, and engineering concepts in the design project. Again, the evaluations in the Onshape group (71%) were higher than those in the Solid Edge group (60%). 71% expressed satisfaction with the workshop. However, the satisfaction of the students who used Solid Edge (80%) was higher than that using Onshape.

## 5   Conclusion

Our research argues that online collaborative design studies should be introduced in technical colleges, and it practically demonstrates a possible approach to incorporate these studies into the college CAD curriculum.

In our approach, learning online collaborative design methods was the main goal of a special workshop dedicated to the topic. In the workshop, the engineering students who already learned the basic CAD course were engaged in team projects in a setting where team members sat apart from each other and performed the design assignment through online collaboration. We divided the teams into two groups: one used the desktop platform Solid Edge and the other the cloud-based platform Onshape.

In this research, we identified several characteristics of learning online collaborative design and showed that the cloud-based platform is better suited to provide them. The students who used Onshape had less difficulty communicating with each other and managing their design projects.

Our data analytics inquiry into the division of work and collaboration within Onshape teams indicated that during the design process, team members actively interacted with each other, alternating between the functions of executing design operations and monitoring their implementation.

The limitations of our research are its small sample size of students majoring in a certain engineering field and the limited time allotted for their learning practice in online collaborative design. Also, we could not compare the effectiveness of our approach to LOCD with alternative approaches, since we did not find in the literature studies proposing such approaches. Therefore, we recommend further research on the development of learning environments and instructional strategies for introducing online collaborative design in college education.

**Acknowledgement.** The study was supported by the Technion Additive Manufacturing Center Dissemination Grant.

# References

1. Jiao, R., Commuri, S., Panchal, et al.: Design engineering in the age of industry 4.0. J. Mech. Design **143**(7), 070801–1–070801–25 (2021)
2. Hirschtick, J.: Introducing Onshape Beta (2015), https://medium.com/onshape-news/introducing-onshape-beta-21098b965549. Last accessed 6 Nov 2023
3. Le, N.: Product Design with Cloud Based and Desktop CAD software: A Comparison between SolidWorks and Onshape. Degree Thesis (2018)
4. Phadnis, V.S., Wallace, D.R., Olechowski, A.: A multimodal experimental approach to study CAD collaboration. Comput.-Aided Des. Appl. **18**(2), 328–342 (2020). https://doi.org/10.14733/cadaps.2021.328-342
5. Siderska, J., Jadaan, K.S.: Cloud manufacturing: a service-oriented manufacturing paradigm. A review paper. Eng. Manag. Product. Serv. **10**(1), 22–31 (2018). https://doi.org/10.1515/emj-2018-0002
6. Maher, M.L., Paulini, M., Murty, P.: Scaling up: From individual design to collaborative design to collective design. In: Gero, J.S. (ed.) Design Computing and Cognition'10. Springer, Netherlands (2011)
7. Paulini, M., Murty, P., Maher, M.L.: Design processes in collective innovation communities: a study of communication. CoDesign **9**(2), 90–112 (2013)
8. Greenholts, M., Verner, I.: A reverse engineering and making approach to enhance learning in school design projects. INTED 2023. IATED (2023)
9. Maher, M.L., Brady, K., Fisher, D.H.: Computational models of surprise in evaluating creative design. Fourth International Conference on Computational Creativity (Vol. 147). University of Sydney, Faculty of Architecture, Design and Planning (2013)
10. Fielding, E.A.S., McCardle, J.R., Eynard, B., Hartman, N., Fraser, A.: Product lifecycle management in design and engineering education: International perspectives. Concurrent Eng. **22**(2), 123–134 (2014). https://doi.org/10.1177/1063293X13520316
11. Hoepfl, M.: Teaching and learning in project-based learning, technology and engineering education, and related subjects. Exemplary teaching practice in technology and engineering education, pp. 1–32 (2016)
12. Seitamaa-Hakkarainen, P., Viilo, M., Hakkarainen, K.: Learning by collaborative designing: technology-enhanced knowledge practices. Int. J. Technol. Des. Educ. **20**(2), 109–136 (2010)
13. Jones, D., Nassehi, A., Snider, C., et al.: Towards integrated version control of virtual and physical artefacts in new product development: inspirations from software engineering and the digital twin paradigm. Procedia CIRP **100**, 283–288 (2021). https://doi.org/10.1016/j.procir.2021.05.121
14. Odell, J., Parunak, H., Brueckner, S., et al.: Changing roles: dynamic role assignment. J. Object Technol. **2**(5), 77–86 (2003)
15. Chitanana, L.: The role of Web 2.0 in collaborative design: an ANT perspective. Int. J. Technol. Des. Educ. **31**(5), 965–980 (2021). https://doi.org/10.1007/s10798-020-09578-x
16. Germani, M., Mandolini, M., Mengoni, M., et al.: Collaborative design system for supporting dynamic virtual enterprises. In: Camarinha-Matos, L.M., Boucher, X., Afsarmanesh, H. (eds.) PRO-VE 2010. IAICT, vol. 336, pp. 577–584. Springer, Heidelberg (2010). https://doi.org/10.1007/978-3-642-15961-9_69

# Education for Industry 4.0: Introducing Engineering Students to Cloud-Based Collaborative Design

D. Cuperman[1,2], I. Verner[1(✉)], and U. Rosen[1,2]

[1] Technion – Israel Institute of Technology, 3200003 Haifa, Israel
dancup@ed.technion.ac.il, ttrigor@technion.ac.il,
rosen@campus.technion.ac.il
[2] Braude College of Engineering, 2161002 Karmiel, Israel

**Abstract.** This paper proposes and explores an approach to learning cloud-based collaborative design in the framework of a new course that introduces engineering students to Industry 4.0. The approach was implemented in the course workshop and was featured by active learning in the FabLab space, the use of the new SaaS CAD platform Onshape, and a challenging collaborative assignment prompting prototype-redesign iterations. We conducted an exploratory case study that evaluated the proposed approach regarding the workshop learning outcomes and the use of the data analytics tools of Onshape. The participants were 18 junior and senior engineering students majoring in mechanical engineering. We collected the educational research data using pre-course and post-workshop questionnaires as well as using the data analytics tools of the Onshape Education Enterprise. Most students highly evaluated the workshop's contribution to fostering their skills in design collaboration, 3D printing, and engineering problem-solving. The data analytics tools of OEE supported the findings from student reflections and provided objective evidence of design, collaboration, and communication processes that occurred in student teams. The researchers considered this study as a step towards the development of the methodology for cloud-based collaborative design education.

**Keywords:** Industry 4.0 · Engineering Education · Cloud-Based Collaborative Design

## 1 Introduction

The technological innovation brought by Industry 4.0 creates a new collaborative ethos that is reshaping various engineering sectors [1]. The spirit of collaboration is particularly relevant in engineering design, rapid prototyping (RP), and additive manufacturing (AM), where cloud-based platforms enable real-time collaboration, resource sharing, and a more cohesive approach to design and digital production tasks [1].

In the current era of computer-aided design, conventional desktop CAD systems have been enriched with cloud-based tools that bolster RP and AM processes [2]. These technological advancements have created new conditions, in which engineering design documents are cloud-stored, allowing for global collaborative access and evaluation of the design process based on data analytics. This access paves avenues for optimizing design solutions and revolutionizes real-time collaborative work, leading to more unified methods of design and digital manufacturing [3].

Industry 4.0 marks a pivotal evolution in the domain of engineering design education. Academic institutions around the world are integrating Industry 4.0 into their curriculum, either by infusing related topics into existing courses [4] or by creating dedicated programs [5].

In light of these developments, it is critical to reassess and modernize educational frameworks [6]. Integrating the topic of cloud-based collaborative design into engineering curricula is imperative to ensure that educational outcomes are in step with the demands of Industry 4.0 [1, 2, 7].

Recently, data analytics has started to be implemented for the analysis of design processes in engineering education. The use of these tools for the exploration of design-based learning processes in CAD environments and the evaluation of their outcomes is very promising but poses significant technological and pedagogical challenges [8, 9].

This research aims to propose and explore an approach to learning online collaborative design in the framework of a new course that introduces students to Industry 4.0. The proposed approach leverages the capabilities of the cloud-based platform Onshape to provide shared access and version control of design documents, as well as data collection and analysis of collective design processes. The approach is implemented in the course workshop dedicated to online collaborative design and 3D printing. Our research questions focused on the evaluation of the proposed approach regarding the workshop learning outcomes and the use of the data analytics tools of Onshape.

## 2 Engineering Courses Dedicated to Industry 4.0

Hernandez-de-Menendez et al. [6], in their review paper, pointed to the key Industry 4.0 technologies that can be used to accelerate the development of engineering education, among them 3D printing, augmented reality (AR), Internet of Things (IoT), artificial intelligence (AI), and cloud computing. The authors surveyed programs, courses, and workshops that academic institutes around the world offered to prepare their students for Industry 4.0. The paper focused on the technologies used, the ways they are taught, and the competencies addressed. They reported that a significant number of institutions already offer such programs and elaborated on three of them. They described MIT's Smart Manufacturing Program, the programs and courses at ETH Zurich's Institute of Virtual Manufacturing, and the master's program in robotic systems engineering at RWTH Aachen University. The latter program includes a course sequence in Computer Vision, Machine Learning, Simulation, and Automation Technology for Production Systems. The authors concluded that educational organizations should keep on adapting their curricula and educational strategies following the rapid advancements of Industry 4.0.

Coskun et al., [10] discussed the need to implement the vision of Industry 4.0 in engineering education while updating existing higher education courses and laboratory exercises. The authors described five programs developed in their university in which some of the courses were updated following this vision, among them Mechanical Engineering, Mechatronic Systems Engineering, and Industrial Engineering. All the programs contain courses in design (Technical Drawing and CAD or Construction of Mechanic Parts) and software engineering projects. The programs also contain courses in industrial robotics and machine learning and a mechatronics project. As a complimentary measure to support the curricula, the authors described plans for a Visual Production Lab that will include computer-aided design and manufacturing (CAD/CAM) and 3D printing facilities.

## 3   The Study

In the study presented in this paper, we developed and evaluated an approach to learning engineering design in the framework of a course that introduces engineering students to the concepts and technologies of Industry 4.0. Our approach focuses learning on experiential practice in collaborative design and rapid prototyping using the cloud-based CAD platform Onshape and 3D printing facilities. When evaluating the learning processes and their outcomes, we used Onshape Education Enterprise (OEE) – a new application for creating educational environments and data analytics for design-based learning processes. Below in this section, we will present the research goal and questions, the research method, the considered course and workshop, the developed learning environment, and the data collection and analysis.

The goal of this research was to evaluate the learning outcomes of our workshop that implemented the proposed approach and explore the use of the new educational environment and data analytics tools for learning cloud-based collaborative design.

The research questions were as follows:

1. How can practice in cloud-based collaborative design contribute to the course that introduces engineering students to Industry 4.0 concepts and technologies?
2. How can using Onshape Education Enterprise and its data analytics tools help to understand and characterize learning through cloud-based collaborative design?

An exploratory case study method was used to analyze and evaluate collaborative learning in a complex innovative technological environment, with the use of new data analytics tools, and in the absence of conditions for more systematic research [11]. The data were collected using the pre-course and post-workshop questionnaires and by Onshape data analytics tools.

The course "Industry 4.0 – the Smart Factory", considered in this study, was developed by the authors (Cuperman and Rosen) and taught at the Braude Academic College of Engineering in the fall semester 2022–2023 as part of the new bachelor program "Advanced Industry in the Data Era" of the Department of Mechanical Engineering. The course consisted of 13 weekly meetings, 5 h each: 2 h of lectures and 3 h of hands-on practice in the FabLab environment. The departmental FabLab is equipped with craft tables and tools, 3D printers, a large laser cutter, CNC milling, and vacuum forming machines, as well as electronics and measurement equipment.

18 engineering students participated in the course. They learned Industry 4.0 concepts and acquired basic competencies in its technologies. The lecture topics related to the historical background, the impact of Industry 4.0, and the new skills required from mechanical and industrial engineers. The hands-on practice in the FabLab included three workshops "The Internet of Things", "Augmented Reality Experiences", and "Collaborative Design and 3D Printing", followed by a mini project.

The "Collaborative Design and 3D Printing" workshop included four consecutive meetings focused on the challenge of online collaborative design. The students used Onshape for the design and produced the models by 3D printing. Onshape is a cloud-based CAD software-as-a-service (SaaS) platform that allows several designers to work simultaneously on the same CAD document and is free for academic use [12]. When using Onshape in the workshop, most students used their own mobile devices to connect to the platform and perform the design, while some used the college's local workstations provided in FabLab.

The learning environment for design activities was set up based on the OEE facilities. OEE includes features for students' collaboration and communication in design, as well as tools to distribute instructional materials, establish virtual classrooms, consult students, monitor their engagement, and analyze performance [9]. The setup was implemented as follows. First, the college Institutional domain was established. Then, the course virtual class was established. All the students were assigned to it and linked to a folder with read-only access, where learning materials were accessible. Subsequently, folders with read-write access were created for the use of student teams. The folders were shared with the course instructors. At the beginning of the workshop, the students registered for the virtual class. As they all already studied CAD with the conventional desktop platform SolidWorks, we did not teach the basics but focused on the key features and cloud-specific functionalities of Onshape.

The workshop assignment was to design a model of a bridge that can carry as much weight as possible under the following constraints: have a given minimal length and width and maximal weight, can be folded or retracted into a given container, and should be manufactured as a print-in-place assembly. The assignment and its constraints were intentionally defined to evoke several design-to-manufacturing iterations and thus enhance the team practice in collaborative design. At the end of the workshop, the students participated in the final contest, in which physical models of the bridges produced by the students were loaded with weight, and the teams were graded according to the maximum load they withstand.

## 4 Findings

From the pre-course questionnaire, most participants of the course (89%) were junior or senior year students majoring in the new Industry 4.0 program. 50% of the students rated their own CAD skills before the course as high, and 72% rated their SolidWorks proficiency as good.

To answer the first research question, the outcomes of learning engineering design in the workshop were evaluated based on the post-workshop questionnaire. Students' reflections were supported by the evaluation of the designed models and by the analysis of student's performance based on OEE data analytics.

As found, most students highly evaluated the workshop's contribution to their engineering practices. Regarding the students' progress in the Industry 4.0 skills, 93% of them stated that the workshop improved their skills in 3D printing and design for additive manufacturing. 87% noted that it facilitated the development of their cloud-based design collaboration skills. 80% noticed an improvement in their Onshape skills. Regarding the generic skills needed in the era of Industry 4.0, 73% reported their progress in engineering design and problem-solving skills. 60% acknowledged that the workshop contributed to their teamwork skills and creativity. From the students' reflections:

*Due to the rigors of the task, I had to think about each size and design the bridge with the utmost care.*

*In this assignment, we, for the first time, worked collaboratively in the cloud and it was interesting.*

The analysis of OEE data supported the evaluations expressed by the students.

All the teams completed the bridge design assignment, produced physical models, and participated in the in-class competition. Figure 1 presents two students' models using different concepts: one used a telescopic mechanism and the other a folding mechanism. While weighing less than 10 g and printed from PLA material, the models succeeded in carrying unexpectedly high loads of more than 8 kg.

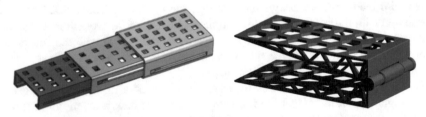

**Fig. 1.** Two CAD models of bridges designed by the student teams.

To answer the second research question, we analyzed the data of students' design performances, collaboration, and communication collected using OEE data analytics tools. This analysis indicated several characteristics of learning cloud-based collaborative design based on the proposed approach.

Significant differences were identified in the distribution of work in student teams consisting of two students, compared to those of three students. In teams of two, both team members made a major contribution to the design process. However, in teams of three, two team members did most of the work, while the third team member made a small contribution. The diagrams in Fig. 2, which were generated by OEE analytics upon a query request, graphically illustrate the relative time spent by each of the team members in the examples of a team of three and of a team of two.

In the diagrams, the blue circles symbolize the CAD documents, and their size represents the relative total time spent on them by the teams. The green circles represent the students. The solid lines connect the students and the CAD documents on which they worked. The width of each line represents the relative time spent by the student on the document. One can see in Fig. 2A that the line connecting Student 1B to the document is narrower than that of Students 1A and 1C, indicating that this student worked less than his peers.

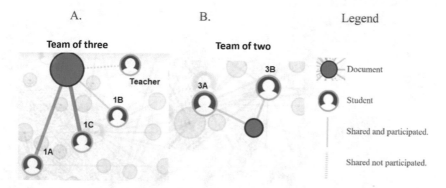

**Fig. 2.** Relative work time of students: A. In a team of three; B. In a team of two.

Another characteristic of the collaborative design indicated in the study was the large amount of design operations of different types used by the team members when performing the exercise, as presented in Table 1.

**Table 1.** The number of CAD operations used by the members of one of the teams.

| Number | Operation | | | | | | |
|---|---|---|---|---|---|---|---|
| | Design | Edit | Import/Export | Share | View | Other | **Total** |
| Student 4A | 1177 | 378 | 26 | 11 | 518 | 176 | **2286** |
| Student 4B | 821 | 165 | 2 | 4 | 391 | 81 | **1464** |

The table indicates that most operations done by the students are related to design. The large number of operations done by both team members and the presence of sharing operations (15 in total) show that the students worked in collaboration.

One more characteristic relates to the evolution of CAD models and teamwork dynamics. Figure 3 presents the time spent per week by one of the teams (2 students) when performing the design exercise. The evolution of the 3D printed CAD models developed during each week is shown under the diagram near the corresponding bars. As the diagram indicates, it took the team about one hour in the first week to come up with a design concept. After validating the concept by 3D printing, a 9-h effort was made by the team in week 2 to design a model that answered most constraints and had holes to reduce weight. In week 3, the students spent 4 h trying to improve the design by answering the width constraint and changing the number and size of the holes. In week 4, the students found out that the weight can be reduced if the model consists of two and not three parts. They spent a lot of time (9 h) to implement this idea and redesign the model. In week 5, the students found a way to increase the model's strength by making it higher and spent 6 h redesigning the model.

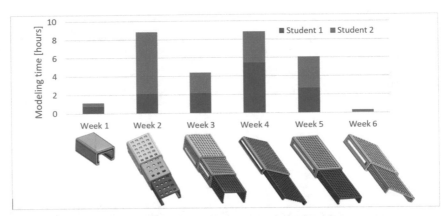

**Fig. 3.** The weekly time spent by the team members and the evolution of the CAD models.

In week 6 the students spent only half an hour on just cosmetic changes of the model. It is interesting to follow up on the dynamics of the time contribution of the team members. From the diagram, almost each week the major time contribution was made by a different team member.

## 5 Conclusion

The first-year experience of the new bachelor program "Advanced Industry in the Data Era" was successful. The subject of Industry 4.0 attracted the high interest of the students and led to full enrollment for the next year's program. The course "Industry 4.0 – the Smart Factory" was designed as a sequence of lectures and workshops, in each of which the students learned a set of emerging digital technologies and applied them to perform challenging assignments. We found that such design has significant advantages: its modular structure allows modification or replacement of workshops in a modular way and concentrates on the advanced issues based on the students' engineering background.

The workshop "Collaborative Design and 3D Printing", considered in this study, introduced the students to the advanced issues of design for additive manufacturing and cloud-based design, based on their background in desktop CAD. The approach implemented in the workshop is featured by learning in the FabLab, using the new cloud-based platform Onshape, and performing a challenging collaborative assignment prompting prototype-redesign iterations.

The educational study that accompanied the workshop acknowledged its effectiveness in learning online collaborative design and additive manufacturing technologies of Industry 4.0. The Onshape Education Enterprise was found suitable for the development of the learning environment for students' practice in collaborative design. The data analytics tools of OEE supported the findings from student reflections and provided objective evidence of design, collaboration, and communication that occurred in student teams.

Our research was a pioneer exploratory case study conducted with a limited number of students and in the absence of a control group. The authors consider it as a first step in the development of the methodology for teaching cloud-based collaborative design. Based on the promising research results, we plan to continue the study and recommend extending research in this direction.

**Acknowledgment.** This research was supported by the grants of the Technion Additive Manufacturing Center and the Technion Gordon Center for Systems Engineering.

# References

1. Schuster, K., Groß, K., Vossen, R., Richert, A., Jeschke, S.: Preparing for industry 4.0–collaborative virtual learning environments in engineering education. Engineering Education 4.0: Excellent Teaching and Learning in Engineering Sciences, pp. 477–487 (2016)
2. Wu, D., Thames, J. L., Rosen, D.W., Schaefer, D.: Towards a cloud-based design and manufacturing paradigm: looking backward, looking forward. In: International Design Engineering Technical Conferences and Computers and Information in Engineering Conference, Vol. 45011, pp. 315–328. American Society of Mechanical Engineers (2012)
3. PwC EU Services: Skills for industry Curriculum Guidelines 4.0: future-proof education and training for manufacturing in Europe. Publications Office of the European Union, Luxembourg (2020)
4. Eppes, T.A., Milanovic, I., Jamshidi, R., Shetty, D.: Engineering curriculum in support of industry 4.0. Int. J. Onl. Eng. **17**(01), 4 (2021). https://doi.org/10.3991/ijoe.v17i01.17937
5. Alhloul, A., Kiss, E.: Industry 4.0 as a challenge for the skills and competencies of the labor force: a bibliometric review and a survey. Science **4**(3), 34 (2022). https://doi.org/10.3390/sci4030034
6. Hernandez-de-Menendez, M., Escobar Díaz, C.A., Morales-Menendez, R.: Engineering education for smart 4.0 technology: a review. Int. J. Interact. Des. Manuf. **14**(3), 789–803 (2020). https://doi.org/10.1007/s12008-020-00672-x
7. Richert, A., Shehadeh, M., Willicks, F., Jeschke, S.: Digital transformation of engineering education – empirical insights from virtual worlds and human-robot-collaboration. Int. J. Eng. Ped. **6**(4), 23 (2016). https://doi.org/10.3991/ijep.v6i4.6023
8. Cuperman, D., Verner, I.M., Levin, L., Greenholts, M., Rosen, U.: Focusing a technology teacher education course on collaborative cloud-based design with Onshape. In: Auer, M.E., Hortsch, H., Michler, O., Köhler, T. (eds.) ICL 2021. LNNS, vol. 390, pp. 465–477. Springer, Cham (2022). https://doi.org/10.1007/978-3-030-93907-6_49
9. Olechowski, A., Deng, Y., DaMaren, E., Verner, I., Rosen, U., Mueller, M.: All's not Fair in CAD: an investigation of equity of contributions to collaborative cloud-based design projects. Comput. Aided Des. Appl. **20**(3), 574–583 (2023)
10. Coşkun, S., Kayıkcı, Y., Gençay, E.: Adapting engineering education to industry 4.0 vision. Technologies **7**(1), 10 (2019). https://doi.org/10.3390/technologies7010010
11. Yin, R. K.: Case study research design and methods. Canadian J. Program Eval. 1–5 (2014)
12. Onshape Homepage: https://www.onshape.com/en/. Last accessed 30 Nov 2023

# Predictive Maintenance and Production Analysis in Smart Manufacturing

B. Kalyan Ram[1]($\boxtimes$), Nitin Sharma[2], Abhishek S. Joshi[1], and Advik Vermani[3]

[1] Indxo AI Pvt Ltd, 3/11, Chanduri, II Cross, I Main, Abhayadhama Road, Whitefield, Bangalore, Karnataka 560066, India
{kalyan,abhishek.joshi}@indxo.ai

[2] Birla Institute of Technology and Science, K.K Birla Goa Campus, Bypass, Road, Zuarinagar, Sancoale, Goa NH 17B403726, India
nitin.sharma@goa.bits-pilani.ac.in

[3] West Windsor-Plainsboro High School South, 346 Clarksville Road, West Windsor, NJ 08550, USA

**Abstract.** Predictive maintenance and production analysis are pivotal in smart manufacturing, leveraging advanced technologies such as machine learning, and real-time data analytics to provide insights into the manufacturing efficiency and ensuring better productivity and sustainability. Predictive maintenance, driven by these key digital transformation tools, anticipates and mitigates equipment failures proactively, minimizing downtime and optimizing maintenance schedules for efficient operations.

Production analysis, emphasizing aspects such as data-driven decision making, efficiency and resource utilization, involves real-time monitoring and evaluation of manufacturing processes. It utilizes data from sensors, IoT devices, and production systems to enhance operational efficiency, ensure product quality, and optimize resource allocation. The integration of Overall Equipment Effectiveness (OEE), a critical parameter, refines production analysis by evaluating equipment efficiency based on availability, performance, and quality.

Digital transformation in manufacturing refers to the integration of digital technologies and data-driven processes throughout the entire manufacturing value chain to enhance efficiency, productivity, and overall business performance. It aims to create a more connected, agile, and efficient ecosystem that can adapt to changing market dynamics and customer demands. It involves a holistic approach, encompassing technology adoption, process optimization, and a cultural shift towards embracing innovation and digital capabilities.

This synergistic integration of predictive maintenance, production analysis, and OEE forms a comprehensive strategy for smart manufacturing. It embodies responsiveness, efficiency, and cost-effectiveness, positioning industries at the forefront of digital transformation.

**Keywords:** Industry 4.0 · Connected Machines · M2M communication · Digital Transformation · Smart Manufacturing · MES · OEE · Cyber Physical Systems

M. E. Auer et al. (Eds.): STE 2024, LNNS 1027, pp. 234–244, 2024.
https://doi.org/10.1007/978-3-031-61891-8_23

# 1  Introduction

## 1.1  Manufacturing Types

Manufacturing can be broadly categorized into two main types: Discrete Manufacturing and Process Manufacturing. These classifications are based on the nature of the production processes and the characteristics of the products being manufactured. A brief description of each types is provided herewith.

**Discrete Manufacturing**
*Characteristics:*

- Involves the production of distinct and separate items or units.
- Units are individually identifiable and can be counted.
- Often characterized by assembly line production, where components are assembled in a sequential manner to create finished products.

  *Examples:*

- Electronics: Manufacturing of smartphones, laptops, and other electronic devices.
- Automotive: Assembly of cars, trucks, and their components.
- Aerospace: Production of aircraft and related components.
- Consumer Goods: Manufacturing of appliances, furniture, etc.

**Process Manufacturing**
*Characteristics:*

- Involves the production of goods through a continuous or batch processing system.
- Products are typically homogeneous and cannot be easily distinguished as individual units.
- The production process is continuous, with materials flowing through various stages.

  *Examples:*

- Chemical Industry: Manufacturing of chemicals, petrochemicals, and pharmaceuticals.
- Food and Beverage: Production of beverages, dairy products, and processed foods.
- Oil and Gas: Refining of crude oil and production of refined petroleum products.
- Paper and Pulp: Manufacturing of paper and pulp products.

**Hybrid Manufacturing**
In some cases, a hybrid approach is used, combining elements of both discrete and process manufacturing.

This approach is common in industries where certain components are produced through discrete processes, and then these components are combined or processed further in a continuous or batch manner.

## 1.2 OEE – A Key Parameter for Smart Manufacturing

Overall Equipment Effectiveness (OEE) is a pivotal metric in manufacturing, representing the efficiency of equipment by evaluating availability, performance, and quality. It provides a comprehensive view of operational performance, helping organizations identify and eliminate inefficiencies to optimize production processes.

APQ, on the other hand, stands for Availability, Performance, and Quality, the three key components that collectively contribute to OEE. Availability assesses the actual production time compared to the planned time, Performance evaluates the speed at which the equipment operates, and Quality measures the ratio of good output to the total output.

In the context of smart manufacturing, integrating OEE and APQ is fundamental. Smart manufacturing leverages advanced technologies, like IoT, AI, and data analytics, to enhance efficiency, agility, and decision-making in real-time. This digital transformation not only improves production processes but also contributes to a sustainable future by minimizing waste, optimizing resource utilization, and fostering environmentally conscious practices. The integration of OEE, APQ, and smart manufacturing technologies is a key step towards achieving a sustainable and efficient manufacturing ecosystem. The Fig. 1 below is an example of typical OEE representations in discrete manufacturing scenarios.

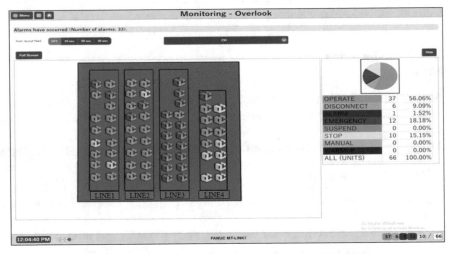

**Fig. 1.** MT-LINKi Machine status dashboard with its OEE

## 2 Objectives

### 2.1 Predictive Maintenance in Discrete Manufacturing

Predictive maintenance in discrete manufacturing involves using data, analytics, and technology to predict when equipment or machinery is likely to fail, allowing for timely maintenance to prevent unexpected downtime and optimize maintenance schedules. This proactive approach helps manufacturers reduce costs, enhance equipment reliability, and improve overall operational efficiency.

**Condition Monitoring**

- Implementing sensors and IoT devices to continuously monitor the condition of critical equipment.
- Collecting real-time data on factors such as temperature, vibration, pressure, and other relevant parameters.

**Data Collection and Analysis**

- Gathering and storing data from various sensors and sources to create a comprehensive dataset.
- Utilizing advanced analytics and machine learning algorithms to analyze historical and real-time data for patterns and anomalies.

**Predictive Modeling**

- Developing predictive models that use historical data to forecast when equipment is likely to experience failure.
- Models may consider factors such as equipment usage, environmental conditions, and production cycles.

**Historical Data Analysis**

- Analysing historical maintenance records to identify patterns and trends in equipment failures.
- Incorporating lessons learned from past incidents into predictive models.

**Kaizen (Continuous Improvement)**

- Adopting Kaizen methodologies (a continuous improvement mindset) by regularly refining and updating predictive maintenance models based on new data and insights.
- Seeking feedback from maintenance teams to enhance the accuracy of predictions.

## 2.2   Production Analysis in Discrete Manufacturing

The optimization of production efficiency is a critical challenge in the discrete manufacturing industry. To remain competitive, manufacturers must constantly explore and implement innovative operations and supply chain management strategies that enhance production efficiency and reduce costs.

Through this analysis, the paper seeks to identify the most effective strategies for enhancing production efficiency and reducing costs. The research will focus on the latest developments in the industry and will provide a comprehensive overview of the challenges and opportunities faced by manufacturers in enhancing production efficiency. This paper will serve as a valuable resource for manufacturers looking to improve their operations and supply chain management strategies for optimal production efficiency.

This outcome offers key information, including the current operational status of each machine, such as whether it is in operation, disconnected, in an alarm state, facing an emergency, suspended, stopped, in manual mode, or undergoing a warm-up process. Furthermore, it allows users to quickly assess the status of each machine in the plant, facilitating efficient monitoring and management of the manufacturing processes by featuring a machine selection option, that enables users to choose a specific machine and access detailed results.

The need for real-time monitoring and analysis of machine status is paramount for ensuring operational efficiency and minimizing downtime. This paper introduces a sophisticated framework that captures and analyzes critical data related to machine states, encompassing connectivity, emergency situations, alarms, and stoppages.

## 3   System Implementation and Features

### 3.1   Predictive Maintenance

By offering real-time visual alerts and facilitating proactive maintenance, the dashboard emerges as an invaluable tool for optimizing productivity and ensuring the efficient operation of industrial machines. The nuanced insights into individual machine behavior, coupled with the ability to customize analyses, empower operators to make informed decisions, ultimately contributing to the overall efficiency and reliability of industrial processes.

The dashboard reverses the color scheme for leakage parameters. Here, red indicates critical values, yellow represents warning states, and blue signifies normal conditions. Warning states for leakage occur between 40 and 20. The integration of date and machine filters enhances the precision of data analysis, allowing users to tailor their focus for targeted insights into specific time frames and machines. This customization feature is particularly useful for identifying patterns and trends associated with temperature and leakage parameters.

The below Fig. 2 represents one such parametric detection and prediction definition in the form of Insulation resistance (leakage resistance) in Servo motors and Spindle motors of CNC machines.

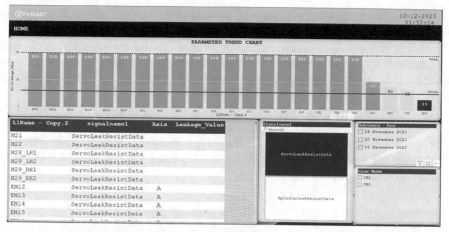

**Fig. 2.** Insulation Resistance (Leakage Resistance) detection and analysis

This study introduces an advanced monitoring dashboard designed to comprehensively assess the temperature and leakage parameters crucial to the productivity of industrial machines. The dashboard's primary focus lies in monitoring two key aspects: servo and spindle temperatures, and spindle and servo leakage. Each parameter is governed by predefined threshold values that determine warning and critical states.

The color-coded visual representation employed for temperature parameters enhances interpretability, with red indicating critical values exceeding 80, yellow denoting a warning state between 60 and 80, and blue representing normalcy below 60. Notably, this dynamic color scheme adjusts based on the specific axis of the parameter, providing a nuanced understanding of individual machine behavior (Fig. 3).

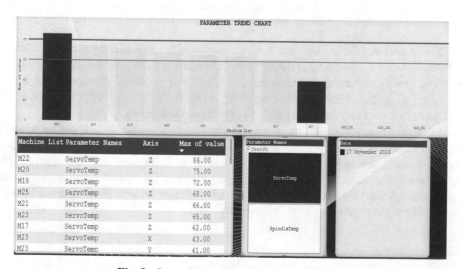

**Fig. 3.** Servo Temperature Analysis dashboard

The need to monitor and understand the trends of various parameters is paramount. Parameter trend graphs serve as indispensable tools, offering a visual representation of how critical factors evolve over time. This paper explores the application of trend charts to parameters like servo temperature and spindle temperature, pulse code temperature, shedding light on their capacity to unveil patterns, anomalies, and potential areas of improvement.

The research employs a systematic approach, utilizing historical data from manufacturing processes to generate parameter trend graphs. By focusing on specific metrics, such as temperature-related parameters, the study aims to showcase the effectiveness of trend charts in capturing nuanced variations and providing a comprehensive view of the operational landscape (Fig. 4).

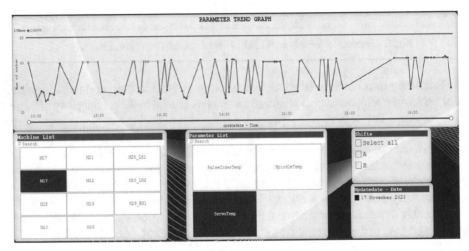

**Fig. 4.** Parameters Trend Graph

The results emphasize the clarity and interpretability afforded by parameter trend graphs. The visual representation of data allows operators and decision-makers to discern patterns, anticipate issues, and optimize processes. For instance, monitoring the trend of servo and spindle temperatures enables proactive maintenance, reducing the risk of equipment failures and enhancing overall operational efficiency.

In conclusion, parameter trend graphs emerge as invaluable tools in the manufacturing domain, offering a user-friendly and insightful way to interpret the trajectory of critical parameters over time. By enhancing visibility and facilitating proactive decision-making, these graphs contribute significantly to the optimization of processes and the mitigation of potential issues, ultimately fostering a more efficient and resilient manufacturing ecosystem.

### 3.2 Production Analysis

As one of the three pillars of the APQ measurement, it is essential to have a dashboard solely dedicated to the analysis of the production offered by the machines in a given data set. The dashboard exists as a useful device in pattern identification and statistical analysis of machine production, focusing on the total production at a given moment in time, during a specific shift, for a certain machine, as selected per the user.

Due to the nature of the data, without tools such as this dashboard, it can be difficult to identify values over only a certain period of time or for a certain machine. However, with the drill-down options facilitated by the dashboard, the user can easily view the production over a period of time or shift. This is represented as the actual value of the machine over the period, in contrast to a historical running total of production count. Moreover, this dashboard presents both a bar graph and a line graph, with the former utilizing the actual value over a period of time, and the latter utilizing the total sum over time (Fig. 5).

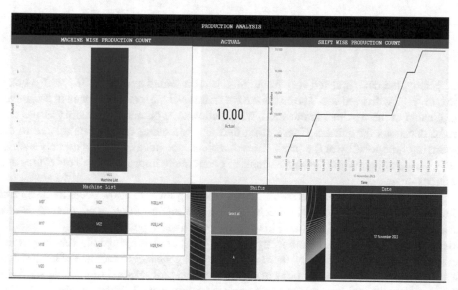

**Fig. 5.** Production Analysis dashboard

The dashboard's ability to present data based on certain modifiable factors makes it a vital tool for any user set on improving the production of the machines, given that many patterns and methods of analysis are only available after filtering data for a certain subset. These insights improve the operators' abilities to make statistically intelligent decisions in the highest favor of improving the production aspect of APQ, with the end goal of improving the OEE in mind.

Here is one such Dashboard that provides the status of machines in a single view with several useful options for line-wise, operation-wise, shift-wise filtering along with appropriate tooltips (Fig. 6).

When managing a large network of machines, as is often the case in modern industrial settings, it is critical that there exists a method to effectively oversee a wide variety of Boolean parameters associated with the machines. For such purposes, a dashboard, consisting of a complete visualization of the statuses and an advanced filtering system, was created.

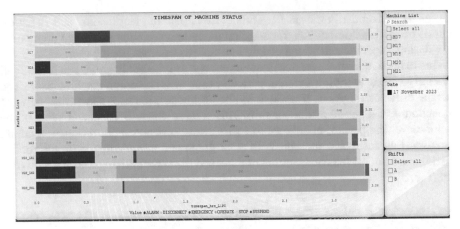

**Fig. 6.** Machine status dashboard

Firstly, the data, updated very frequently, is represented as either 'OK' with a green box, or 'NG' with a red box. The boxes are the statuses of the machines for each parameter applicable to the machine in question, which allows the operator to identify issues and make decisions far quicker. Considering that the data being collected is related to the operating functionality of the production machines, specifically certain aspects such as their fans or spindles, the dashboard enables operators to improve overall efficiency and reduce outages.

The interface possesses an assortment of filtering tools to allow users to customize their analysis based on factors such as machine name, signal (Boolean parameter) names, as well as the status of the signal. This allows for extensive possibilities, such as rapidly identifying a list of problems simply by clicking 'NG' or identifying long standing trends in individual machines. Finally, there exists a logging system to indicate when the data was last collected all combinations of machine and signal selected in the data set. With the complete "signal to machine" grid representation as well as the capability to filter by a number of different factors, this dashboard stands as an essential tool in the pursuit of the end goal of improving operational efficiency.

The utilization of Boolean parameters proves instrumental in expediting problem res-olution. Through real-time monitoring of CNC battery levels and fan statuses, operators can promptly identify and address issues, ensuring the continuous and optimal function-ality of critical equipment. The color-coded visual cues enhance the interpretability of complex operational data, empowering operators to make informed decisions swiftly. The impactful application of Boolean parameters, exemplified by CNC battery low, APC battery low, and CNC fan status, to streamline problem-solving processes in manufac-turing operations. Boolean parameters, inherently binary in nature, delineate conditions as 'true' or 'OK' for positive statuses and 'false' or 'NG' for negative ones. Leveraging this inherent simplicity, a visual feedback system is proposed, employing color-coded indicators where green signifies positive states and red indicates negative ones.

This not only underscores the significance of Boolean parameters in manufacturing but also demonstrates their practical implementation in solving operational challenges.

By facilitating a clear and immediate understanding of equipment statuses, Boolean parameters contribute to a proactive problem-solving approach, minimizing downtime and optimizing manufacturing efficiency (Fig. 7).

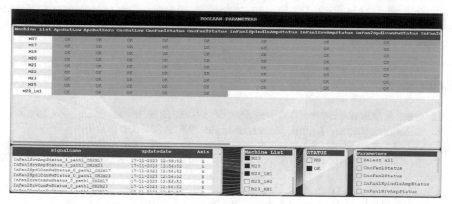

**Fig. 7.** Boolean Parameters dashboard

# 4   Results and Discussion

The activity started with Data collection on the operational status of each individual machine, including whether it is connected, disconnected, in an emergency state, in an alarm state, or stopped. This information is gathered at high-frequency intervals, allowing for a granular understanding of machine behavior.

It further translated into the framework being equipped with powerful filtering mechanisms, enabling users to customize their analysis based on date ranges, shifts, and specific machines. This feature facilitated a targeted examination of machine status data, supporting more nuanced and context-specific investigations.

The results of our analysis demonstrate the effectiveness of the framework in providing a detailed understanding of industrial machine behavior. By examining the duration of each operational state, operators can identify patterns, anticipate issues, and optimize machine performance.

# 5   Future Scope

Digital Transformation in Discrete manufacturing is an incredible tool to ensure efficient, effective and sustainable manufacturing. The scope to enhance Digital Transformation is currently seeming infinite. With opportunities to enhance quality, productivity and maintainability of the manufacturing ecosystem, it ensures sustainability and profitability too. The future scope includes integration to Business Intelligence framework, establishing Digital Twin and facilitating a safe and secure Cyber Physical System.

## 6  Conclusion

This research introduces a robust framework for real-time monitoring and analysis of industrial machine status, emphasizing the importance of temporal analysis and customizable filtering options. The insights gained from this system empower operators to make informed decisions, thereby enhancing overall operational efficiency and reducing unplanned downtime.

## References

1. Salonen, A., Gopalakrishnan, M.: Practices of preventive maintenance planning in discrete manufacturing industry. J. Qual. Maintenance Eng. **27**(2), 331–350 (2020)
2. Jun, H.-B.: A review on the advanced maintenance approach for achieving the zero-defect manufacturing system. Front. Manuf. Technol. (2022). https://doi.org/10.3389/fmtec.2022.920900
3. Continuous Improvement and Data Analytics in Discrete and Process Manufacturing, kaizen.com/insights/discrete-process-manufacturing
4. Optimizing Production Efficiency Through Data-Driven approach from metrology.news
5. Seven key guidelines for optimizing your manufacturing from www.calsoft.com
6. Manufacturing Capacity Planning: Optimizing Production from www.machinemetrics.com/blog/manufacturing-capacity-planning

# Automated Quality Control of 3D Printed Tensile Specimen via Computer Vision

Rizwan Ullah[1]($^{\boxtimes}$), Silas Gebrehiwot[1], Thumula Madduma Patabendige[1], and Leonardo Espinosa-Leal[2]

[1] School of Engineering, Culture and Wellbeing, Arcada University of Applied Sciences, Jan-Magnus Janssons Plats 1, 00560 Helsinki, Finland
`{ullahriz,gebrehis,maddumat}@arcada.fi`
[2] Graduate School and Research, Arcada University of Applied Sciences, Jan-Magnus Janssons Plats 1, 00560 Helsinki, Finland
`espinosl@arcada.fi`

**Abstract.** This research explores the integration of a robotic arm using computer vision for automated quality control for sorting 3D printed tensile specimens. The study, conducted, focuses on utilizing a Niryo NED-2 robotic arm with a vision system. The robotic arm captures cross-sections of tensile specimen, and a Python program processes vision feeds, filtering images based on 2D contours. Tensile samples were manufactured using Fused Deposition Modeling (FDM) with PLA material, incorporating known offsets (both positive and negative). Their dimensions were predicted and compared with the actual geometrical measurements. Experimental results showcase the system's accuracy in measuring specimen dimensions, demonstrating low error rates. The study highlights the potential for automated quality control in additive manufacturing, presenting a valuable tool for Industry 4.0. The robotic arm's vision system proves effective in enhancing efficiency and reliability in 3D printing quality inspection processes.

**Keywords:** Machine vision · Industry 4.0 · Additive manufacturing · Robots

## 1 Introduction

The four industrial revolution or Industry 4.0 is the term that defines the industrial change caused by the union of different technonologies such as artificial intelligence, robotics, additive manufacturing, virtual reality, among others. Developments created under the combination of these have stimulated the creation of new autonomous tools able to disrupt the classical industrial processes [1]. Quality control in manufacturing has been one of these that has shown the use of machine learning as a way to improve both the speed and the selection performance [2]. In additive manufacturing the quality of the workpieces manufactured depend on the capability of the device to print the workpieces within the desired parameters [3]. Several advances to ensure high quality have been done in the last years, however some issues remain. In this context, robotic arms integrated with camera devices can use advanced computer vision libraries to create reliable pipelines to ensure automatic quality control of manufactured pieces.

© The Author(s), under exclusive license to Springer Nature Switzerland AG 2024
M. E. Auer et al. (Eds.): STE 2024, LNNS 1027, pp. 245–252, 2024.
https://doi.org/10.1007/978-3-031-61891-8_24

This paper presents a prototype that uses a robotic arm for quality control of 3D printed PLA pieces. We Analyse the capabilities as well as the limitations of the systems and discuss the possible implications for the future of quality control in additive manufacturing.

## 2   Methodology

In the context of this research, the primary goal is to see the integration of robotic arm to sort "good" and "not-good" tensile specimen based on the computer vision method as means of quality measurement. In this research, emphasis is placed to tune the program that processes the robotic-arm vision feed (of images) and filter them based on their 2-dimensional (2d) cross section of contours.

### 2.1   Experimental Setup

The vision system [4] of a Niryo NED-2 robotic arm is used in this research to capture cross-section of tensile specimen as depicted in Fig. 1.

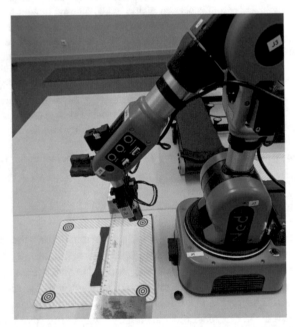

**Fig.1.** Niryo NED-2 vision system measuring the cross section of a tensile specimen

Regular measuring scale and metal pieces are used to place the specimen against a fixed reference. We present the Niryo's Ned-2 Camera's specifications of in Table 1.

**Table 1.** Niryo NED-2 Camera specifications

| Quality parameters | Specifications |
|---|---|
| Lens size | 2.1 mm |
| Maximum resolution | 1080p |
| Resolution | 640 × 480 px |
| Frame Rate | 30fps |

## 2.2 Sample Design Variables

Integrating additive manufacturing (AM) with recent 3D technologies like scanning and digital image correlation (DIC) enables rigorous and automated quality inspection processes. In relation to this, our objective focuses on scanning samples' contour using machine vision technology that is integrated with our Niryo robot arm. The samples are designed according to the ISO 527-2 standard for specimen type 1B. We used the fused filament fabrication (FFF) additive manufacturing (AM) technology to produce the samples using Polylactic Acid (PLA) as material [5–7]. The selection of the sample geometry is related to our routine procedures in sample design for material testing. Our current objectives focus on accurate contouring of the various sample width and length dimensions. The 2-D cross section of a standardized tensile specimen is shown in Fig. 2. The tensile specimens were manufactured with the intended deviations from the nominal size on positive and negative offsets, see Table 2.

**Table. 2.** Positive and negative offsets to nominal size of standardized tensile specimen

| Nominal sample dimensions [mm] | Positive dimensional offset from nominal [mm] | Negative dimensional offset from nominal [mm] |
|---|---|---|
| Width W = 20 Length, L = 150 | +0.5 | −0.5 |
| | +1 | −1 |
| | +2 | −2 |
| | +3 | −3 |
| | +4 | −4 |

As indicated in Table 2, the category of samples is created based on equidistance positive and negative offsets of the nominal 1B sample dimension. The positive offsets are made in 0.5 mm and 1 mm increments whereas the negative offsets are made in 0.5 mm and 1 mm decrements of the nominal contour. A range of dimensional offsets between +4 mm and −4 mm are made resulting in a total of 11 sample variants. Figure 2 shows the studied sample variants.

The specimens were manufactured using fused deposition modeling (FDM) using Creality K1 Max AI Fast 3D Printer and shown in Fig. 3.

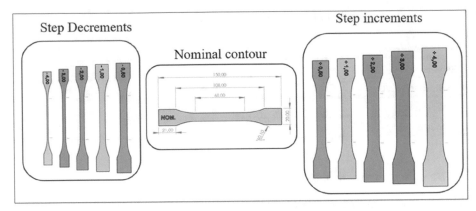

**Fig. 2.** Sample variations for the machine vision contouring experiments.

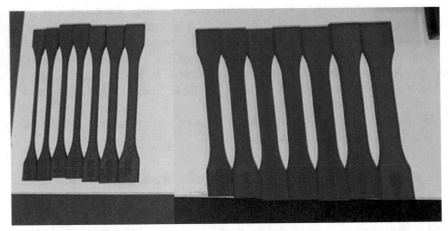

**Fig. 3.** Tensile specimen manufactured using FDM according to the offsets enlisted in Table 2.

## 2.3 Program for Capturing Images

The imaging capturing system of Niryo NED-2, employed for image acquisition, relies on a Python program designed for capturing images of tensile specimens.

In the python program, the Niryo NED-2 robotic system is employed to capture images of tensile specimens in a controlled environment. The Python program utilizes the Pyniryo library for interfacing with the robotic system, as well as the OpenCV library for image processing tasks. The code begins by defining color thresholds (minimum and maximum values) for blue tone. It then connects to the Niryo robot, calibrates its position, and captures images continuously.

The captured images undergo a series of image processing steps. Initially, the image is undistorted using camera intrinsic, followed by noise reduction using bilateral filtering. A mask is created to isolate the blue color channel, and edge detection is performed using the Canny algorithm [8]. Morphological transformations, specifically erosion followed by dilation, are applied to enhance the specimen's contours. The contours are then identified, and a minimum area bounding rectangle is fitted to the specimen. The program calculates the dimensions of the specimen, considering rotation and scaling factors.

Furthermore, the code includes a module named 'scaling_module,' which is imported at the beginning. This module contains a function named 'calculate_scaling_factor,' responsible for determining the scaling factors based on a reference image of a known length. The calculated scaling factors are subsequently utilized to adjust the dimensions of the specimen. The processed images, along with relevant information i.e., height and length of tensile specimen, are saved, and the data is appended to a CSV file for further analysis.

## 3   Results and Discussion

### Contouring Results

The scans are made for the sample categories and presented in Fig. 4 (nominal case), 5 (positive increment of 0.5 mm) and 6 (negative increment of 0.5 mm). We classified the contouring into width and length dimensions, running 22 measurements. The sample variants are classified based on the dimension offsets considered in our study. Table 3 presents the results of the measurements.

**Table. 3.** The nominal sample dimensions and results of contour (width, length) measurements using the robot machine vision.

| Sample category Based on offsets (±mm) | Nominal dimensions | | Measured dimensions | |
|---|---|---|---|---|
| | Width [mm] | Length [mm] | Width [mm] | Length [mm] |
| −4 | 12 | 142 | 11.76 | N/A |
| −3 | 14 | 144 | 12.61 | 144.74 |
| −2 | 16 | 146 | 14.73 | 146.54 |
| −1 | 18 | 148 | 16.83 | 149.13 |
| −0.5 | 19 | 149 | 18.05 | 150.00 |
| 0 | 20 | 150 | 18.95 | 150.81 |
| +0.5 | 21 | 151 | 19.37 | 152.09 |
| +1 | 22 | 152 | 20.21 | 152.96 |
| +2 | 24 | 154 | 23.17 | 155.13 |
| +3 | 26 | 156 | 25.25 | 156.86 |
| +4 | 28 | 158 | 27.38 | 159.44 |

Based on the results obtained, we calculated the percent error of the measurement. All percentage errors in the length contours are found below. A maximum of 0.91% error for the +4 mm, and a minimum of error for the −2 mm sample variants is obtained in length measurements. The width contour showed increments in all calculated errors. A maximum of 9.92% error for the −3 mm and a minimum of 2% for the −4 mm samples are observed. The variation of calculated errors for the length and width are due to the dimension scaling effect. The higher nominal dimension results in the lower calculated error, see Table 4.

**Table 4.** Errors of measurement data

| Sample category (± mm) | Errors in Width [%] | Errors in Length [%] |
|---|---|---|
| −4 | 2.00 | – |
| −3 | 9.92 | 0.52 |
| −2 | 7.96 | 0.37 |
| −1 | 6.47 | 0.76 |
| −0.5 | 4.98 | 0.68 |
| 0 | 5.26 | 0.54 |
| +0.5 | 7.76 | 0.72 |
| +1 | 8.12 | 0.63 |
| +2 | 3.48 | 0.73 |
| +3 | 2.88 | 0.55 |
| +4 | 2.20 | 0.91 |

Even the nominal tensile specimen has about 5% error in predicting width and compared to this error does not grow much for other offsets as well. This can be solved either by adding some compensation to the program or training it to use filtration based on an acceptable tolerance. Although in this research, capability of vision system of a low-cost robot arm is checked, based on the lesson learned a Python based program can

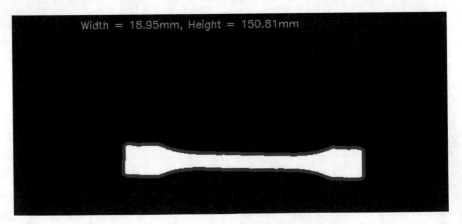

**Fig. 4.** Contours as captured for nominal size

be used to filter "good" vs "not-good" tensile specimen and hence this system can have the potential to integrate within Industry 4.0 (Figs. 5 and 6).

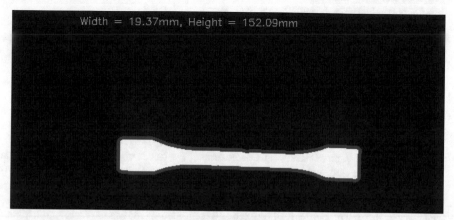

**Fig. 5.** Contours as captured for + 0.5 mm offset

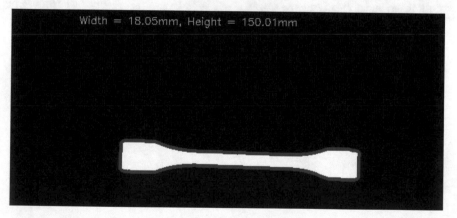

**Fig. 6.** Contours as captured for –0.5 mm offset

## 4  Conclusions

In conclusion, the NED-2 vision system, integrated with a robotic arm, shows great potential for automated quality control in the 3D printing of tensile specimens. The system's accuracy in contour measurements and low error rates make it a valuable tool for enhancing the efficiency and reliability of quality inspection processes in additive manufacturing, particularly in the context of Industry 4.0. Future research will be focused on improving the hardware and the recognition accuracy. Moreover, we plan to create a full pipeline where the robot arm can perform automatic quality control of the pieces in real time using a conveying belt attached to the 3D printer.

# References

1. Espinosa-Leal, L., Chapman, A., Westerlund, M.: Autonomous industrial management via reinforcement learning. J. Intell. Fuzzy Syst. **39**(6), 8427–8439 (2020)
2. Magar, V.M., Shinde, V.B.: Application of 7 quality control (7 QC) tools for continuous improvement of manufacturing processes. Int. J. Eng. Res. General Sci. **2**(4), 364–371 (2014)
3. Wu, H.C., Chen, T.C.: Quality control issues in 3D-printing manufacturing: a review. Rapid Prototyping J. **24**(3), 607–614 (2018)
4. "Niryo Vision Set - User Manual — Vision Set User Manual v1.0.0 documentation." Accessed 3 Jan 2024. https://docs.niryo.com/product/vision-set/v1.0.0/en/index.html
5. Gebrehiwot, S.Z., Espinosa-Leal, L., Linderbäck, P., Remes, H.: Optimising the mechanical properties of additive-manufactured recycled polylactic acid (rPLA) using single and multi-response analyses methods. Int. J. Adv. Manuf. Technol. **11**, 1–6 (2023)
6. Gebrehiwot, S.Z., Espinosa-Leal, L.: Characterising the linear viscoelastic behaviour of an injection moulding grade polypropylene polymer. Mech. Time-Depend. Mater. **26**(4), 791–814 (2022)
7. Gebrehiwot, S.Z., Espinosa Leal, L., Eickhoff, J.N., Rechenberg, L.: The influence of stiffener geometry on flexural properties of 3D printed polylactic acid (PLA) beams. Prog. Addit. Manuf. **6**, 71–81 (2021)
8. Rong, W., Li, Z., Zhang, W., Sun, L.: An improved CANNY edge detection algorithm. In: 2014 IEEE International Conference on Mechatronics and Automation, pp. 577–582. IEEE (2014)

# Learning in Virtual Environments

# Japanese High School Students' Attitudes and Ways of Learning with Technology Outside of School After the Pandemic

Yutaro Ohashi(✉)

Shibaura Institute of Technology, 307 Fukasaku, Minuma-ku, Saitama City, Saitama, Japan
yohashi@shibaura-it.ac.jp

**Abstract.** After the pandemic, the learning environment surrounding students have changed dramatically in terms of technology, and there has been insufficient research on what kind of ICT environment high school students actually use to study and what kind of attitudes they have toward school and learning. The author conducted a nationwide online questionnaire survey on how high school students changed their learning methods and who (or what) do they trust in thinking about their learning and career path. The author found that there has been little progress in the informatization of education in schools, and that the use of ICT in pupils' learning was already widespread before pandemic. Schoolteachers are still trusted by students to some extent, however, they are not as trusted as cram school teachers. Twenty per cent of respondents trusted information from nonhuman sources. In terms of the types of people and things that students trust, pupils were categorized as school-oriented, relationship-oriented, rationality-oriented, nonhuman, and school-distrusting, revealing a picture of students that schools may not have envisaged before.

**Keywords:** Students' view of learning · Learning method · Trust · Learning app · Juku (private tutoring school)

## 1 Introduction

### 1.1 Impacts of COVID-19 on Schools

Due to the COVID-19, it was reported that 1.5 billion children across the globe have been affected by school closures as of March 2020 [1]. Like schools around the world, many schools in Japan had been forced to close for long periods of time and many children have lost the opportunity to learn. In Japan, the request to close schools came on February 2020, and local governments had to quickly decide if they were going to close schools from the beginning of March 2020 and 98.5% of schools were closed as of March 2020. According to a survey conducted by Ministry of Education, Culture, Sports, Science and Technology in Japan (hereinafter MEXT), 99% of public schools (elementary, lower secondary, upper secondary schools and special needs education schools) which were closed due to COVID-19 restarted in June 2020 [2]. This was

© The Author(s), under exclusive license to Springer Nature Switzerland AG 2024
M. E. Auer et al. (Eds.): STE 2024, LNNS 1027, pp. 255–266, 2024.
https://doi.org/10.1007/978-3-031-61891-8_25

a long-awaited recommencement; however, it also marks the beginning of starting to explore a 'new way of life' in schools. Schools are taking measures including staggering students' attendance, regular disinfection and better air ventilation, and spacing of seats and desks. Education boards, schools, and teachers are in the midst of exploring new ways of teaching.

During the school closure, the government informed local authorities about the need to strengthen online education. However, a survey carried out at the beginning of the 2020 school year showed that only 5% of local authorities had implemented online interactive education [3], with many relying on self-study using old-fashioned textbooks and workbook, which raised the issue of ensuring the quality and equality of education. The issues such as pupils' disorder of daily life during the closure period [4], and the possible growing gap in learning progress between pupils [5] were pointed out.

### 1.2　Growing Interest in Technology-Based Education During the Pandemic

In the face of this lack of opportunity to learn, new online learning services have attracted attention [6]. In addition to these, *juku* (private tutoring school or cram school) is also important educational service providers. According to a survey conducted by Benesse, the largest provider of education services, 49.1% of primary school students, 57.0% of junior high school students, and 36.3% of high school students attend private tutoring schools [7]. Private tutoring schools are often seen as symbols of a 'cold meritocracy', characterized by endless re-education and drills, the business aspect of after-school schools, and competition for admission to the 'right' universities [8]. Currently, tutoring schools are regarded as important educational service providers by parents and students who feel that they cannot rely on schools in the wake of the COVID-19. In response to the prolonged closure of schools, tutoring schools have responded in a variety of ways, using the teaching methods they have developed over the years. The major tutoring schools have opened their online classes to the public in a variety of ways.

### 1.3　Focus of the Study

The author's focus in this article is not only on the way students learn, but also on the way they think about learning in the face of these major changes. It goes without saying that trust between teachers and students is important in education. However, as the pandemic has disrupted teaching and learning in schools, author wondered if students are building new trust outside of school. There has been no large-scale research on this point, and I thought it worthwhile to study. In this study, I decided to conduct a nationwide online survey of high school students. High school students were chosen as the target group because I believe that high school students are at an age when they are able to establish their own learning methods and attitudes, even though they are under the care of their parents. Two research questions were raised in this paper.

(1) How have high school students' learning methods changed (or did not change) after COVID-19 pandemic, especially their use of ICT, private tutoring schools, distance learning or learning applications?
(2) Who (or what) do high school students trust in thinking about their learning and career path?

## 2 Methods

With the cooperation of an internet research company, an online questionnaire survey consisting of the following 10 questions was conducted among high school students.

Q1 Type of school (public/private) and grade (single answer question).
Q2 Type of study guidance during school closure (single answer question).
Q3 Career path you are currently considering (multiple choice question).
Q4 Use of cram schools, correspondence courses, learning apps, etc. (multiple choice question).
Q5 What devices do you use to study (multiple choice question).
Q6 Advantages of using Q5 for study (multiple choice question, free description).
Q7 Changes in life and learning under the COVID-19 situation (multiple choice question, free description).
Q8 Person (or thing) you trust with your studies (single answer question).
Q9 Person (or thing) you trust with your future career (single answer question).
Q10 Reasons for Q8 and 9, or concerns about your own study, career path, or life (free description).

For the optional answers, the total was calculated and, in some cases, comparisons were made by attribute. Responses to open-ended questions were post-coded using a case-code matrix. This approach to the analysis of qualitative data is based on grounded theory [9], and identifies general patterns for a specified case by fully considering both individuality and materiality [10]. Word counting was not used in this study, as it does not always capture the significance of a word and may decontextualize its description. The analytical procedure was as follows.

A. Codes or keywords were assigned to each description using the following sub-procedures:
    A-1. Open-ended coding to freely assign a code to a description.
    A-2. Focused coding to assign a more abstract code to a description.
B. Codes were structured to attain an overall balance.
C. Steps B and C were repeated to ensure consistency across all codes.

## 3 Results

The survey was conducted in August 2020 and responses were received from 542 respondents. The types of school of the respondents (Q1) were public (349, 64%), private (186, 34%) and unknown (7, 1%). The career paths that respondents are considering (Q3) are as follows (Table 1). This is very close to the national average as of 2019 (53.7% for universities, 4.4% for junior colleges and 23.8% for vocational schools) [11], and is therefore considered to be an appropriate sample for exploring national trends.

### 3.1 Changes in the Way High School Students Learn After the Pandemic

With regard to the type of study instruction provided by the school during the period of closure due to the COVID-19 (Q2), the most common type of instruction overall was

**Table 1.** Career paths being considered by respondents (Q3).

| Type of career | Number of respondents (%) |
|---|---|
| University | 316 (58.3%) |
| Junior college | 36 (6.6%) |
| Technical school | 127 (23.4%) |
| Employment | 117 (21.6%) |
| Other | 1 (0.2%) |
| Undecided | 36 (6.6%) |

'study instruction using textbooks, reference books, etc. (59%) (Table2).' In May 2020, MEXT conducted a survey of school establishers (local governments in prefectures and municipalities where schools are located), which covered only public schools. This survey is novel in that it surveyed both public and private schools at the same time.

**Table 2.** Types of study guidance during the school closure (Q2) (The classification is based on the one used in the MEXT survey [12]).

| | Overall (542) | Public schools (349) | Private schools (186) |
|---|---|---|---|
| Study guidance using textbooks and reference books | 320 (59%) | 234 (67%) | 84 (45%) |
| Teaching using learning apps on smartphones or tablets | 164 (30%) | 92 (26%) | 70 (38%) |
| Simultaneous interactive online teaching | 151 (28%) | 69 (20%) | 80 (43%) |
| Teaching learning through non-TV videos | 121 (22%) | 74 (21%) | 46 (25%) |
| Teaching learning through television broadcasting | 23 (4%) | 15 (4%) | 7 (4%) |
| Study guidance over the phone | 10 (2%) | 7 (2%) | 3 (2%) |
| Other | 40 (7%) | 32 (9%) | 7 (4%) |

In terms of the use of ICT for learning outside school (Q5), 70% of the respondents used a smartphone and 80% used either a smartphone, a tablet or a computer (Table 3). It should be noted that, in principle, Japanese high schools regulate or do not allow the bringing in of smartphones, so this question assumes a situation in which students are studying at home or in tutoring schools. The use of any device was higher among those who planned to go to university, with nearly 90% using some kind of device. Even among non-university-bound students, about 70% use some kind of device, suggesting that ICT-based learning is now commonplace.

**Table 3.** Use of ICT in learning (Q5).

| | Overall (N = 542) | Students wishing to go to college (N = 316) | Non-college applicants (N = 226) |
|---|---|---|---|
| Smartphones | 387 (71.4%) | 247 (78.2%) | 140 (61.9%) |
| Computers (desktop and laptop) | 114 (21.0%) | 79 (25.0%) | 35 (15.5%) |
| Tablets | 112 (20.7%) | 85 (26.9%) | 27 (11.9%) |
| I don't use any of the above | 111 (20.5%) | 39 (12.3%) | 72 (31.9%) |

In terms of the use of cram schools, distance learning or learning apps (Q4) (Table 4), more than half of the respondents who wanted to go to college used some kind of educational services. On the other hand, about 80% of the respondents who were considering other career paths did not use such services.

**Table 4.** Use of cram schools, distance learning and learning apps (Q4).

| | Overall (N = 542) | Students wishing to go to college (N = 316) | Non-college applicants (N = 226) |
|---|---|---|---|
| Using distance learning (no face-to-face instruction) | 59 (10.9%) | 45 (14.2%) | 14 (6.2%) |
| Attending a private school (face-to-face teaching only) | 59 (10.9%) | 50 (15.8%) | 9 (4.0%) |
| Paid learning apps | 42 (7.7%) | 35 (11.1%) | 14 (6.2%) |
| Free apps | 41 (7.6%) | 30 (9.5%) | 7 (3.1%) |
| Attending a private school (both face-to-face and online teaching) | 29 (5.4%) | 23 (7.3%) | 6 (2.7%) |
| Attending a private school (online tutoring only) | 25 (4.6%) | 21 (6.6%) | 4 (1.8%) |
| Other | 4 (0.7%) | 1 (0.3%) | 3 (1.3%) |
| I do not use any of the above | 331 (61.1%) | 152 (48.1%) | 179 (79.2%) |

The results for what is important to them in learning with ICT (Q6) are as follows (Table 5). It can be seen that students tend to focus on efficiency and learning effectiveness, such as being able to learn at their own pace and learn effectively.

**Table 5.** What is important in ICT-based learning (Q6).

|  | Number of responses (N = 431) |
|---|---|
| Study at your own pace, whenever you want | 187 (43.4%) |
| Study more efficiently in less time | 131 (30.4%) |
| You'll get the answer soon | 121 (28.1%) |
| You can manage your study plan and schedule | 116 (26.9%) |
| There is a video | 112 (26.0%) |
| You can focus on the areas and subjects you want to study | 94 (21.8%) |
| Shows the history of your studies | 80 (18.6%) |
| No need to commute | 74 (17.2%) |
| No need to meet people | 72 (16.7%) |
| Large number and variety of questions and sample answers | 70 (16.2%) |
| It is easy to understand what is not taught at school and what is difficult to understand | 63 (14.6%) |
| There are audio materials | 55 (12.8%) |
| You can get detailed advice | 54 (12.5%) |
| You will find learning fun | 52 (12.1%) |
| Beautiful picture and sound quality | 41 (9.5%) |
| You don't need a camera to see your face or a microphone to speak | 38 (8.8%) |
| No need to write or take notes | 36 (8.4%) |
| You can study according to the tendency of the examinations of the school you want to enter | 33 (7.7%) |
| You can ask questions about what you don't understand and it's easy to ask questions | 31 (7.2%) |
| No data too large, no interruptions in communication | 27 (6.3%) |
| No need to read the textbooks yourself | 25 (5.8%) |
| The duration of each video or audio is short | 25 (5.8%) |
| Large letters and pictures are easy to read | 24 (5.6%) |
| Ability to interact with other students (chat, talk, ask questions, etc.) | 13 (3.0%) |
| You can study with other students | 12 (2.8%) |
| Many people have used the same study methods to get into the school of their choice | 8 (1.9%) |
| Other | 10 (2.3%) |

As for the changes in their life and study after the pandemic (Q7), they spent less time and less often with their friends (50.2%), spent more time and more often with their

**Table 6.** Changes in living and learning as a result of the spread of new coronaviruses (multiple responses).

| | Number of responses (N=542) |
|---|---|
| I see my friends less and less often. | 272 (50.2%) |
| I spend more time on my smartphone or use it more often. | 262 (48.3%) |
| I wake up later in the morning. | 208 (38.4%) |
| I spend more time playing games or play them more often. | 183 (33.8%) |
| I go to bed later at night. | 176 (32.5%) |
| I have started to use my smartphone late in the night. | 134 (24.7%) |
| I spend more time and communicate more often with my friends on my smartphone. | 107 (19.7%) |
| I've come to enjoy being at home. | 81 (14.9%) |
| I started to find it hard to stay at home. | 79 (14.6%) |
| I spend more time other than playing games. | 62 (11.4%) |
| I spend less time other than playing games. | 58 (10.7%) |
| I have more fights or quarrels with family members. | 53 (9.8%) |
| I wake up earlier in the morning. | 47 (8.7%) |
| I fall asleep faster at night. | 28 (5.2%) |
| I spend less time communicating with my friends on my smartphone. | 27 (5.0%) |
| I have changed/am considering changing my career path. | 18 (3.3%) |
| I have less fights or quarrels with my family. | 15 (2.8%) |
| I play games for less time or less often. | 11 (2.0%) |
| I've started using new paid learning apps. | 10 (1.8%) |
| I spend less time on my smartphone and use it less often. | 9 (1.7%) |
| I've started to use free apps for learning. | 7 (1.3%) |
| I see my friends more often or for longer periods of time. | 6 (1.1%) |
| I have changed/am considering changing the entrance examination method for my desired career path. | 5 (0.9%) |
| I started to use distance learning and/or cram school. | 4 (0.7%) |
| Other | 3 (0.6%) |
| No particular change | 97 (17.9%) |

smart phones (48.3%), woke up later in the morning (38.4%), spent more time and more often playing games (33.8%), and went to bed later at night (32.5%) (Table 6). This was in line with the author's expectation. However, for example, 'I came to think that being at home is enjoyable' and 'I came to think that being at home is hard' were almost equal (15%), and 'I spent more time or increased the frequency of other than playing games' and 'I spent less time or decreased the frequency of other than playing games' were also almost equal (about 10%). It should also be noted that the negative effects are not always obvious.

On the other hand, changes in learning, such as started to use paid learning apps (1.8%), started to use free apps for learning (1.3%), and started to use distance learning and cram schools (0.7%), were found to be less frequent than other changes in life (shaded area in Table 6). Therefore, learning using smartphones and other devices was already widespread to some extent before the pandemic, and the pandemic did not seem to have changed the way students learn. In addition, 18% of the respondents answered that there was no change.

## 3.2 Someone (or Something) High School Students Trust in Thinking of Their Studies and Career Path

In terms of the someone (or something) they trust about their learning (Q8), school teachers were the most common (43.9%), followed by friends (12.5%) and parents and family (10.9%) (Table 7). Twenty per cent of the respondents said that they trust information from people other than those they know directly (results of tests and mock exams (10%), study apps (4%) and online information (4%)), i.e. nonhuman information. It is interesting to note that, when compared by whether or not they attend cram schools, respondents who attend cram schools are more likely to trust their cram school teachers than their school teachers (26% trust school teachers and 47% trust cram school teachers).

**Table 7.** Someone (or something) you trust with your learning (Q8).

| Item | Overall (N = 542) | Cram school students (N = 106) | Non-cram school students (N = 436) |
|---|---|---|---|
| School teachers | 238 (43.9%) | 28 (26%) | 210 (48%) |
| A friend (a friend you have actually met) | 68 (12.5%) | 8 (8%) | 60 (14%) |
| Parents, family and relatives | 59 (10.9%) | 6 (6%) | 53 (12%) |
| The results of tests and mock examinations | 52 (9.6%) | 6 (6%) | 46 (11%) |
| Cram school teachers | 50 (9.2%) | 50 (47%) | - |
| Learning apps | 21 (3.9%) | 2 (2%) | 19 (4%) |
| Online information | 21 (3.9%) | 1 (1%) | 20 (5%) |
| Online videos (other than those provided at school or by a teacher) | 14 (2.6%) | 3 (3%) | 11 (3%) |
| Friends (you have never met in person but only communicate online) | 5 (0.9%) | 1 (1%) | 4 (1%) |
| Other | 14 (2.6%) | 1 (1%) | 13 (3%) |

The results of the question about whom to trust regarding the career path (Q9) were broadly similar to those of Q8 (Table 8). Overall, schoolteachers were the most commonly trusted (47.6%), followed by parents/family (18.1%) and cram school teachers (10.7%). As in the previous question, students who use cram schools said that they trust their cram school teachers (55%) more than their schoolteachers (28%) in thinking of their career path.

**Table 8.** Someone (or something) you trust in thinking of your career path (Q9).

|  | Overall (542) | Cram school students (106) | Non-cram school students (436) |
|---|---|---|---|
| School teachers | 258 (48%) | 30 (28%) | 228 (52%) |
| Cram school teachers | 58 (11%) | 58 (55%) | - |
| The results of tests and mock examinations | 49 (9%) | 4 (4%) | 45 (10%) |
| Online videos (other than those provided at school or by a teacher) | 5 (1%) | 0 (0%) | 5 (1%) |
| Learning apps | 11 (2%) | 2 (2%) | 9 (2%) |
| A friend (a friend you have actually met) | 24 (4%) | 1 (1%) | 23 (5%) |
| Friends (you have never met in person but only communicate with online) | 7 (1%) | 1 (1%) | 6 (1%) |
| Online information | 15 (3%) | 0 (0%) | 15 (3%) |
| Parents, family and relatives | 98 (18%) | 9 (8%) | 89 (20%) |
| Other | 17 (3%) | 1 (1%) | 16 (4%) |

The last question (Q10) asked about the reasons for the questions about people (or things) they trust (Q8, 9) or about their own concerns about their studies, career and life. After excluding 'nothing in particular' answers and statements that did not make sense, 268 responses were counted as determinable answers. After applying the analysis method mentioned above, five categories were generated: 'school-oriented,' 'relationship-oriented,' 'rationality-oriented,' 'nonhuman,' and 'school-distrusting.'

**School-oriented Type (N = 101).** This type of respondents basically puts their trust in schools and teachers. This was the most common type of respondents, with 'schoolteachers' being the most common answer for both Q8: whom they trust with their studies and Q9: whom they trust with their career. Some examples are shown below.

'Because I feel like my teacher is watching me study the most.' (Second year in a public school, wishing to go to university.)

However, some respondents were reluctant to do so because they had no one else to rely on. There were also some respondents who trusted schools and teachers but were dissatisfied with the lack of distance education coordination. Therefore, it seems that not all respondents in this category trust schools and teachers equally. 'Because I have no one to turn to but my teacher at school.' (Third year in a private school, wishing to go to vocational school).

**Relationship-oriented Type (N = 71).** This type of respondents trusts their family, friends and sometimes their long-time tutor the most. It does not necessarily mean that they do not trust the school or teachers.

'(My friends are) easy to ask because they are in the same class. (My family) knows me best and can give me good advice.' (Third year in public school, wishing to go to work, choosing 'friends' for Q8 and 'parents, family and relatives' for Q9).

**Rationality-oriented Type (N = 45).** This type of respondents changes who (or what) they trust about their learning and who (or what) they consult about their career, depending on their objectives (mainly higher education or job hunting).

'Because the kind of person you can trust depends on the kind of study you do.' (First year in a public school, considering university, junior college, or work, choosing 'schoolteacher' for Q8 and 'cram school teacher' for Q9.)

**Nonhuman Type (N = 26).** This type of respondents does not trust schools or other people, and basically only trusts information from nonhuman sources. This type is similar to the school-distrusting type (see below), but differs in that they have no or few people they trust in thinking of their learning and career path.

'(Online information is) good because you can get information from people who have actually attended the university, and it is easier to understand. It is comparable with other universities and it can avoid studying in the wrong way.' (Third year in a public school, wishing to go to university, choosing 'online information' for both Q8 and Q9.)

**School-distrusting Type (N = 25).** This type of respondents showed a strong distrust of schools and teachers after or even before the pandemic. Basically, they do not agree with teachers or school policies. Similar to the nonhuman type above, but basically different in that there is someone to trust, whether online or offline.

'Because I feel I can take things more lightly when I haven't met someone in person.' (Third year in a private school, wishing to go to vocational school, choosing 'online friends' for both Q8 and Q9.)

## 4   Discussion

### 4.1   Changes in the Way High School Students Learn After the Pandemic

The results of the survey show that during the school closure, schools were mainly using old-fashioned learning methods, such as using textbooks and reference books, and were not promoting online learning or learning with ICT devices. In the aforementioned survey conducted by MEXT in May 2020, only 5% of respondents said they were using 'simultaneous interactive online instruction', but in this survey, the figure was 28%. This suggests that online teaching has spread somewhat extent in August when the survey was conducted. Even in this context, however, online and interactive learning instruction from schools was not sufficient.

In contrast to schools that have been slow to develop their ICT environment, the use of ICT (mainly smartphones) for learning at the individual level is now commonplace, highlighting the gap between schools and students. The study also found that the school closure did not lead to a shift in the way students learn digitally, but that the use of smartphones and tablets was already prevalent among students to some extent before the pandemic. In their answers to the question 'what is important in learning with ICT (Q6)', many students emphasized the importance of efficiency and learning effectiveness, which can be seen as a sign that they are trying to establish their own individual optimization of learning, which schools are currently unable to provide. It has to be said that schools are lagging behind the educational service providers and even the use of individual level when it comes to teaching with ICT.

## 4.2 Someone (or Something) High School Students Trust in Thinking of Their Studies and Career Path

Although schoolteachers were the most favored person to be trusted in thinking of students' study and career path, the percentage of those who went to cram school trusted their cram school teachers more than their schoolteachers. In fact, this result was unexpected. Twenty per cent of the respondents said that they trusted nonhuman information, although this was less than the proportion who said that they trusted people they knew directly (i.e., schoolteachers, family, friends, etc.). Whether this number is too much or too little is open to debate. Since the survey was conducted after the pandemic, it is not possible to compare how trust has changed. It is therefore not easy to foresee what the future holds, but the survey shows that, partly because of the effects of pandemic, schools and teachers do not enjoy absolute trust as teaching agents. There is no doubt that the school will continue to be an important place of learning, but it can be inferred that the weight of its importance will not be the same as before. The emergence of the 'rationality-oriented,' 'nonhuman' and 'school-distrusting' categories in students' attitude suggests that students may be more dry-minded than teachers think, and see school as just one of the places to learn. If teaching is not the sole domain of the teacher, what happens to the teacher's professionalism? This question is by no means a criticism of teachers. We are all aware of the significance of schools and teachers through the school closure. The essential elements such as caring, protecting, nurturing, and looking after children, which cannot be easily achieved through technology or commercial educational services, must remain. I believe that schools and teachers are at a crossroads in rethinking their role.

## 5  Conclusion

In face of the huge impact of the COVID-19 on education and students' learning, the author conducted a nationwide online questionnaire survey to find out (1) how high school students' learning practices, especially their use of ICT, cram schools, distance learning and learning apps, have changed after the pandemic, and (2) who (or what) they trust in thinking of their learning and their career path. With regard to (1), author found that there has been little progress in the informatization of education in schools, that the use of ICT (mainly smartphones) in learning at the individual level by pupils was already widespread to some extent before the pandemic, and that the pandemic did not directly lead to a digitization of learning methods. Regarding (2), it is clear that although schoolteachers are still trusted to some extent by students, they are not as trusted as cram school teachers. Twenty per cent of respondents said they trust information from non-human sources, such as test and mock exam results, learning apps and online information, suggesting that schools and teachers are not absolutely trusted. Students' way of trust were categorized as 'school-oriented,' 'relationship-oriented,' 'rationality-oriented,' 'nonhuman,' and 'school-distrusting,' revealing a picture of students that schools may not have envisaged before. These results suggest that the gap in attitudes between schools and students may lead to further problems.

For me, the results of this survey raised new questions, such as what is the role of schools and teachers in the first place, and what is the educational role of educational service providers such as cram schools, and technologies such as learning apps. I hope to use this survey as a springboard for further research.

# References

1. Washington post: https://www.washingtonpost.com/education/2020/03/26/nearly-14-billion-children-around-globe-are-out-school-heres-what-countries-are-doing-keep-kids-learning-during-pandemic/. Last accessed 1 Dec 2021
2. MEXT. https://www.mext.go.jp/content/20200513-mxt_kouhou02-000006590_2.pdf. Last accessed 1 Dec 2021
3. MEXT, https://www.mext.go.jp/content/20200424-mxt_kouhou01-000004520_8.pdf. Last accessed 1 Dec 2021
4. Save the Children Japan: https://www.savechildren.or.jp/jpnem/jpn/pdf/kodomonokoe202005_report.pdf. Last accessed 1 Dec 2021
5. Takahashi, W., Takahashi, S.: A Study of Corona Shocks and Educational and Economic Disparities (in Japanese). Staff Report, No.20-SR-10, pp. 1–9 (2020)
6. Japan Education News. https://www.kyoiku-press.com/post-213719/. Last accessed 1 Dec 2021
7. Benesse: https://berd.benesse.jp/up_images/research/2017_Gakko_gai_tyosa_web.pdf. Last accessed 1 Dec 2021
8. Pattersen, L.: Japan's "Cram Schools." Educ. Leadersh. **50**(5), 56–58 (1993)
9. Glaser, B.G., Strauss, A.L.: The Discovery of Grounded Theory: Strategies for Qualitative Research. Aldine Publishing Company, Chicago (1967)
10. Sato, I.: Qualitative Data Analysis Methods (In Japanese). Shinyosha, Tokyo (2008)
11. MEXT: https://www.mext.go.jp/content/20200603-mxt_kouhou01-000004520_4.pdf. Last accessed 1 Dec 2021
12. MEXT, https://www.mext.go.jp/content/20200421-mxt_kouhou01-000004520_4.pdf. Last accessed 1 Dec 2021

# Avatars in Immersive Virtual Reality Education: Preliminary Insights and Recommendations from a Systematic Review

Ines Miguel-Alonso[1], Anjela Mayer[2](✉), Jean-Rémy Chardonnet[3],
Andres Bustillo[1], and Jivka Ovtcharova[2]

[1] Department of Computer Engineering, Universidad de Burgos, Burgos, Spain
`imalonso@ubu.es`
[2] Institute for Information Management in Engineering, Karlsruhe Institute
of Technology, 76133 Karlsruhe, Germany
`anjela.mayer@kit.edu`
[3] Arts et Metiers Institute of Technology, LISPEN, HESAM Université,
71100 Chalon-sur-Saône, France

**Abstract.** Immersive Virtual Reality (iVR) has gained popularity in education for its ability to engage students and enhance learning. With the growing use of avatars in collaborative iVR environments, there is a need for structured guidelines to optimize their effectiveness. This preliminary systematic literature review aims to synthesize key findings from relevant studies, investigates the potential of avatars in iVR education and proposes best practices to promote positive learning outcomes. Gathering data from various studies, the review provides initial insights that will aid future researchers and developers in effectively integrating iVR and avatars, optimizing them for the educational setting, and ensuring their effectiveness in enhancing learning outcomes.

**Keywords:** Immersive Virtual Reality · Avatars in Education · Collaborative Learning

## 1 Introduction

In recent years, Immersive Virtual Reality (iVR), often referred to as the interactive Virtual Reality, has witnessed substantial growth, particularly in the realms of learning and training. iVR simulators serve as potent tools, not only for replicating real-world scenarios but also for crafting engaging experiences that enhance learning and training outcomes. Advancements in hardware capabilities have facilitated collaborative environments where multiple users interact. A prevalent practice involves user identification through avatars, distinct from

---

I. Miguel-Alonso and A. Mayer—Both authors contributed equally to this work.

© The Author(s), under exclusive license to Springer Nature Switzerland AG 2024
M. E. Auer et al. (Eds.): STE 2024, LNNS 1027, pp. 267–274, 2024.
https://doi.org/10.1007/978-3-031-61891-8_26

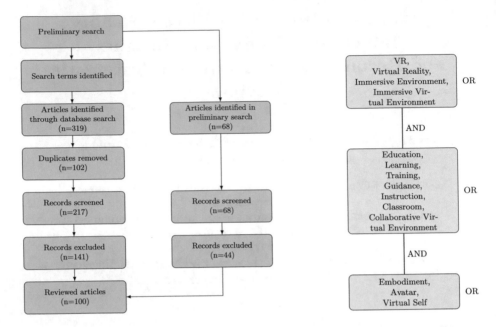

**Fig. 1.** Adapted PRISMA methodology (left) and structure of the constructed search term (right) used for the literature review.

agents as clarified by [12]. Avatars represent players in collaborative settings, while agents depict non-player characters capable of interacting within the virtual environment. The term *Hybrid Agents-Avatars* introduced by [12] refers to avatars with additional computer-controlled features for improved communication. This preliminary review focuses on avatars within the context of learning, education, and training, with the primary goal of offering valuable insights to researchers and developers for the effective utilization and implementation of avatars in iVR environments, emphasizing their effectiveness in enhancing educational outcomes.

While avatars are widely used in both non-iVR and iVR video games, the existing literature varies in its approach. This preliminary review focuses on avatars within the context of learning, education, and training. The primary goal is to offer valuable insights to researchers and developers for the effective utilization and implementation of avatars in iVR environments.

## 2   Methodology

This literature review was conducted following an adapted version of the Preferred Reporting Items for Systematic Reviews and Meta-Analyses (PRISMA) methodology, as outlined in [16] and illustrated in Fig. 1. Initially, a broad search using *Google Scholar* was conducted to identify key terminology within the field.

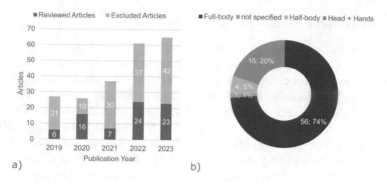

**Fig. 2.** a) Overview of found articles by the year of publication, b) identified avatar types in reviewed articles.

Complementary, *Research Rabbit*[1] was utilized to uncover related papers not captured by the initial search terms. This phase yielded 68 relevant articles.

Transitioning to a more focused approach, the search terms were refined, adhering to the boolean expression scheme shown in Fig. 1. Subsequently, a systematic search was conducted across selected digital libraries, including IEEE Xplore, ACM, ERIC, Web of Science, and Scopus. This phase, targeting open access and peer-reviewed articles, conference papers, and book chapters published in English from January 2019 to November 2023, resulted in the identification of an additional 319 articles.

After the removal of 102 duplicate records, a total of 285 articles ($n = 217$ from the structured search and $n = 68$ from the preliminary search) underwent screening. During the screening process, the following exclusion criteria were applied:

- Article is a review, meta-analysis, or survey.
- Application or experiment does not include user-controlled avatars.
- Technology utilized does not encompass iVR (VR, AR, MR).
- Field of application is outside learning, education, or training.
- Focus on machine learning and not on motion rehabilitation or training.

After the exclusion, 100 articles remained to be reviewed ($n = 76$ from the structured search and n = 24 from the preliminary search). Figure 2a) depicts the distribution of found, included and excluded articles by the year of publication.

## 3   Analysis

In this section the most relevant findings about avatar representation, collaborative environments, embodiment, presence and co-presence are summarized, providing guidelines and useful insights for the implementation of avatars in iVR for education.

---

[1] Research Rabbit: https://researchrabbitapp.com/.

## 3.1    Avatar Representation and Embodiment

In virtual environments, the representation of users significantly influences behaviour, shaping both how users are perceived by others and their own experiences as they control avatars. While the integration of full-body avatars is an evolving field, the majority of articles in this preliminary review (74%; $n = 56$) employ full-body avatars, with tracking for head and hands feasible through Head-Mounted-Display (HMD) systems. Approaches to lower body animation vary, including inverse kinematics (IK), motion tracking with additional sensors, or motion capture, as discussed by [2,15]. Some articles 5% ($n = 4$) use half-body avatars, and others 20% ($n = 15$) visualize only the head and hands, underscoring the diversity in avatar representation.

Despite their prevalence, the impact of avatars on learning outcomes is often unexplored in the reviewed articles. However, some studies investigate user representation through avatars. Notably, [5] found that minimal customization enhances learning, especially for users perceiving lower competence and higher similarity and presence. In a virtual classroom setting, [7] observed increased focus with realistic avatars, while [9] cautioned against task-irrelevant realism due to heightened cognitive workload. Facial animation's role is highlighted by [17], showing that realistic movements and lip-sync enhance presence, knowledge retention, and enjoyment. In contrast, [8] noted shifts in attention based on avatar styles, emphasizing the need for context-aware considerations.

Beyond avatar representation, the concepts of Embodiment and Sense of Embodiment (SoE) play crucial roles in shaping immersive virtual reality experiences. SoE involves feeling present in a virtual body, while Embodiment signifies user connection with a virtual body. The Sense of Agency (SoA) relates to controlling actions in the environment. Co-embodiment involves shared embodiment experiences among users. Studies, like [11], reveal co-embodiment's impact on learning, with increased Sense of Agency in co-embodied conditions. [4] explores the relationship between embodiment, empathy, and behaviour change, emphasizing the interconnectedness of these concepts. Additionally, [10] finds increased focus with realistic avatars in a virtual classroom.

## 3.2    Collaborative Environments

IVR Collaborative Environments (iVRCE) can be defined as a virtual environment shared among multiple users, irrespective of their physical locations or the involved time frame. IVRCE is not yet a ubiquitous phenomenon; in fact, the majority of articles analysed in this review do not involve any form of collaboration. However, given that this review encompasses topics related to learning/education and training, those articles that do incorporate collaboration in their applications can be classified within these fields.

Firstly, within the learning domain, [22] proposes an iVRCE shared by both AR and iVR users in a museum exploration. To successfully integrate these technologies in the same context, the researchers suggest using spatial anchors on the doors of the rooms to prevent tracking loss and provide feedback to the user regarding the alignment precision. Additionally, they propose indicating the relative directions and paths of users. Even if users have different devices, they should be capable of performing the same actions in a human-like manner, including natural gestures. On the other hand, [13] found that incorporating an avatar to guide the users using facial expressions was not as successful as expected. Users did not pay attention to these expressions while performing tasks, resulting in no significant differences. [14] demonstrated that incorporating avatars related to the lesson improved the motivation, immersion, and students' comprehension, despite study limitation. Furthermore, [10] succeeded in showing that co-embodiment with weight adjustment prevented performance decline after assistance. [9]'s research indicated that realism in avatars and environment is not always necessary for learning; in fact, excessive realism could increase cognitive load. Emphasis should also be given to the interaction of roles and instructions, not just the tasks users perform. Other notable studies include [1,15], which present interesting applications in a shared learning environment. In the context of Social VR, experiences are implemented to enable users to socially interact and participate in learning activities together. This is evident in [6,19,25]. [25] emphasized the importance of avoiding distractions in the environment enhances learning, and the importance of the avatar of the teacher, which students rated higher than their own avatar.

## 3.3   Presence and Co-presence

Presence refers to the profound sensation of being physically present within a given environment, particularly in the context of a virtual setting. Despite users physically inhabiting the real world, the immersive nature of iVR can make them feel as though they are truly situated within the virtual environment, blurring the lines between the two spaces. This seamless integration of the iVR environment into their perceived reality is a testament to the effectiveness of avatars in enhancing the sense of presence. Co-presence, on the other hand, delves into the ability to share the virtual environment with other users, transcending physical boundaries. Even when physically separated, users can experience a shared space, thanks to the capabilities of avatars. As evidenced by [25], avatars play a pivotal role in both presence and co-presence, particularly in scenarios where iVR is the sole condition for interaction. However, the work of [26] sheds light on a critical aspect—the visual quality of avatars. Not only does it impact the sense of presence, but if the avatars' expressions and gestures lack realism, the overall sense of presence can diminish. [17] reinforces this idea, emphasizing the importance of incorporating lip synchronization to further amplify the sense of presence. To bolster co-presence, [23] propose integrating gesture cues, making the interface more intuitive for users. Customization of avatars' appearances, as illuminated by [5]'s study, emerges as a significant factor influencing both presence and learning outcomes. Contrary to expectations, [20]'s research challenges the notion that embodied controls directly increase presence. Instead, it suggests that presence, in itself, can be a facilitator for learning. [7]'s study further nuances this by revealing that the stylistic representation of avatars doesn't yield statistically significant differences in terms of presence. [3] take a pragmatic approach by opting for a single-item oral measure to gauge presence in iVR. This decision is rooted in the potential loss of VR-induced phenomena upon headset removal, making multiple surveys less accurate. Interestingly, [24] demonstrate that an assistance condition guiding users' movements in reality can paradoxically decrease presence, even without a proper avatar in the iVR environment. They stress the relevance of haptic feedback in enhancing presence during object manipulation tasks, emphasizing the need for coherence between haptic feedback and visual perception. [21]'s research unveils the impact of avatar collisions on co-presence, showcasing that removing collisions diminishes the sense of co-presence. However, providing feedback during avatar collisions in Social VR can counter-intuitively increase the sense of presence.

Additional studies by [8, 14, 18] contribute to the broader understanding of presence within the realm of iVR, enriching the discourse surrounding avatars and their implications for training and learning in virtual environments.

## 4   Conclusion and Outlook

This preliminary review of the literature provides essential insights into the use and implementation of avatars in immersive Virtual Reality (iVR) for educational purposes. The most popular form of user representation in iVR is the full-body avatar. Thereby, visual quality and realistic behaviour seem to improve the sense of presence, engagement and the learning outcome. Nevertheless, task-irrelevant realism should be avoided since it can add to the cognitive workload and distract users from the learning task. Collaborative environments are yet not well represented in educational iVR research, although they play a significant role in group work and virtual classrooms. In collaborative iVR avatars can enhance motivation, immersion and comprehension and thus are positively supporting educational group activities. The concepts of embodiment and presence are highlighted as central to the user experience in iVR, with a significant impact on learning outcomes. These findings are instrumental in guiding the development of effective educational iVR systems with avatar user representations. iVR is especially suitable to simulate dangerous environments and situations for training. When utilized for spatial guiding in real or virtual environments, the avatars' expressiveness seem not to play a significant role. An improved feeling of presence and co-presence plays a crucial role in iVR and affects learning outcomes. However, it is not clear whether the avatars visual appearance or its realistic behaviour and embodiment has a greater impact on presence. In any case, the coherence between different feedback mechanisms, like the visual and haptic feedback, is benefiting the sense of presence.

This review emphasizes on the importance of avatars for learning outcomes in iVR and summarizes important aspects which are instrumental in guiding the development of effective educational iVR systems with avatar user representations. An extended review will be published in the near future, encompassing all reviewed literature as well as detailed analysis results to provide a deeper understanding of these aspects.

# References

1. Al-Suwaidi, F., Agkathidis, A., Haidar, A., Lombardi, D.: Immersive technologies in architectural education: a pedagogical framework for integrating virtual reality as the main design tool in a fully virtualised architectural design studio environment. In: Proceedings of the 41st Conference on Education and Research in Computer Aided Architectural Design in Europe (eCAADe 2023) (2023)
2. Braun, P., Grafelmann, M., Gill, F., Stolz, H., Hinckeldeyn, J., Lange, A.-K.: Virtual reality for immersive multi-user firefighter-training scenarios. Virtual Reality Intell. Hardw. (2022)
3. Czub, M., Janeta, P.: Exercise in virtual reality with a muscular avatar influences performance on a weightlifting exercise. In: Cyberpsychol.: J. Psychosoc. Res. Cybersp. 15(3), Article 10 (2021)
4. Elzie, C., Shaia, J.: A pilot study of the impact of virtually embodying a patient with a terminal illness. Med. Sci. Educ. 31, 665–675 (2021)
5. Fitton, I. Clarke, C., Dalton, J., Proulx, M., Lutteroth, C.: Dancing with the avatars: minimal avatar customisation enhances learning in a psychomotor task. In: CHI 2023 - Proceedings of the 2023 CHI Conference on Human Factors in Computing Systems, Hamburg, Germany, vol. 714. Association for Computing Machinery, New York (2023)
6. Friston, S., et al.: Ubiq: a system to build flexible social virtual reality experiences. Comput. Sci.: Hum.-Comput. Interact. (2021)
7. Gao, H., Bozkir, E., Hasenbein, L., Hahn, J.-U., Göllner, R., Kasneci, E.: Digital transformations of classrooms in virtual reality. In: Proceedings of the 2021 CHI Conference on Human Factors in Computing Systems, CHI 2021. Association for Computing Machinery, New York (2021)
8. Hasenbein, L., et al.: Learning with simulated virtual classmates: effects of social-related configurations on students' visual attention and learning experiences in an immersive virtual reality classroom. Comput. Hum. Behav. 133, 107282 (2022)
9. Huang, W., Walkington, C., Nathan, M.J.: Coordinating modalities of mathematical collaboration in shared VR environments. Intern. J. Comput.-Support. Collab. Learn. (2023)
10. Kodama, D., Mizuho, T., Hatada, Y., Narumi, T., Hirose, M.: Effect of weight adjustment in virtual co-embodiment during collaborative training. In: Proceedings of the Augmented Humans International Conference 2023, AHs 2023, pp. 86–97. Association for Computing Machinery, New York (2023)
11. Kodama, D., Mizuho, T., Hatada, Y., Narumi, T., Hirose, M.: Effects of collaborative training using virtual co-embodiment on motor skill learning. IEEE Trans. Vis. Comput. Graphics 29(5), 2304–2314 (2023)
12. Kyrlitsias, C., Michael-Grigoriou, D.: Social interaction with agents and avatars in immersive virtual environments: a survey. Front. Virtual Reality (2022)
13. Le Tarnec, H., Augereau, O., Bevacqua, E., De Loor, P.: Improving collaborative learning in virtual reality with facial expressions. Association for Computing Machinery (2023)
14. Liebermann, A., Lente, I., Huth, K.C., Erdelt, K.: Impact of a virtual prosthetic case planning environment on perceived immersion, cognitive load, authenticity and learning motivation in dental students. Eur. J. Dent. Educ. 28, 9–19 (2023)
15. Liu, J., Zheng, Y., Wang, K., Bian, Y., Gai, W., Gao, D.: A real-time interactive Tai Chi learning system based on VR and motion capture technology. Procedia Comput. Sci. 174, 712–719 (2019)

16. Page, M.J., et al.: The PRISMA 2020 statement: an updated guideline for reporting systematic reviews. BMJ **372** (2021)
17. Peixoto, B. Melo, M., Cabral, L., Bessa, M.: Evaluation of animation and lip-sync of avatars, and user interaction in immersive virtual reality learning environments, pp. 1–7 (2021)
18. Porssut, T., Blanke, O., Herbelin, B., Boulic, R.: Reaching articular limits can negatively impact embodiment in virtual reality. PLoS ONE **17**(3), e0255554 (2022)
19. Rante, H., Zainuddin, M.A., Miranto, C., Pasila, F., Irawan, W., Fajrianti, E.D.: Development of social virtual reality (SVR) as collaborative learning media to support Merdeka Belajar. Int. J. Inf. Educ. Technol. **13**, 1014–1020 (2023)
20. Ratcliffe, J., Tokarchuk, L.: Evidence for embodied cognition in immersive virtual environments using a second language learning environment, pp. 471–478 (2020)
21. Reinhardt, J., Wolf, K.: Go-through: disabling collision to access obstructed paths and open occluded views in social VR. In: Proceedings of the Augmented Humans International Conference, AHs 2020. Association for Computing Machinery, New York (2020)
22. Schott, E., et al.: UniteXR: joint exploration of a real-world museum and its digital twin. Association for Computing Machinery (2023)
23. Wang, P., Wang, Y., Billinghurst, M., Yang, H., Xu, P., Li, Y.: BeHere: a VR/SAR remote collaboration system based on virtual replicas sharing gesture and avatar in a procedural task. Virtual Reality **27**, 1409–1430 (2023)
24. Wenk, N., Jordi, M.V., Buetler, K.A., Marchal-Crespo, L.: Hiding assistive robots during training in immersive VR does not affect users' motivation, presence, embodiment, performance, nor visual attention. IEEE Trans. Neural Syst. Rehabil. Eng. **30**, 390–399 (2022)
25. Yoshimura, A., Borst, C.W.: A study of class meetings in VR: student experiences of attending lectures and of giving a project presentation. Front. Virtual Reality **2**, 648619 (2021)
26. Zacarin, M.R.J., Borloti, E., Haydu, V.B.: Behavioral therapy and virtual reality exposure for public speaking anxiety. Temas Psicologia **27**(2), 491–507 (2019)

# Enhancing Educational Outcomes Through Learning Pathways and Artificial Intelligence-Driven Learning Analytics

Martin Trommer[✉], Susanne Franke, Tobias Teich, Ralph Riedel,
Sven Hellbach, and Daniel Franke

University of Applied Sciences Zwickau (WHZ), Kornmarkt 1, 08056 Zwickau,
Germany
{Martin.Trommer,Susanne.Franke}@FH-Zwickau.de
https://www.fh-zwickau.de

**Abstract.** This paper explores the implementation of personalized learning pathways using Learning Analytics to enhance educational quality. It addresses the challenges posed by learner heterogeneity, varying learning speeds, and different learning styles, emphasizing the need for individualized learning approaches to optimize learning outcomes. Especially in the context of e-learning, the integration of Learning Analytics and Artificial Intelligence (AI) allows enhancing educational quality through the definition and dynamic adaption of learning pathways. This is achieved by adapting learning content, questions and difficulty levels in real-time based on the students' performance and behaviors. The paper includes the design of a platform architecture for data acquisition, preparation, and utilization in dynamically adjusting course content.

**Keywords:** Learning Analytics · xAPI · Learning Platform
Architecture

## 1 Introduction

The University of Applied Sciences Zwickau (WHZ) has embarked on an ambitious educational journey to meet the dynamic demands across various professional domains. This includes research not only projects in the field of e-learning, but especially also projects in cooperation with companies operating in car manufacturing, healthcare or steel industry in order to individualize learning. Learning pathways, also known as personalized learning journeys, are a contemporary educational approach specifically designed to meet the unique needs, interests, and capabilities of individual learners. This adaptive strategy has gained prominence in education as it seeks to optimize learning experiences and the learners' outcomes in an increasingly diverse and technology-driven learning environment.

The development of learning concepts focusing on the individualization of learning and knowledge transfer (learning pathways) is highly relevant in view of

M. E. Auer et al. (Eds.): STE 2024, LNNS 1027, pp. 275–284, 2024.
https://doi.org/10.1007/978-3-031-61891-8_27

the many problems that exist in the education sector, e.g., differentiated learning levels, varying learning speeds and different learning styles of the individual learners require personalized learning material. More specifically, this results in the following challenges:

1. Learner heterogeneity manifests itself in significant differences in terms of individual cognitive performance and levels of knowledge.
2. Learners have varying learning speeds, which can lead to frustration and inefficiency in a rigid curriculum-based system.
3. Learners have different preferences regarding learning styles (visual, auditory, kinaesthetic), learning location and learning time.

Thus, the individualization of learning contributes to equal opportunities in the educational system and enables every learner to fully exploit their individual potential. As a result, the development and implementation of flexible learning concepts tailored to the needs of learners is of eminent importance for educational research and practice. An essential core element is the development of methodologies for the creation and implementation of individualized learning pathways:

1. The implementation of individualized learning approaches ensures that each learner is challenged and supported according to his or her current level of knowledge, which in turn leads to increased motivation and self-efficacy by avoiding under- or overchallenging.
2. Individualized learning enables learners to determine their own learning pace and thus makes their learning processes more effective and efficient.
3. The integration of diverse teaching methods and resources in individualized learning concepts leads to optimized knowledge transfer.

This contribution explores the application of learning pathways by using Learning Analytics as a strategic instrument to enhance the educational quality. The main goal is to systematically capture, evaluate and improve individual learning progression by the dynamic adaptation of learning pathways and course content based on Artificial Intelligence (AI) functionality to meet specific educational prerequisites. It is discussed how a basic platform architecture should be designed to be able to capture learning data, how data can be prepared for Learning Analytics and how this data can be used to dynamically adapt learning content.

## 2    Theoretical Background: Individualization of Learning

Individualized learning experiences have become vital in today's education due to the diverse needs of students. Each learner brings unique strengths, learning styles, and backgrounds to the table, calling for tailored teaching methods. In this context, personalized learning improves engagement, motivation and knowledge retention. Individualized learning also empowers learners by fostering independence and self-organized work. By adapting teaching methods and assessments to suit each learner's pace and preferences, educators create inclusive environments that helps bridging achievement gaps and allows every learner to thrive. Tools like adaptive technologies and flexible teaching strategies ensure that each learner gets the support he or she needs [4]. According to [11], embracing personalized learning does not only include meeting different learning needs - it also focuses on preparing learners to take charge of their learning journeys. In order to dynamically adapt learning content based on learning pathways, the use of Learning Analytics is an essential precondition.

Learning Analytics is defined as *"the measurement, collection, analysis and reporting of data about learners and their contexts, for purposes of understanding and optimising learning and the environments in which it occurs"* [10]. It uses theories and methods from machine learning and data science, pedagogy, cognitive psychology, statistics, computer science, neuroscience, and social and learning sciences. With a clear quantitative focus, the challenge is to process a large amount of data (in real-time). Learning Analytics is an essential tool to improve learning pathways. Here, learning pathways are defined as a sequence of intermediate steps from learner's initial knowledge state to a desired knowledge state [9].

To measure learning performance and progress, suitable indicators must be identified. Hereby, a distinction is made between measuring the degree of target achievement, i.e. comparing the result with the learning target/minimum requirement, and measuring change, i.e. comparing the achieved results of learners over time [6]. There are two main approaches for measuring, subjective (e.g., self-evaluation) and objective (e.g., open or closed questions on the learning content like free text answers or multiple choice). They are used to evaluate the individual learning progress in each learning pathway's step and, hence, to improve the learning progress by providing suitable learning content and giving suggestions on which next step is the most suitable one.

According to [5], two approaches for developing tasks to measure learning progress are distinguished: curriculum sampling (i.e., constructing representative curriculum samples) and developing robust indicators (i.e., measures to map a broad performance spectrum). Our use case in the context of university teaching focuses on specific modules, which represents curriculum sampling.

Numerous scientific results exist particularly in the field of mathematics. An overview is given in [3]. The author concludes that most studies focus on static indicators (mostly at the end of the learning process) and do not combine them with continuous progress monitoring.

The effectiveness of learning highly depends on the thorough design of the learning pathways and the choice of suitable implementation options [2]. In the last years, e-learning platforms as one implementation option have developed rapidly. Learning management systems like Moodle provide predefined components like content pages, forums, quizzes and provide interfaces to external sources. The components can be used to create courses and have to be filled with content. Additionally, rule-based learning paths can be implemented by defining access conditions for the components such that, e.g., certain other components have to be viewed or solved.

Many of these platforms automatically include data storage and provision. This data however has to be evaluated and analysed to provide useful results. An approach for an expedient usage of time data automatically stored by Moodle is given in [8]. This enables a deeper analysis and evaluation of the learners' performance. However, the questions remains on how to include these results in the learning pathways.

In summary, learning management systems provide an efficient way to provide learning content. However, this content is usually implemented in a linear way with little to no adaption with respect to the individual learner's skills and progress. As already mentioned, it is our goal to present a possibility to enhance learning management systems by individualising the learning experience using Learning Analytics. This leads to the following demands:

1. The usage of a learning management system with open interfaces to acquire detailed learner-relevant data and include analysis results, the identification of and connection with other activity providers (here, the selection of these activity providers is context-dependent),
2. Enabling data storage in a suitable database management system and enabling analysis with suitable software and specifically tailored programming tools,
3. The implementation and integration of thorough analysis of the learners' data as basis for the adaptation of the learning pathways.

## 3   Approach WHZLearnXP

Referring to the requirements explained in Sect. 2, individualizing the learning process through adaptive learning pathways enables the improvement of educational success. For this reason, a Learning Experience Architecture called "WHZ LearnXP" was established to provide a strong foundation for AI-based individual learning. The basic structure if depicted in Fig. 1.

WHZ LearnXP, in its current expansion stage, which is referred to as "Basic Analytics", consists of a Moodle instance (activity provider) and a Learning Record Store (LRS), which is a database based on a MongoDB (LRS endpoint), as well as an analytics component which consists of an AI server and analytical tools.

The first demand defined in Sect. 2 is met by using a learning management system (LMS) as activity provider with open interfaces. Within the realm of learning experiences, an activity provider can be defined as an entity or source that delivers structured and purposeful learning data such as LMS, mobile apps, serious games and simulations, virtual reality (VR) and augmented reality (AR) environments or in-classroom response devices [12]. Moodle as an LMS and activity provider is built on a modular and flexible framework, designed to support online learning environments. The Moodle architecture consists of a frontend

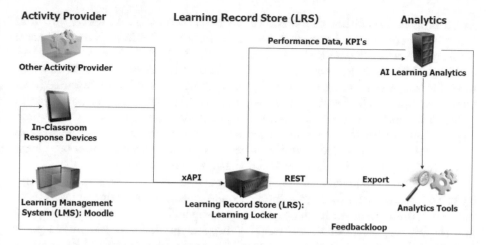

**Fig. 1.** WHZ LearnXP - Platform Architecture

interface, backend components such as relational databases (e.g., MySQL), modules as well as plugin functionality. An xAPI plugin was integrated within the Moodle LMS, which transmits learning data from the activity provider to the LRS in the form of a specific annotation (learning statements). Communication between the entities takes place via the xAPI interface (inbound). WHZ LearnXP also allows communication with other activity providers such that the learners' data can be acquired. These are, for example, response devices used during classroom teaching or other software the learners use to solve certain tasks.

The second demand, addressing the data storage, is covered by the usage of databases that can collect learning data and provide first analysis functionality. In that context, the LRS serves as a repository for storing and retrieving data generated by various activities providers. The primary purpose of an LRS is to collect, store, and provide access to data related to learning experiences, see [10]. The core component is a database (MongoDB) optimized for handling large volumes of learning records efficiently. The LRS is able to receive learning statements via the xAPI plugin in Moodle. External systems can use API Endpoints in order to enable the retrieval and submission of learning data. The LRS frontend additionally provides simple Learning Analytics functionality based on a Data Processing and Query Engine. It is capable of handling various types of queries and data analysis. This allows educators or administrators in an LRS dashboard to retrieve specific learning data, generate reports or perform analytics on learner activities [13]. The connection to the analysis tools (outbound) is established via a file interface or the Representational State Transfer (REST).

The third demand, addressing the analysis, is met by the integration of the third part of the architecture, which comprises an Analytics environment and thus covers the demand for the connection of the data that were collected from the activity providers. AI Learning Analytics functionality uses an AI server, which is available within the WHZ IT infrastructure. The AI server provides processing performance by using a 1.8 x Nividia A100 GPUs with 640 GB total memory, 2.6 x Nvidia NVSwitches, storage OS: 2x 1.92TB M.2 NVME drives and SuperMicro MP Superserver ($16 \times 15.36$TB SSD). A development environment for the programming language Python 3.9 with the relevant libraries (e.g., pandas, numpy, scikitlearn, tensorflow, pytorch) is installed on the server in order to be able to carry outAI-based Learning Analytics. Besides the Python environment, other analytics tools such as Microsoft Power BI or KNIME can be integrated. The activity providers as well as the LRS are executed on a local server that is provided in flexibly scalable Docker containers.

In the current expansion stage of WHZ LearnXP, Moodle acts as main activity provider. Activities (e.g., quiz questions, feedback, glossary) or resources (e.g., files, URLs) are assigned to the topics, in which the learning content is mapped and the learning progress measurement is implemented. From a technical point of view, defined user activities are transmitted to the LRS via the xAPI interface in the form of xAPI statements in asynchronous learning process while the learning pathway is being actively worked on. The statements are transformed into an XML structure and prepared for being transmitted. This can be explained using the example of a learner who has worked on the content of a particular chapter, as it is depicted in Fig. 2.

**Fig. 2.** Statements

The learning statements are sent to the LRS in a continuous data stream or as groups and then processed for initial analysis purposes in both statistical and graphical form. In addition to other elements, the data extract contains the user, a timestamp, the specific statements made by the users, their decisions and various meta information, such as identification numbers, information about the activity provider and connection information.

Based on the received learning statements, the LRS offers basic data evaluation functionalities, such as the number of incorrectly answered quiz questions, in order to map the learning behavior of groups or individual participants. This functionality gives trainers the opportunity to obtain evaluations that can be interpreted immediately in real-time.

All implementations were carried out in compliance with the applicable data protection regulations, so that the use of Experience Learning functionality is data protection compliant. This also includes anonymization and pseudonymization (such that learners' identification is not possible for the data analysts), secure data storage or access control and permissions (this is implemented by user roles with respective rights).

Based on the developed platform architecture, feedback loops can be implemented to adapt the learning content in the LMS in real-time. In order to prepare the mentioned WHZ LearnXP platform for full feedback loop functionality, another major step of data processing is essential.

## 4  Data Processing and AI Integration

Educational Data Mining has become a pivotal element in enhancing the learning experience, leveraging analytics tools to evaluate data within LMS. In this section, we present an approach to optimize learning pathways and individualize student analysis using data mining techniques.

Considering the architecture depicted in Fig. 1, it becomes obvious that data stored in the LRS originates in varying sources, which will even be extended during the future usage of WHZLearnXP. The main part stems from the LMS, where the learning content is provided. Initially, rule-based learning pathways are implemented within the LMS. They allow control on access to various components based on conditions such as quiz scores or forum participation. Thereby, it is crucial to develop a comprehensive concept for these pathways prior to deployment to ensure consistency throughout the learning phase, as any midcourse alterations could adversely affect the comparability of the results.

A more complex analysis is possible using the anonymized user data from the LMS such that the movement through the learning pathways are tailored to the learner's previous actions and needs; for instance, a student may be prompted or required to revisit content (which helped other students with similar learning problems) or explore it in a different format (e.g., text instead of a video) to reinforce understanding. This can only be realized with an algorithm learning from the students' past behaviour.

We distinguish student-specific and cohort-specific analysis. The former one sheds light on individual learning progress, styles, and behaviors. These insights are valuable not only for educators but also for learners, who can refine their study habits based on their personal data. Moreover, educators can use this analysis to assess the cohort's overall learning patterns and preferences, pinpoint challenging content for further exploration, gauge the appropriateness of assessment difficulty, and recognize students' strengths, which could be the basis for the formulation of topics for bachelor and master theses.

To enable this analysis, a robust conceptual framework of the learning pathways is necessary, encompassing the structuring of the learning content, the development of learning pathways, and the creation of a model for measuring progress. Learning progress is measured by evaluating the students' results in diverse tasks (a ratio of achievable points and actually achieved points) in combination with a classification of the respective task's difficulty (e.g., according to Bloom's taxonomy [1]). Additional measures include the duration students need to solve the tasks. The conception of the tasks should be aligned with the definition of learning goals, where the learning goals themselves reflect the relevant knowledge and competences the students acquire in this course. The database also includes the definition of thresholds which define the movement through the learning pathway. These values are parameterized and defined by the respective teacher. Additionally, external activity providers specifically tailored to certain disciplines can be used (for example, software development environments, simulation or planning software). Hence, considering the technical prerequisites (see Fig. 1), data integration plays an essential role. In practice, especially the data provided by Moodle must undergo preprocessing to eliminate superfluous details and to format it for analysis. The data is provided in a json-style format with nested information (for example, the current page is stored in object, definition, name, en). Unnecessary information has to be deleted (e.g., IDs of the tuples or data on URLs where the learning content is provided), and timestamps are used to derive durations. Questions in Moodle are stored in xml format and have to be integrated as well. Relevant information like question type, correct answer, given answer and points have to be extracted. AI allows a further enhancement of the analysis. While closed question can include difficult tasks for the students to solve, the predetermined solutions bear the risk of learners correctly guessing the answer without actually having the necessary knowledge. Natural language processing allows the real-time evaluation of complex tasks using open questions, where no hints are given at all. Furthermore, using historical data, learner-specific patterns can be derived and help to refine the learning pathways. This includes learning styles to present specific content and predictions on the future learning progress as basis for, e.g., regular reminders for learners at risk to invest more time in the learning process. From an educator's perspective, the evaluation of content and tasks by difficulty levels allows a better understanding of the whole learner cohort and also enables the refinement of learning pathways. This can be achieved in two ways. The first possibility is to use unsupervised methods and create clusters of tasks with similar difficulty. Secondly, supervised learning uses already labeled tasks (with, e.g., the levels *beginner*, *advanced* and *expert*) and identifies features leading to this classification. Clustering methods can also be used to identify learners with similar learning styles and competence levels, which allows a customized suggestion regarding the next steps along the learning pathway.

Ultimately, subjective learners' feedback serves as a critical component, guiding the continuous improvement of the educational framework. This kind of feedback should be requested after each learning section and should include questions regarding the easiest and most difficult content, an evaluation of the type of content provision, see for example [7]. This iterative feedback loop serves as valuable input for the teacher, who can, after a critical evaluation of the feedback, adjust or extend the content accordingly. This process ensures that the learning platform evolves to meet the dynamic needs of learners, providing a tailored and effective educational journey.

# 5 Summary and Outlook

The concept of learning pathways and individualized learning journeys tailored to each learner has emerged as a crucial strategy in contemporary education. It addresses the challenges posed by learner heterogeneity in terms of cognitive performance, varying learning speeds, diverse learning styles, and preferences.

Personalized learning not only ensures equal opportunities but also allows learners to fully utilize their potential. Implementing flexible learning concepts that cater to individual needs becomes very important for both research and practical applications in education. This paper provided the application of learning pathways using Learning Analytics as a strategic tool to enhance educational quality. This involves the measurement, analysis, and optimization of learning and learning environments, utilizing methodologies from various disciplines such as data science and AI. The paper also highlighted the significance of e-learning platforms like Moodle and emphasized the need for data evaluation and analysis. For this reason, the WHZ LearnXP architecture was introduced, aiming to leverage AI-based Learning Analytics for maximal educational success.

The future goal involves deeper analysis of learner activities to understand individual behaviors and preferences. As part of the next steps to establish a WHZ LearnXP architecture, more activity provides will be included in the architecture in order to meet the specific needs from different courses. Additionally, the analytics tools will be expanded significantly to enable the system to generate even more personalized feedback loops according to the learner's proficiency in real-time.

The implementation of concepts like the Learning Experience Platform WHZ LearnXP with adaptation to the targeted asynchronous deepening of learning content allows teachers the opportunities to better assess the learning progress of learners and to adapt the learning material and learning methodology as required. The concept is being fine-tuned as the project progresses so that further promising potential can be leveraged.

**Acknowledgments.** The project on which this report is based was funded by the Federal Ministry of Education and Research under grant number 16DHBKI063. The authors are responsible for the content of this publication.

# References

1. Bloom, B.S., Engelhart, M.D., Furst, E.J., Hill, W.H., Krathwohl, D.R.: Taxonomy of educational objectives: the classification of educational goals. In: Handbook I: Cognitive domain. David McKay Company, New York (1956)
2. Branch, R.M., Dousay, T.A.: Survey of Instructional Design Models, 5th edn. Association for Educational Communications and Technology AECT (2015)
3. Foegen, A.: Progress monitoring measures in mathematics. A review of the literature. J. Spec. Educ. **41**(2), 121–139 (2007)
4. Freitas, S.D., Yapp C.: Personalizing Learning in the 21st Century. Network Educational Press (2005)

5. Fuchs, L.S.: The past, present, and future of curriculum-based measurement research. Sch. Psychol. Rev. **33**, 188–192 (2004)
6. Makransky, G., Terkildsen, T., Mayer, R.: Role of subjective and objective measures of cognitive processing during learning in explaining the spatial contiguity effect. Learn. Instruct. (2019)
7. Nicol, D., Thomson, A., Breslin, C.: Rethinking feedback practices in higher education: a peer review perspective. Assess. Eval. High. Educ. **39**(1), 102–122 (2014)
8. Rotelli, D., Monreale, A.: Processing and understanding moodle log data and their temporal dimension. J. Learn. Anal. **10**(2), 126–141 (2023)
9. Scott, P.H.: Pathways in Learning Science: a case study of the development of one student's ideas relating to the structure of matter. In: Duit, R., Goldberg, F., Niedderer, H. (eds.) International Workshop on Research in Physics Learning: Theoretical Issues and Empirical Studies. University of Bremen (1991)
10. Srinivasa, K.G., Muralidhar, K.: A Beginner's Guide to Learning Analytics. Springer, Cham (2021). https://doi.org/10.1007/978-3-030-70258-8
11. Tadlaoui, M.A., Khaldi, M.: Personalization and Collaboration in Adaptive E-Learning, pp. 1–33. Business Science Reference, Hershey (2019)
12. Torrance, M., Houck, R.: Making Sense of xAPI. Learning technologies, vol. 4. Association for Talent Development, Alexandria (2019)
13. Watershed: Learning Record Store. https://www.watershedlrs.com/. Accessed 25 Nov 2023

# The Development an IoT-Based Virtual Model of Power Grid System with Renewable Energy Sources for a Laboratory Practicum of Control Systems

Maryna Stupak ⓘ, Hanna Telychko(✉) ⓘ, Hlib Stupak ⓘ, and Valerii Potsepaiev ⓘ

Donetsk National Technical University, 2, Shybankova Square, Pokrovsk 85300, Donetsk, Ukraine
hanna.telychko@donntu.edu.ua

**Abstract.** Today, Ukraine faces challenges that significantly impact the educational process in higher education institutions. Initially, the COVID-19 pandemic and later the war have hindered students' ability to develop hard skills effectively. Approximately 50% of students in technical universities in Ukraine lack access to laboratories for practical training. To compensate for educational losses and enhance the quality of practical training, it is essential to focus efforts on virtualizing laboratories and ensuring remote continuous access to them.

Modern technologies and cloud services enable the creation of digital twins of technological processes that can replace physical equipment, conduct simulation modeling of various systems, create interactive systems, and allow students to research and test solutions they have developed in a safe environment. This realization method ensures uninterrupted access to educational content for learners, regardless of their actual location and at a convenient time in secure conditions.

The result of this work is a prototype IoT system created using databases, Node-RED, and protocols such as MQTT, HTTP/HTTPS, among others. The developed IoT system familiarizes education seekers with examples of building geographically distributed computer-integrated networks for automation and monitoring of control object statuses using Industry 4.0 and IoT. The virtual model of the IoT system represents a power grid consisting of consumers, power plants, and renewable energy sources. All model components interact in a unified information space, allowing students to gain practical experience in virtual environments with networking technologies, programming, data analysis, and control algorithm development. This work was supported by the European Commission [grant number ENI/2019/413–664 "EDUTIP"] and Edunet World Association (EWA).

**Keywords:** Industry 4.0 · Virtual Laboratory · Green Transition · Cloud Technologies · IoT/IIoT

# 1  Introduction

The practical experience gained during education is an integral part of the skills students need to acquire at the university. This is particularly relevant for students in technical specialties, undergoing training in areas such as automation and computer-integrated technologies, robotics, electrical engineering, IoT, and IIoT. These fields are rapidly advancing within the framework of Industry 4.0, requiring corresponding hard and soft skills.

The aim of this work is to provide students with the opportunity to enhance their hard skills in the fields of automated control systems, IoT/IIoT, and Industry 4.0 using cloud environments in a remote mode.

To compensate for educational losses and improve the quality of practical training, it is essential to concentrate efforts on virtualizing laboratories and ensuring secure remote continuous access to them. Therefore, the task at hand is to create a virtual space with the capability of acquiring practical skills corresponding to Industry 4.0 technologies.

The choice of a virtual environment as the primary tool for acquiring practical skills is justified by several factors. Firstly, students physically cannot attend the university and work in laboratories due to rocket attacks and the need to adhere to personal safety requirements. This is particularly crucial for participants in the educational process studying in universities near active conflict zones. Secondly, the issue of blackouts and power outages. The experience of the past year demonstrated that even higher educational institutions located in relatively safe regions of Ukraine struggled to conduct laboratory practice and practical training for students in the absence of electricity.

Thirdly, the outdated material and technical base and technical equipment pose a challenge. Although this factor is not directly related to the first two, it remains quite relevant. Even with the autonomy of universities, funds for updating laboratory equipment are not always sufficient. The decision to use virtual environments is based on these three challenges faced by institutions of higher technical education.

Modern technologies and cloud services allow the creation of digital twins of technological processes that can replace physical equipment, conduct simulation modeling of various systems, create interactive systems, and enable students to research and test solutions they have developed in a safe environment. This implementation method will ensure uninterrupted access to educational content for learners, regardless of their actual location and at a convenient time in secure conditions.

Additionally, another significant advantage of using cloud environments for acquiring practical skills is that students will have the opportunity to expand their knowledge in related fields, particularly those associated with information technologies.

# 2  Main Part

The development of a virtual IoT laboratory model is based on the use of a simulation approach. This development involves the utilization of massive datasets obtained through simulation rather than experimental means. Simulation is considered as an "indirect mediated method of scientific investigation of objects of cognition (the direct study of which, for certain reasons, is impossible, complicated, or impractical) by studying their models" [1]. Modeling is seen as an epistemological category that denotes a way of understanding an object based on constructing and investigating a model of the object, with subsequent transfer of the acquired knowledge to the actual object [2].

There are various types of modeling, including:

- Natural modeling (conducting research on the real object with subsequent processing of experimental results);
- Simulation modeling (building models that describe processes as they would occur in reality);
- Physical modeling (experimental study of physical processes and phenomena based on their physical similarity, using physical models);
- Mathematical modeling (establishing a correspondence between a given object and a mathematical object, a mathematical model, and studying this model to obtain characteristics of the considered real object).

The main stages of modeling include:

1. Problem formulation, description of the studied system, and identification of its components and relationships.
2. Formalization (creation of a mathematical model representing a system of equations reflecting the essence of the studied object).
3. Algorithm development, the implementation of which solves the formulated task.
4. Planning and execution of calculations and obtaining results.
5. Analysis and interpretation of results [3].

Through these types of modeling, the most accurate forecast or reproduction of a system can be achieved. Simulation modeling is a specific case of mathematical modeling. For objects without analytical models or solution methods, mathematical models are replaced by simulators or simulation models. A simulation model is a logical-mathematical description of an object that can be used for experimentation on a computer for the purpose of design, analysis, and evaluation of the object's functioning.

The planned outcome is a virtual model of an IoT power grid with renewable energy sources, flexible for simulation, research, and laboratory practical work. The chosen base model for investigation is an electrical network with both renewable and traditional energy sources. The selection of an electrical network with renewable energy sources emphasizes the importance of green transition, transformation, efficient production, and in-depth knowledge in the field of electrical power engineering for students in non-specialized specialties.

Furthermore, this specific field offers broad opportunities and relative simplicity in creating a simulation model compared to technological processes. The basic model includes flexibility, the ability to influence linear and nonlinear factors and processes, feedback, energy storage elements, and the factor of territorial distribution. In conclusion, in the context of considering management and automation tasks, interdisciplinary student training expands, and they acquire practical skills in various areas.

For the Internet of Energy (IoE), various detailed architectures have been proposed, and key components for creating an automated and flexible power distribution system or Energy Internet have been defined in [7, 8] (see Fig. 1).

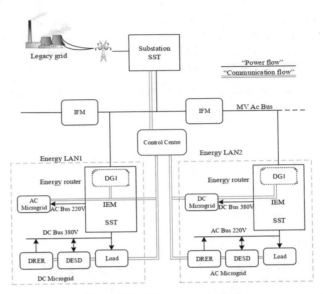

**Fig. 1.** Key Components of IoE

The main element in this system is the "energy router" or Intelligent Energy Management (IEM) block, which includes a unified interface for connection and Distributed Grid Intelligence (DGI) components—a distributed intelligent network. The Intelligent Fault Management (IFM) block isolates the local energy network in case of electrical failure or during scheduled disconnection. IEM and IFM must exchange data through a reliable and secure communication network. The primary functions of the energy router include information processing, electricity dispatching, and voltage transformation. Its role is to enhance the reliability, efficiency, and safety of the energy system, as well as optimize energy usage by balancing supply and demand. The energy router can receive, process, and transmit information about the network's current generating power and demand. According to the IoE concept, energy routers will be placed at various points in the energy network, similar to transformers in today's networks or in consumer metering devices with the capability to connect to intelligent consumption/generation devices. The device is responsible for communicating with numerous distributed sources and loads, as well as for local demand and supply balancing. When local supply exceeds local consumption, the energy router transfers excess energy to the network. The energy router also communicates with other routers for broader demand and supply balancing on a larger scale [9]. In essence, this implementation aims to realize a distributed IoT system.

In creating the IoT system, the use of cloud services and Node-RED is envisioned to facilitate interaction between the main components of the model. This approach will also enhance students' knowledge in the fields of information technology, programming, and telecommunications, providing a foundation for building an IoT system at the technology and protocol levels.

One major challenge that needs to be overcome in creating a realistic simulation environment is the adequacy of time series data forecasting methods. Forecasting will be used to implement data arrays on model components.

The process of analyzing existing methods for time series forecasting involves a sequence of values describing the progression of a process over time, measured at sequential time points, usually at regular intervals. These data characterize the generation of electrical energy. From the most common methods for time series forecasting, autoregression, decision tree methods, and artificial neural networks can be highlighted [4, 5].

Autoregression, like any method, has its advantages and disadvantages in time series forecasting. Traditionally, for electricity consumption forecasting, autoregressive methods are used, based on predicting future values using recent past data. Other factors influencing electricity generation are indirectly considered through their impact on previous consumption values. This approach yields good results under stable conditions when there are no abrupt changes in climate [6]. However, in unexpected and abrupt changes in external parameters, such an approach cannot successfully predict the situation. For example, in sudden severe cold weather, even with an approximate weather forecast, the method is unable to account for this change. Since the task of forecasting electricity consumption involves many factors, and this method becomes unsuitable for this research in case of unexpected and abrupt changes in external parameters.

Another well-known method is the Decision Tree method. If the dependent variable in a time series has discrete values, the decision tree method is used for classification forecasting tasks. In cases where the dependent variable is continuous or constant over a specific time interval, decision tree branches help establish the dependence of this variable on independent variables. This method was first introduced by Holte and Hunt in the late 1950s [4, 5]. In its simplest form, a decision tree represents a hierarchical flowchart structure where rules are expressed as a sequence of "Yes" or "No" answers to various questions.

Recently, experts studying time series forecasting, particularly electricity consumption time series, have increasingly focused on using artificial neural networks. The primary advantage of using artificial neural networks for electricity consumption forecasting lies in the ability to use various input parameters, such as historical data on electricity generation, weather conditions, weather forecasts, time of day, season, type of day, and many others. The impact of each input parameter on the output can be very complex and nonlinear. Artificial Neural Networks (ANN) are analytical systems where formulated tasks may not be very clearly defined. The uncertainty in task formulation is compensated by the ability of ANN to learn and recognize complex patterns in data. An important feature of ANN is their ability to adapt to changes in the external environment and modify their behavior and knowledge.

Research has shown that the most promising approach is the use of neural networks, as electricity generation and consumption depend on many parameters such as working or non-working days, seasons, temperature, weather conditions, and others.

To forecast the level of generation in systems based on renewable energy sources, we will use ANN. The generation of electricity by photovoltaic converters varies significantly depending on the climatic conditions in which the installation operates. The power of photovoltaic panels changes according to the amount of sunlight reaching their working surface. The arrival of solar radiation on Earth varies depending on weather conditions and has pronounced daily and seasonal fluctuations. It is also important to consider that solar elements have a lower limit of sensitivity to illumination below which they cease to produce energy. For silicon crystalline solar modules, this is approximately 150–200 W/m$^2$. For thin-film modules, it is slightly lower, within the range of 100–200 W/m$^2$. The nominal power of solar panels is produced at an illuminance of 1000 W/m$^2$ and a panel temperature of 25 °C [5].

Temperature fluctuations and overheating of solar elements in solar modules adversely affect the functioning of photovoltaic elements on solar panels, as the photoeffect and power generation efficiency decrease.

In the study on creating an artificial neural network for forecasting electricity generation, a feedforward neural network (perceptrons) configuration with open-loop training using the error backpropagation method based on the Nonlinear AutoRegressive with eXogenous inputs (NARX) architecture is employed [5, 10, 11].

Figure 2 provides examples of the standard NARX architecture.

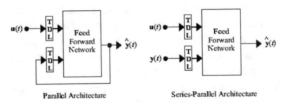

Parallel Architecture          Series-Parallel Architecture

**Fig. 2.** Standard NARX Architecture (1 - Direct access; 2 - Sequential-parallel architecture)

All training occurs in an open loop with a sequentially-parallel architecture. The typical workflow involves creating the network in an open loop, and only after successful training, which includes verification and testing stages, it transforms into a closed loop for multi-stage forecasting (see Fig. 3).

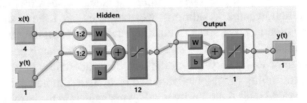

**Fig. 3.** Structure of the open loop in the ANN architecture.

To utilize the formed structures of neural network models, a program was developed in the MATLAB software environment – an artificial neural network with backpropagation, employing the Levenberg-Marquardt training function with data processing organized into 1000 iterations [10] (see Fig. 4).

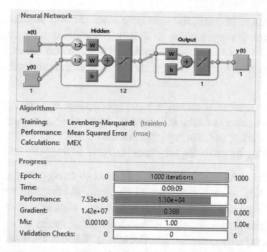

**Fig. 4.** Obtained results of neural network formation after training.

The electricity consumption forecasting model, based on the developed artificial neural network, demonstrates favorable results, especially for average daily values. However, for hourly forecasting, the utilization of neural networks requires modification, specifically employing a dedicated neural network for each hour, as implemented in the subsequent step.

It is noteworthy that the same sequential-parallel architecture of the neural network is utilized for the implementation and construction of electricity consumption forecasts. However, for network training, the dataset from [12] served as the foundation.

The Household Power Consumption dataset is a multivariate time series dataset that describes the electricity consumption for a single household over four years.

The data was collected between December 2006 and November 2010 and observations of power consumption within the household were collected every minute.

It is a multivariate series comprised of seven variables (besides the date and time); they are:

- global_active_power: The total active power consumed by the household (kilowatts).
- global_reactive_power: The total reactive power consumed by the household (kilowatts).
- voltage: Average voltage (volts).
- global_intensity: Average current intensity (amps).
- sub_metering_1: Active energy for kitchen (watt-hours of active energy).
- sub_metering_2: Active energy for laundry (watt-hours of active energy).
- sub_metering_3: Active energy for climate control systems (watt-hours of active energy).

The obtained forecasting results of the neural network coincide by 99% with the values used for network training, serving as reference benchmarks. This holds true for both consumption levels and the forecasted generation levels from renewable energy sources.

With the data arrays in hand, the next step involves implementing a distributed IoT system, whose primary purpose lies in creating a unified information infrastructure. The Information and Communication System (ICS) in the energy sector defines the technical and software tools that facilitate the collection and transmission of information, specifically:

– Microprocessor devices for measurement, processing, and storage.
– Communication channels as the information transmission environment.
– Servers for processing and storing information.
– Software technologies providing information processing, transmission, and storage.

In practice, such a system is implemented based on energy routers, and the main task to be addressed in this work is how to realize physical connections in the information space. Essentially, we have implemented two neural networks that generate arrays of data, which are intended to be transmitted to the energy router through IoT means. The energy router, in turn, will make certain decisions. We can represent the energy router as a system for transmitting information flows. The tool used for this purpose is Node-RED – a programming tool for connecting hardware devices, APIs, and online services. It provides a browser-based editor that simplifies the connection of data streams using a wide range of nodes. Each node has a clearly defined purpose; it receives some data, performs some operation on it, and then passes on this data. The network is responsible for the flow of data between nodes. Node-RED consists of a runtime environment based on Node.js, configured on a web browser for access to the flow editor. For model development, the following packages will be used:

– node-red-contrib-modbus – for working with the data transmission protocol Modbus;
– node-red-dashboard – nodes for building the user interface;
– node-red-contrib-mssql-plus – nodes for connecting to a database MS SQL Server.

Figure 5 depicts a segment of the information flow diagram of the Node-RED system implemented in the Azure cloud environment.

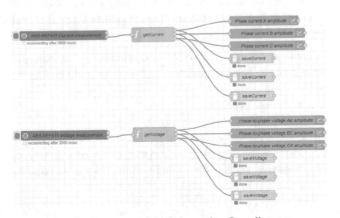

**Fig. 5.** Fragment of the information flow diagram

Figure 6 presents the results of monitoring this system.

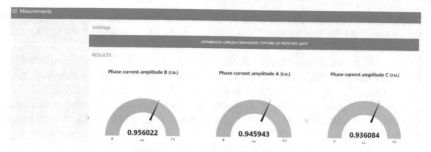

**Fig. 6.** Measurement Dashboard

## 3 Conclusion

The result of this work is a prototype IoT system created in Microsoft Azure using databases, Node-RED, and protocols such as MQTT, HTTP/HTTPS, and others. The developed IoT system should familiarize education recipients with modern examples of building geographically distributed computer-integrated networks for automation and monitoring of control objects using industrial communication networks and Industry 4.0 and IoT technologies. The virtual model of the implemented IoT system represents a power supply network consisting of consumers, power stations, and renewable energy sources. All model components interact in a unified information space. The developed IoT system will allow students to gain practical experience in network technologies, programming, data analysis, and the development of control algorithms and hypotheses testing in a safe simulation environment.

## References

1. Бажан, М. П.: Українська радянська енциклопедія: у 12 томах/голов. редкол. К.: Головна редакція УРЕ, vol. 7, p. 526 (1982)
2. Ядровская, М.В.: Моделирование как метод обучения информационным технологиям. Вестник ДГТУ. **4**(65), 121–128 (2012)
3. Майер, Р.В.: Компьютерное моделирование: учебно-методическое пособие для студентов педагогических вузов. Глазов. гос. пед. ин-т, Глазов (2015). 24,3 Мб. http://maier-rv.glazov.net/Komp_model.htm
4. Стогній, Б.С., Кириленко, О.В., Денисюк, С.П.: Інтелектуальні електричні мережі електроенергетичних систем та їхнє технологічне забезпечення. Техн. Електродинаміка **6**, 44–50 (2010)
5. Автоматизоване проектування гібридних сонячних енергетичних систем (2021). https://dspace.nau.edu.ua/handle/NAU/48977
6. Денисюк, С.П., Бєлоха, Г.С., Чернещук, І.С., Лисий, В.В.: Світові тенденції впровадження відновлюваних джерел енергії та особливості їх реалізації при відновленні економіки України // Енергетика: економіка, технології, екологія, № 4 (2022)
7. Huang, A.Q., et al.: The future renewable electric energy delivery and management (FREEDM) system: the energy internet. Proc. IEEE **99**(1), 133–148 (2011)

8. Chen, D., et al.: Distributed and autonomous control of the FREEDM system: a power electronics based distribution system. In: 2014-40th Annual Conference of the IEEE Industrial Electronics Society, IECON. IEEE (2014)

9. Ступак, М.В., Ступак, Г.В.: КОНЦЕПЦІЯ INTERNET OF ENERGY ДЛЯ УПРАВЛІННЯ ОБ'ЄКТАМИ ЕЛЕКТРОМЕРЕЖ. Науковий вісник ДонНТУ **1**(8)-2(9) (2022). https://doi.org/10.31474/2415-7902-2022-1(8)-2(9)-152-161. ISSN 2415-7902

10. Barrera, J.M., Reina, A., Maté, A., Trujillo, J.C.: Solar energy prediction model based on artificial neural networks and open data. Sustainability **12**, 6915 (2020). https://doi.org/10.3390/su12176915

11. Jlidi, M., et al.: An artificial neural network for solar energy prediction and control using Jaya-SMC. Electronics (2023). https://doi.org/10.3390/electronics12030592

12. Hebrail, G., Berard, A.: Individual household electric power consumption. UCI Machine Learning Repository (2012). https://doi.org/10.24432/C58K54

# Augmented and Virtual Reality

# Effectiveness Study of an Augmented Reality App as Preparation Tool for Electrical Engineering Laboratory Courses

Mesut Alptekin[✉] and Katrin Temmen

Paderborn University, 33098 Paderborn, Germany
{mesut.alptekin,katrin.temmen}@upb.de
https://td.upb.de

**Abstract.** An Augmented Reality Application (AR App) called PEARL has been developed at the Paderborn University to prepare electrical engineering students for the compulsory laboratory work. The App enables students to familiarize themselves with laboratory equipment such as oscilloscope in an intuitive and realistic way, independent of time and location constraints. An effectiveness study with pre and post-test design has been conducted with students from two universities in order to investigate the effects of PEARL concerning the knowledge and motorical skill gain. To do so, the students were divided into two groups, an intervention group (IG) using PEARL and a control group (CG) having only access to classical preparation materials. The results show that both, the IG and CG have growth in all skill levels that were measured by the online test. However, there is no significant difference between IG and CG. This is also true for the questionnaire regarding students anxiety before lab work, experiment-related self-concept and general experimental interest. Both groups have a significant improvement on these affective components at the respective times, but no group difference could be measured.

**Keywords:** Augmented Reality · Laboratory Work · Electrical Engineering · Case Study · Effectiveness Study

## 1 State of Research

Numerous publications highlight a high didactic potential to Augmented Reality (AR) technology in educational context [1], e.g. by improving learning comprehension or spatial imagination. However, there are currently only a few projects investigating this potential regarding preparation for laboratory work. Current projects are limited to linking real preparatory materials with videos and sources to other websites [2], require special technical requirements or can only be used on-site in real laboratories [3]. Furthermore, the quality of AR scenarios as a supplement to a real laboratory, as in [3], cannot be derived on the basis of

M. E. Auer et al. (Eds.): STE 2024, LNNS 1027, pp. 297–304, 2024.
https://doi.org/10.1007/978-3-031-61891-8_29

questionnaires about motivation. Overall, the general potentials have so far only been insufficiently proven. The aim of this paper is therefore to examine several aspects of the effectiveness of AR with regard to preparation for electrical engineering laboratory practicals, in particular the use of laboratory equipment. This includes the different taxonomy levels according to Bloom as well as the influence on the experimental self-concept, the interest in experimental topics and the feeling of being overwhelmed.

## 2    Design

### 2.1    Research Design

The research study consists of a pre-post-test design with a control group comparison as shown in Fig. 1.

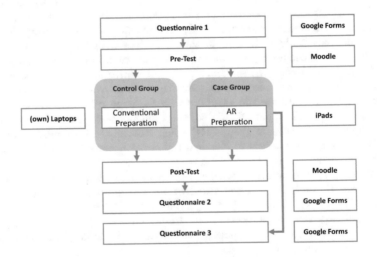

**Fig. 1.** Research Design with pre-post-test design with control group comparison.

The study was conducted at two universities with electrical engineering students in Germany, namely TU Dortmund (conducted at 05.05.2023) and RWTH Aachen (conducted at 16.05.2023). At the beginning, the participating students were asked to complete the online test items in the Moodle platform of the respective university (pre-test). This was followed by an approx. 15-minute online questionnaire. Besides the sociodemographic questions, there were also questions regarding students' previous experience with laboratory equipment and experimentation. Further questions refer to technical and experimental self-concept as well as students' technical and experimental interests. Students were then randomly divided into two groups; an intervention group (IG) and a control group (CG). Both groups were given 30 min to familiarize themselves with oscilloscopes by using the provided materials. After preparation, students had to answer a post-test which is identical to the pre-test. In addition, IG were asked to fill out a second questionnaire after the test, which refers to the user interface (UI) and the perceived satisfaction and user experience (UX) of the App.

## 2.2  Participants

Table 1 represents the number of subjects who participated in the test and surveys from each university and at each time point.

**Table 1.** Number of participants at time points t1 and t2 of the respective university.

|  | TU Dortmund | | | RWTH Aachen | | |
|---|---|---|---|---|---|---|
|  | t1 | t2 | t1 & t2 | t1 | t2 | t1 & t2 |
| Test | 41 | 34 | 31 | 41 | 40 | 39 |
| Questionnaire | 25 | 29 | | 47 | 28 | |
| SUS & UEQ | - | 14 | - | - | 17 | - |

All students who participated in only one test or who could not be assigned due to incorrect/missing anonymous codes were excluded from further investigation. This results in a total of $N = 70$ participants, with $N_{tu} = 31$ from TU Dortmund and $N_{rwth} = 39$ from RWTH Aachen. When subdividing the participants, 28 are allocated to the IG ($N_{IG} = 28$) and 42 to the CG ($N_{CG} = 42$).

# 3  Materials

## 3.1  Measuring Instruments

The Moodle course includes a 21-question test on oscilloscopes, covering general understanding, time base, sensitivity, and trigger functions.

Two similar questionnaires were employed in the study. The first questionnaire, which had to be completed twice by students, focused on socio-demographics, prior laboratory experience, experimental self-concept, general experimental interest, and feelings of overwhelm and anxiety. The second questionnaire, exclusive to the IG, collected data on AR App experience and questions regarding System Usability Scale (SUS) and User Experience Questionnaire (UEQ); both measuring user experience (UX) and user interface (UI).

The detailed analysis of the second questionnaire is reserved for a separate publication, covering overall app evaluation across different versions and time periods.

## 3.2 Intervention

To standardize the study, IG students were equipped with 2018 iPad models running iOS version 14.1 and AR App version 0.17.7.

Students were instructed to bring a laptop/tablet and headphones for CG to use materials provided on Moodle. This included a video, co-developed with E-Labor (students from electrical students council), explaining oscilloscope functions. The user manual, transcriptions, and structured explanations were also provided. Both groups received nearly identical content but on different media.

## 4 Statistical Evaluation

### 4.1 Descriptive Statistics on Subjects

The study focused on second-semester undergraduates with limited laboratory experience. The first section of questionnaire 1 assessed prior experience, revealing that a total of 51 students reported knowledge of laboratory work, either through school (23), study (26) or apprenticeship (2) (see Fig. 2). 26 students stated out that they have already worked with an oscilloscope, of which 2 participants are even very familiar with it (see Fig. 2).

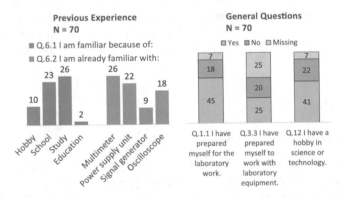

**Fig. 2.** General questions about prior knowledge regarding laboratory work and devices and question with yes/no answers.

Of the 70 participants, 45 prepared for lab work, with 25 specifically preparing for equipment use. Additionally, 41 students reported a hobby in scientific or technical fields, and 13 in the IG had experience with AR Apps (see Fig. 2).

Regarding general preparation for laboratory work during their studies, 40 students stated that they use the provided scripts and 30 also used experimental instructions. Only 8 students prepared specifically for using oscilloscopes; using materials are distributed into simulations (12), internet (12) or with a real device (7) (see Fig. 3).

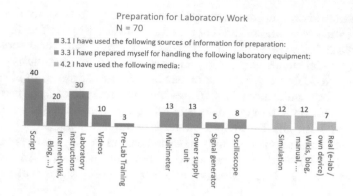

**Fig. 3.** Descriptive statistics: General question about preparation.

## 4.2   General Linear Model (GLM)

**Moodle Test.** From the online Moodle test, results of 58 subjects could be used for further analysis. Of these, 34 participants were in the CG and 24 in the IG.

Figure 4 shows the descriptive statistics consisting of the means and standard deviation for each taxonomy level and each group.

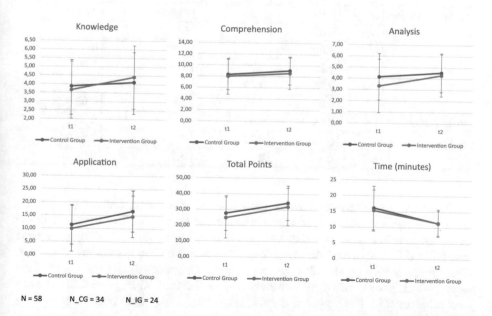

**Fig. 4.** Descriptive statistics: Mean and standard deviation for each taxonomy level, total test result and duration for each group.

The results show improvement in both groups compared to the pre-test, evident in total scores and individual taxonomy levels (Fig. 4). The IG generally performs weaker than the CG in all taxonomy levels and total scores. Both lines show parallel trends, suggesting no observable group differences due to the intervention. Notably, very high performance gains are observed in both groups.

To assess the significance of knowledge increase, further analysis is conducted with tests for within and between subject effects, presented in Tables 2 F-values are compared with the F-distribution table (compare table in [4]), revealing significant effects in taxonomy levels (comprehension (4.85), analysis (4.15), and application (29.93)) and the total score across time points (31.15 and 32.24). The duration for completing Moodle tasks is also highly significant (F-value of 20.71).

**Table 2.** Results for within-subject and between-subject tests for each taxonomy level in Moodle test.

|  |  | Tests of within-subjects effects - sphericity assumed | | Tests of between-subject effects | |
|---|---|---|---|---|---|
|  | Source | Time Point | Time*Group | Constant Term | Group |
| Knowledge | F | 3.97 | 1.08 | 476.35 | 0.01 |
|  | Sig. | 0.05 | 0.3 | 0 | 0.91 |
|  | $\eta^2$ | 0.07 | 0.02 | 0.89 | 0 |
| Comprehension | F | 4.85 | 0.06 | 626.04 | 0.35 |
|  | Sig. | 0.03 | 0.8 | 0 | 0.56 |
|  | $\eta^2$ | 0.08 | 0 | 0.92 | 0.01 |
| Analysis | F | 4.15 | 0.77 | 370.51 | 1.54 |
|  | Sig. | 0.05 | 0.39 | 0 | 0.22 |
|  | $\eta^2$ | 0.07 | 0.01 | 0.87 | 0.03 |
| Application | F | 29.93 | 0.08 | 190.55 | 0.85 |
|  | Sig. | 0 | 0.77 | 0 | 0.36 |
|  | $\eta^2$ | 0.35 | 0 | 0.77 | 0.01 |

However, most taxonomy levels show a small effect size, reflected in partial eta square ($\eta^2 <= 0.10$). But application, total scores, and duration indicate a very large effect size ($>= 0.2$; compared to table in [5]). This suggests that time point differences explain over 30% of the variability in dependent variables. *Time*group* shows no significance, confirming no group difference across time points, which is evident by all F-values (below the critical $F_{1,55} = 4.016$) (compare table in [5]). Both groups significantly improved their performance to a similar degree.

**Questionnaire 1.** Looking at the descriptive statistics for the questionnaire, the mean values for the scale experimental self-concept, general experimental interest, as well as overload/anxiety (see Fig. 5), results in similar interpretation as derived in the Moodle test. Here too, the standard deviation is very high due to the small number of participants and ranges from 0.7 for experimental self-concept and interest to 1 for anxiety.

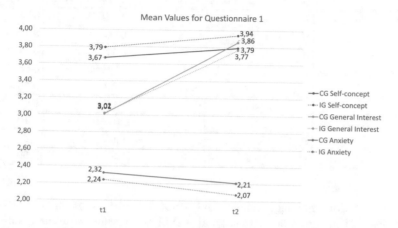

**Fig. 5.** Descriptive statistics: Mean for each scale and for each group in questionnaire 1.

Both, CG and IG show better performance after preparation. That is, the exp. self-concept and the general exp. interest increase for both groups similarly (parallel run of lines), while anxiety decreases for both. The slope, i.e., the rate of change for general exp. interest is strongest, indicating that the intervention independent of material, could increase students interest for laboratory work.

Table 3 shows the tests from the within-subject and between-subject tests for the respective scales of the questionnaire.

It can be seen that all scales have a high F-value ($>=$ 4.016) and thus a very high significance level. That is, the intervention in both groups over the two time points has a significant effect on both, exp. self-concept and general exp. interest, as well as on the students' overload/anxiety level. The effect is medium large for the exp. self-concept ($\eta^2 = 0.068$) and very large for general exp. interest (0.561) and anxiety (0.124). However, similar to test results in Moodle, no difference between the groups (time*group) is evident. Both, the significance and the F-value are very low for group and time point.

## 5   Summary and Further Research

For this study, students from TU Dortmund and RWTH Aachen participated in pre- and post-test assessments to gauge the impact of the AR App on their understanding and skills in using oscilloscopes. Overall, the study indicated that the use of the AR App positively impacted students' knowledge and practical

**Table 3.** Results for within-subject and between-subject tests for the respective scales of the questionnaire.

| | | Tests of within-subjects effects - sphericity assumed | | Tests of between-subject effects | |
|---|---|---|---|---|---|
| | Source | Time Point | Time*Group | Constant Term | Group |
| Exp. Self -concept | F | 4.27 | 0.04 | 1850.54 | 0.58 |
| | Sig. | 0.04 | 0.85 | 0.00 | 0.45 |
| | $\eta^2$ | 0.07 | 0.00 | 0.97 | 0.01 |
| General exp. Interest | F | 75.48 | 0.30 | 2108.83 | 0.08 |
| | Sig. | 0.00 | 0.59 | 0.00 | 0.78 |
| | $\eta^2$ | 0.56 | 0.01 | 0.97 | 0.00 |
| Overload/ Anxiety | F | 8.33 | 0.32 | 309.62 | 0.19 |
| | Sig. | 0.01 | 0.57 | 0.00 | 0.67 |
| | $\eta^2$ | 0.12 | 0.01 | 0.84 | 0.00 |

skills, but it does not significantly outperform typical materials (if they are well-prepared) in terms of knowledge and skill acquisition. Similar results were observable when evaluating the questionnaires; the survey revealed increases in experimental self-concept and general experimental interest among students in both groups and reduces the feeling of overload and anxiety before laboratory work. Future research should focus on long-term retention to determine the lasting knowledge and skills acquired through the AR App over a longer period. More comprehensive studies with larger sample sizes needs to be conducted to better understand the impact of AR technology in laboratory work.

# References

1. Akcayir, M., Akcayir, G.: Advantages and challenges associated with augmented reality for education: a systematic review of the literature. Educ. Res. Rev. **20**, 1–11 (2017)
2. Akcayir, M., Akcayir, G., Pektas, H.M., Ocak, M.A.: Augmented reality in science laboratories: The effects of augmented reality on university students' laboratory skills and attitudes toward science laboratories. Comput. Hum. Behav., Bd. 57, Nr. Supplement C, S. 334–342 (2016)
3. Martín-Gutiérrez, J., Fabiani, P., Benesova, W., Meneses, M.D., Mora, C.E.: Augmented reality to promote collaborative and autonomous learning in higher education. Comput. Hum. Behav. **51**, 752–761 (2015)
4. Lenhard, W., Lenhard, A.: Berechnung von Effektstärken Psychometrica. (2022). https://www.psychometrica.de/effektstaerke.html
5. Stocker, H.: Methoden der Empirischen Wirtschaftsforschung. Fakultät für Volkswirtschaft und Statistik. Universität Innsbruck (2023). https://www.uibk.ac.at/econometrics/einf/tab_f_stat.pdf
6. Alptekin, M., Temmen, K.: Design concept and prototype for an augmented reality based virtual preparation laboratory training in electrical engineering. In: 2018 IEEE Global Engineering Education Conference (EDUCON), pp. 963–968. IEEE (2018)

# Enhancing the Immersion of Augmented Reality Through Haptic Feedback

Joshua Grodotzki(✉) , Benedikt Tobias Müller , and A. Erman Tekkaya

Institute of Forming Technology and Lightweight Components, TU Dortmund
University, 44227 Dortmund, Germany
Joshua.Grodotzki@iul.tu-dortmund.de

**Abstract.** Mobile multi-actoric force feedback gloves (bHaptics Tactgloves) have been coupled with a mobile holographic display device (HoloLens 2). The communication is realized using open programmable interfaces, while the gloves are paired with a laptop. The holographic device, which must be connected to the internet, sends information via a network protocol to the cloud. This information is read by a program on the laptop and converted into actions sent to the gloves. The intensity and duration of the force-feedback can be modulated depending on the action the user is currently performing. Clicking a holographic button triggers a low-intensity, short-duration pulse at the fingertip while moving the entire hand into an area of potential danger causes all fingers and wrist motor to vibrate at maximum intensity as long as the user does not remove the hand from that area. First tests have shown that for novice users and users familiar with Augmented Reality, the additional sensory feedback through haptics offers a significantly higher level of immersion. While previously holograms could only be interacted with by hands without any haptic feedback, they now report that they have a sense of being able to touch and feel the holograms. The feedback gloves are viewed to be bulky and stiff, hence further development is needed on the hardware side, similar to the holographic devices themselves. Overall, connecting force-feedback gloves with Augmented Reality shows large potential in the fields of manufacturing, maintenance and education, as the level of immersion is further increased.

**Keywords:** Haptic Feedback · Augmented Reality · Immersion · Immersivity

## 1 Introduction

In robotics and digital animation, research [1] often refers to the *uncanny valley* when artificial objects are neither too abstract nor realistic enough to please the user and have them naturally enjoy the interaction with these objects. As the level of active participation is drastically increased when switching from robots and animations to user-centered extended reality applications (i.e. Augmented, Virtual & Mixed Reality), the perceived level of immersion becomes a decisive factor for widespread dissemination of such technologies. Immersion (or also immersivity) in this case refers the quality or effectiveness of a hardware-software combination, for instance a learning application

M. E. Auer et al. (Eds.): STE 2024, LNNS 1027, pp. 305–312, 2024.
https://doi.org/10.1007/978-3-031-61891-8_30

running on an augmented reality device, to be perceived as immersive by the user. In such a case, the user does not think about using the combination as such but focuses on the task at hand [2]. In simple terms, if users naturally immerse in the application without complaining about the equipment or the quality of the application itself, it is referred to as a highly immersive experience.

As immersivity is improved with incorporating multi-sensorial interaction forms [3, 4], research on enhancing the immersivity in extended reality has often focused on adding haptic feedback. Most virtual and augmented reality devices provide a sufficiently high quality of visual and acoustic feedback, but naturally lack the ability to provide haptic feedback. In some cases, especially for virtual reality, the user controls the environment via game controllers. Those may have vibration feedback included, which is known from smartphones vibrating.

A more elaborate form of haptic feedback is achieved using gloves which incorporate multiple actuators. Enhancing virtual training in construction and safety training was found to benefit from such a combination [5]. When it comes to the acquisition of psycho-motoric skills, as it is required in surgery, through virtual reality training, haptic feedback can improve results, especially in early stages of the learning process [6]. If the hands must remain free of actuator devices, the feedback can be induced in different positions of the body, such as the wrist [7] or the cubital joint [8], through which more complex feedback situations, such as lifting virtual weights, are enabled. Regarding the immersivity of virtual reality in general, good visual and aligned haptic feedback offer a greater sense of being immersed in the application than any of the two modes alone [9].

In augmented reality (AR), the great benefit over virtual reality is to interact with the actual environment through transparent displays. Additionally, latest generations of AR devices, such as the Microsoft HoloLens 2, offer a controller-free interaction with the holograms, which is found to be advantageous for learning [10]. First steps towards combining AR with haptic or force feedback were done as early as 1999, where a static feedback device was coupled with a static AR device [11]. Ai et al. [12] developed a new approach for the force actuation by using microfluidic actuators, which offers nuanced feedback based on the current action performed by the hands.

In a more general survey, Bermejo and Hui [13] explored the potential of adding haptic feedback to AR. Even though they attribute many benefits to such a combination, they conclude that no affordable and portable solution exists despite multiple available AR and haptic devices. Here, this work aims to close the gap.

## 2   Aim

The goal of this work is to establish a robust communication between a mobile holographic device (Microsoft HoloLens 2) and lightweight multi-actoric force feedback gloves (bHaptics Tactgloves). As this enlarges the commonly known level of immersion of Augmented Reality by adding haptical feedback, a training application will be developed, in which users can familiarize with the additional sensory information. Secondly, an established AR-application [10] will be enhanced by the new feedback system. Both applications will be tested with users of different ages and background in a first survey to investigate the effect of haptic feedback in comparison to conventional audio-visual feedback. Recommendations will be derived, where such a novel combination of AR and haptic feedback can further improve the perceived immersion of AR in manufacturing training and education.

## 3 Implementation

As haptic feedback technology for augmented and virtual reality is still in its infancy, widespread support for an easy connection of the two systems is still lacking. To circumvent device incompatibility, a bridge was developed to offer an open API which can be used by a wide range of devices. The gloves, here the multi-actoric force feedback gloves Tactgloves from bHapctics, must be paired with a Windows computer (via Bluetooth), preferably a laptop to maintain mobility of the proposed solution. However, the laptop only runs a small receiver program to relay any desired action from the user's device to the SDK of the gloves. The main holographic mobile device (Microsoft HoloLens 2), which shall be in communication with the gloves, sends information packages via MQTT, which is a robust and reliable industry network protocol, to a broker. The information must be formatted according to the glove's specifications, to access all functionalities correctly. The connection between HoloLens and laptop can be established either in a peer-to-peer fashion (via local WiFi or Bluetooth) or via a cloud-based approach through the internet. The latter method has the additional benefit that no additional configuration or pairing between the receiver and sender is required, as both simply connect to the same known MQTT broker. This method is also able to extend the interaction to different types of force-feedback devices, such as vests, wristbands or even mock-tools.

The bHaptics Tactgloves provide a total of six induvial actuators, i.e. force-feedback motors creating a vibrating sensation, per hand, see Fig. 1. One is located on the outside of the wrist and is intended to function as a more general feedback system. The others are located on the inside of the fingertips and are triggered when the corresponding fingers engage in an action that causes the actuators to be activated. Each actuator's vibration can be modulated in terms of duration and intensity. The minimal pulse duration is in the range of a few milliseconds whereas the maximum is a continuous vibration. The intensity can be varied from 1% to 100% while the maximum is defined by the manufacturer. The slim design of the gloves enables the correct tracking by the HoloLens, necessary for an immersive controller-free experience.

**Fig. 1.** User's perspective on the force-feedback gloves as viewed through the HoloLens. The bHaptics gloves contain a wrist actuator on the outside and for each fingertip one on the inside.

## 4 Application for Training Purposes

As the haptic feedback presents a new addition to the other built-in types of interaction modes (audio and visual), users need to familiarize themselves with the gloves as a device and how to use them. The implementation was done in such a way that the users do not need to change their behavior compared to previous usages of the HoloLens 2. This plays an important role in the perceived immersivity of the gloves themselves. Nevertheless, before starting with actual tasks that now might include haptic feedback as a response, the technology as such should be understood in a controlled training environment. To this end, a training application was designed and tested with 20 people of different ages and backgrounds. Most of them were familiar with operating the controller-free HoloLens 2 based on their direct hand interaction. Figure 2 provides an overview of the training application with all its features.

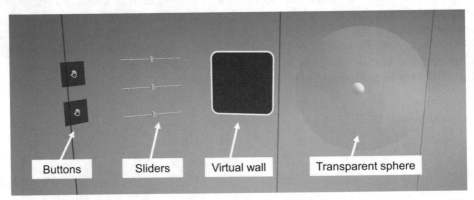

**Fig. 2.** Overview of the training app including the features: buttons, sliders, a virtual wall for understanding dangerous zones and a transparent sphere for haptic-focused feedback.

The app comprises objects such as buttons, sliders, a virtual wall and a grabbing task. In the case of the buttons (Fig. 3a), an interaction is commonly accompanied by a visual 3D highlighting and movement of the button, similar to a real movable button being pushed down. An additional *click* sound typically confirms the correct interaction with the button. In the new version of button interactions, there is a short 20 ms pulse to the fingertip which triggered the button. All nine participants confirmed that a single short pulse feels like the best way to support the cognition of interacting with the button. Further, all agreed that this feature presents a higher level of immersion compared to an interaction without haptic feedback. The lower button in Fig. 3a provides no acoustic feedback, only visual and haptic. Here, the majority agreed that the combination of all three types of feedback, visual, haptic and acoustic, provides the best experience.

For sliders, the configurations (from top to bottom in Fig. 3b) continuously increasing haptic feedback with sound, haptic feedback at the discrete sections of the slider with sound, and discrete haptic feedback without sound were compared. Most of the participants (7 out of 9) stated that, as before, the combination of sound and haptics results in the best perception of the interaction. However, sound and haptics should align, as it was in case two, since the sound appeared only when the slider was in one of the discrete positions.

Lastly the effect of all actuators being active at the same time as a means to discomfort people was tested via the virtual wall. In the beginning the wall appears entirely transparent (in Fig. 2. it is artificially colored black for better visibility). Once the user moves the hand through the frame it turns full red (Fig. 3c). Additionally, the wrist and all finger actuators vibrate continuously at maximum intensity as long as the hand is penetrating the plane of the frame. Compared to previous studies [10], where only a visual-acoustic deterrence was implemented, the majority of the participants, who were also part of the previous study, showed a much stronger and more consistent response: they pulled their hand out of the dangerous area. As this was the intended reaction, it proves that high intensity vibrations can severely support the learning process regarding potential dangers. The participants reported unanimously that the sensation was highly deterring but by no means hurtful.

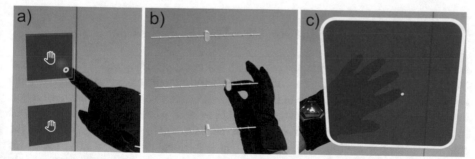

**Fig. 3.** Visualization of the multi-modal interaction with the different features: a) button is highlighted, b) sliders change color and follow either discrete or continuous values and c) visual demonstration of the virtual wall upon full penetration.

Haptic feedback can also provide beneficial information when, for instance, visual feedback is not sufficiently unambiguous. To demonstrate this phenomenon, the fourth feature consists of a ball inside a highly transparent sphere, see Fig. 4. The task here is to grab the ball inside the sphere and place it as close as possible to the surface of the sphere without penetrating it. Without any acoustic or haptic feedback, this is a challenging task, as the visual representation of the transparent sphere might be deceiving, compare Fig. 4a) and b). Especially when the ball shall be placed away from the user, it becomes even more difficult. A new modulated haptic feedback feature was hence implemented which provides the user with information about the ball's position relative to the center of the sphere. The two fingertips involved in grabbing the ball vibrate continuously at highest intensity at the center and with no intensity upon touching the surface. When comparing the users' performance with and without active haptic feedback, a significant improvement was found in the initial studies. The users reported that they have a good feeling about the ball's position inside the sphere even though they would be unsure about it solely based on visual information.

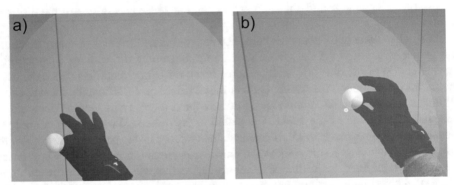

**Fig. 4.** When the ball inside the sphere is moved, the haptic signal to the fingers grabbing the ball is modulated in intensity as a function of the ball's distance to the sphere's center.

## 5 Forming Press Application with Haptic Extension

An existing application for the analysis of an extrusion process, which had been thoroughly investigated in [10], was enhanced by haptic feedback based on the insights gathered from the studies of the training app. Several features were implemented, including slider-based process control and full haptic feedback when reaching into dangerous areas, both Fig. 5. The findings from the previous investigations were confirmed by the analysis of this application. Where adding haptic feedback was found to be useful for the majority of interactions, such as control sliders and disassembling the machine into its individual components, the most significant improvement was achieved in the field of security.

The app includes, similar to the virtual wall in the training app, a holographic shield (Fig. 5b) which appears in red accompanied by a loud warning sound upon touching. In the updated version of the app, all six actuators gradually increase in intensity when moving the hand close to this dangerous zone of the machine. In this way, the haptic

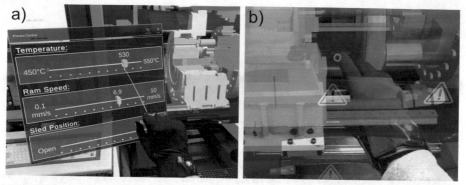

**Fig. 5.** Extrusion press application with additional haptic feedback features. In a) the new slider-based process control window is visible and b) shows the audio-visual warning feedback now coupled with intense haptic response.

feedback precedes the audio-visual warning and improves the learning about potentially hazardous areas.

## 6   Conclusion and Outlook

A robust, wireless communication between the holographic mobile device HoloLens 2 and the multi-actoric force feedback gloves bHaptics Tactgloves has been success-fully established. Through the MQTT protocol and a bridge, information is exchanged between the two devices. A training application with four different features has been developed, such that users can learn to use and work with additional haptic feedback in augmented reality environments. Initial studies among 20 users of different ages and backgrounds showed that in all cases, the additional haptic feedback improved the level of immersion. For simple clicks, only single pulses of short duration (20 ms) should be used. In cases of warnings, continuous maximum intensity vibration can actually make the user stop the motion which triggered the haptic warning. It works significantly better compared to conventional audio-visual warnings. This knowledge was then transferred to an existing manufacturing-related application where the findings were confirmed.

Other more advanced and costly haptic gloves, such as the SenseGlove Nova, include Bowden cables that can restrict the actual movement of the fingers. This can create the additional sensation of grabbing objects. Future research has to investigate if such bulky and cumbersome gloves can be easily connected with the HoloLens and what their benefits and drawbacks regarding educational training are.

**Acknowledgements.** This research was funded through the research and development project KORESIL (German abbreviation for: Concepts for the resource-efficient and safe production of lightweight structures) which is funded by the German Federal Ministry of Education and Research (BMBF) (project number 02P20Z004) and supervised by the Project Management Agency Karlsruhe (PTKA).

# References

1. Mori, M.: The uncanny valley. Energy **7**(4), 33–35 (1970)
2. Ermi, L., Mäyrä, F.: Fundamental components of the gameplay experience: analysing immersion. In: DiGRA Conference proceedings (2005)
3. Brown, E., Cairns, P.: A grounded investigation of game immersion. In: Extended Abstracts on Human Factors in Computing Systems (CHI EA '04). Association for Computing Machinery, New York, NY, USA, pp. 1297–1300 (2004)
4. Bowman, D.A., McMahan, R.P.: Virtual reality: how much immersion is enough? Computer **40**(7), 36–43 (2007)
5. Patil, K., Bhandari, S., Ayer, S.K., Hallowell, M.: Potential for virtual reality and haptic feedback to enhance learning outcomes among construction workers. In: Proceedings of the 18th International Conference on Construction Applications of Virtual Reality (CONVR2018) (2008)
6. van der Meijden, O.A.J., Schijven, M.P.: The value of haptic feedback in conventional and robot-assisted minimal invasive surgery and virtual reality training: a current review. Surg. Endosc. **23**, 1180–1190 (2009)
7. Pezent, E., et al.: Explorations of wrist haptic feedback for AR/VR interactions with tasbi. In: UIST '22 Adjunct. Association for Computing Machinery, New York, NY, USA, Article 79, pp. 1–5 (2022)
8. Suzuki, R., Egawa, M., Yamada, Y., Nakamura, T.: Development of a 1-DOF wearable force feedback device with soft actuators and comparative evaluation of the actual objects and virtual objects in the AR space. In: 2016 14th International Conference on Control, Automation, Robotics and Vision (ICARCV), Phuket, Thailand, pp. 1–6 (2016)
9. Gibbs, J.K., Gillies, M., Pan, X.: A comparison of the effects of haptic and visual feedback on presence in virtual reality. Int. J. Hum. Comput. Stud. **157**, 102717 (2022)
10. Grodotzki, J., Müller, B.T., Tekkaya, A.E.: Enhancing manufacturing education based on controller-free augmented reality learning. Manuf. Lett. **35**(Supplement), 1246–1254 (2023)
11. Vallino, J., Brown, C.: Haptics in augmented reality. In: Proceedings IEEE International Conference on Multimedia Computing and Systems, Florence, Italy, pp. 195–200 (1999)
12. Ai, Q., He, X.: ILab-Gloves–design of AR experimental gloves based on ergonomics and force feedback technology. In: Stephanidis, C., Antona, M., Ntoa, S., Salvendy, G. (eds.) HCI International 2023 – Late Breaking Posters, vol. 1958, pp. 191–197. Springer, Cham. (2024). https://doi.org/10.1007/978-3-031-49215-0_23
13. Bermejo, C., Hui, P.: A survey on haptic technologies for mobile augmented reality. ACM Comput. Surv. **54**(9), 184 (2022)

# Augmented Reality Based Control of Autonomous Mobile Robots

Benedikt Tobias Müller[✉][iD], Joshua Grodotzki[iD], and A. Erman Tekkaya[iD]

Institute of Forming Technology and Lightweight Components,
TU Dortmund University, 44227 Dortmund, Germany
benedikttobias.mueller@tu-dortmund.de

**Abstract.** A new approach to control and interact with autonomous mobile robots via a head-mounted Augmented Reality device is presented. Such robots are often used for highly repetitive tasks, however, in small and medium-sized enterprises, tasks are often non-repetitive. In such cases, the programming of mobile robots is neither time nor cost-efficient. Hence, a direct human-machine interaction was developed, connecting an Omron LD-90 with a Microsoft HoloLens 2 via a cloud-based communication. Users can send inputs to the mobile robot using voice commands or select from a pre-defined set of actions. The robot will execute the task and send feedback to the HoloLens about its position and other status information. The potential of this new communication approach is exemplified using a small shop floor where a forming machine operator has to fulfill a wide range on tasks, where the mobile robot can support the worker.

**Keywords:** Augmented Reality · Autonomous Mobile Robots · Human Machine Interaction · Cloud Communication · Simultaneous Localization And Mapping

## 1 Introduction

Amid an increasing integration of robotics into many industrial fields and the recent rise of cobots (collaborative robots that can safely work with and directly next to humans), the combination with extended reality (XR) technologies offers new innovative opportunities. Currently, mobile robots and extended reality are of limited use for smaller and medium-sized enterprises (SME) as integrating them into the current workflow can be tedious, cost-intense and may require major adjustments to their existing hardware and software infrastructure. Consequently, mobile robots are most commonly found in large warehouses and logistic companies and less in SMEs.

The market for autonomous mobile robots (AMR) has been steadily increasing and is expected to increase even faster in the following years [7]. The share of cobots in this market has also been ever increasing [14]. With the recent release of Spot by Boston Dynamics in 2020, mobile robots are now also capable of navigating through rough terrain. Although mobile robots may already be widely in

M. E. Auer et al. (Eds.): STE 2024, LNNS 1027, pp. 313–323, 2024.
https://doi.org/10.1007/978-3-031-61891-8_31

use by major companies, they are most often used to do pre-planned actions or work only in company with other robots. Robots equipped with SLAM (simultaneous localisation and mapping) technology, which enables continuous tracking and localisation in a dynamically changing environment, are theoretically suitable for use in SMEs. In such a case, the tasks are often one-off operations, making pre-planning inappropriate or uneconomical.

The whole field of extended reality, and in particular augmented reality, is still an emerging technology that is starting to enter the mainstream audience with the upcoming release of the Apple Vision Pro and recent cheaper alternatives such as the Meta Quest 3. Whereas virtual reality is common for private use, such as gaming, augmented reality is already being used more widely in engineering, manufacturing and construction industries aswell as healthcare, with the Microsoft HoloLens 2 being primarily used in industry applications.

Combinations of mixed reality with robotics were previously already investigated in multiple fields. For example a way to use mixed reality and robotics to aid in search and rescue missions was developed in [4]. It uses mixed reality to control a robot in dangerous terrain remotely to keep the user safe. Other implementations researched the use for SLAM-visualization and mapping [8], through displaying detailed sensor data, such as a LIDAR (laser based 3D mapping of the environment), as geometry overlayed in mixed reality. Other research looked into the concept of using mixed reality as a tool for training with cobots virtually [13]. Here the user can train on a digital twin without requiring access to the physical machine. A different group of researches described a way to utilize virtual reality to do safe pre-testing of potential cobot working spaces [1]. This concept aims to avoid any human casualties during testing, which could be caused by testing with robots and humans directly. In the fields of stationary robots and cobots the integration of AR is already studied. In a literature review that analyzed the current state of AR in combination with cobots, positive effects for the worker, such as performance and task awareness were identified [3]. Additional applications have been discussed to increase user safety by visualizing safe areas of robots [2].

However, what is largely missing here is a concept that aims to provide support and help to workers, for example in smaller and medium-sized enterprises, by combining mixed reality and cobotic in a way that puts the cobot in a companion role, aswell as keeping usability and ease of use in mind.

After introducing the aim of the paper in the following chapter, Sect. 3 highlights the specifics of all the technologies involved in this solution of the stated problem. Section 4 discusses the required setup and the application itself. This is followed by a demonstration of a potential use case for this technology combination, before concluding with a summary and an outlook.

## 2   Concept and Aim

The aim of this paper is to demonstrate the integration of augmented reality with a mobile robot in the workplace as a human-robot-companionship using current state of the art hardware and to show the potential for future innovations in this field. This is being done under the assumption that some sort of extended reality, may it be in the form of smart glasses like the HoloLens 2, smartphones, smartwatches or other devices will be used in industrial environments. Users should be able to work with the robot companion in a way that helps them focus on the work that requires their specific domain knowledge and abilities, while alleviating any additional cumbersome tasks, such as material or tool transport. One key aspect of this research is the multi-modal input capability to enable a greater area of application. For example, when a worker has no free hands to control any device, he or she might have to rely on voice control. Based on these key points a few key features are determined to be required for a successful integration of mobile robots into SMEs. Users have to be able to call the robot to their current location, see where the robots current location is and send it to a specific location. Additionally, the ability of the robot to directly follow the user wherever he or she needs to go can be helpful in the way of a moving workbench which carries tools, material or similar. The clarity of the feedback that is given to the user is of critical importance during all of these tasks to allow the worker to keep focus on the work and not actively monitor the actions of the robot companion. An overview of the proposed concept is presented in Fig. 1.

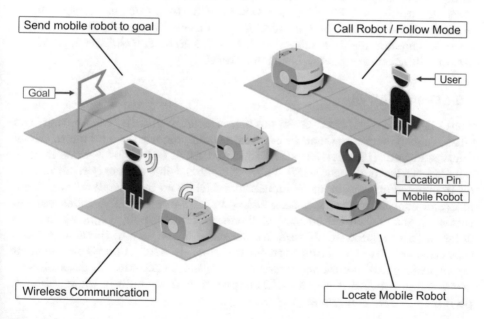

**Fig. 1.** A representation of the desired interaction capabilities of the proposed human-robot-companionship concept

## 3   Devices and Technical Concepts

In the following, definitions and explanations regarding the key technologies that are used in this work are given. Whereas SLAM simply describes a concept that is used in a wide range of products, the Omron LD-90 and the Microsoft HoloLens 2 are merely used as an example of the current state of technology in the respective fields and can be exchanged for any similar or upcoming product. This underlines the intended flexibility and general applicability of the presented approach.

### 3.1   SLAM

Simultaneuous localization and mapping (SLAM) is often found in robotics and describes the concept of creating a map of the surrounding area (mapping) and localizing within that map (localization) at the same time [5]. This is achieved using built-in sensors like cameras, LIDAR (i.e. laser-based distance scanning) or ultrasonic. SLAM is the de-facto standard in mobile robots as it requires no external equipment or setup to be performed. Another field that benefits from SLAM is the field of augmented and virtual reality. Many current virtual reality headsets switched from outside-in tracking, like it was found on the first consumer products HTC Vive and Oculus Rift, to inside-out tracking using cameras. Same goes for the upcoming Apple Vision Pro, mobile devices and all major augmented reality headsets, such as the Microsoft HoloLens 2, which all use, or will use once released, inside-out tracking.

### 3.2   Omron LD-90

The LD-series by Omron is a product range of mobile robots that can autonomously perform assigned tasks while complying with several international safety standards (ISO EN1525, JIS D6802 and ANSI B56.5) for working in close proximity to (several) people [12]. These standards define the safety requirements for unmanned and autonomous industrial vehicles and include all relevant technological aspects to ensure worker safety in all situations. The robot used in this work, the Omron LD-90, is a small chassis model capable of payloads up to 90 kg and a top speed of 1.35 m/s. Autonomous in the case of the mobile robot refers to its ability to determine the optimal route to a target on its own without pre-planned paths and taking dynamic obstacles into account. To determine its position and rotation it uses a SLAM approach with its built-in sonar, LIDAR, position encoders and gyroscope.

### 3.3   Microsoft HoloLens 2

The Microsoft HoloLens 2 is the newest augmented reality headset by Microsoft. The HoloLens features a see-through display that is capable of placing holograms in the real world around the user [11]. Using its cameras, depth sensor, accelerometer, gyroscope and magnetometer it uses SLAM to determine its position and rotation in space as well as creating a map of the surrounding and previously seen areas [10]. Since its launch, Microsoft has released multiple different editions for different work environments, such as the Industrial Edition for clean rooms and hazardous locations (HAZLOC) environments as well as the Trimble XR10 edition for construction sites with an included hard hat [9]. Based on previous research, common negative side effects that might affect long tearm wearing the HoloLens such as motion sickness were not observed while using the HoloLens [6]. In that study, only a handful of participants noted some degree of eye strain after prolonged continuous use, which can be mitigated by the design of the app or by simply flipping up the HoloLens visor when not in use.

## 4    Interaction Concept and Implementation

On the hardware side only the mobile robot itself, a small and inexpensive micro-controller aswell as an end-user device, in this case the HoloLens 2, is needed. The prerequisites and required steps to setup and configure the developed system are laid out in Subsect. 4.1. The following section takes an in-depth look at the resulting application, which was made with Unity, and its use cases.

### 4.1    Setup

Before the system can be used, a couple small tasks have to be completed to properly configure all devices. The autonomous mobile robot requires an initial walk through the facility and all areas in which it should be used in the future to create a base map. This process has to be done only once after unboxing and can be done either with the included controller or a standard tablet. This map in its current form is not in an optimal format for the use with the HoloLens as it is in a specialized format which is tailored to the Omron LD-90. As such, the map file contains a lot of unnecessary information for the intended use case since it stores the main map data as a point cloud. However, on the HoloLens only an image of the map with basic additional information regarding scale and position offset are required. Therefore, a post-processing pipeline was developed that takes the map file and extracts all required information. In this specific case, the developed converter extracts key parameters of the map and creates a reduced visual representation for human use. To create the connection between the two separate SLAM-based maps of the mobile robot and the HoloLens multiple real world features must be known in both systems. For this use case, visual tracking markers, in the form of a special variant of QR-Codes, were identified to deliver the most reliable results. To offer a consistent experience with a known good marker type and design, an app has been developed to create the desired amount of tracking markers with a consistent design and tracking quality.

As the ideal location for those markers are corners on walls as seen in Fig. 2, cabinets or pillars multiple variants per marker are created as such that regardless of the position and available space a marker can be placed and easily identified on the map of the mobile robot. With just a couple of those markers

Fig. 2. Two examples of tracking markers in an industrial environment

scattered around the area on e.g. a wall, fence post or floor, the HoloLens app can determine not only its position in space relative to its starting position, as it is usually the case, but determine its position in the same coordinate system as the robot with a high degree of certainty and accuracy. Finally, predetermined target locations need to be defined, which later will be available as shortcuts in the app. Once all this data is gathered the resulting map configuration is done and ready for use by the HoloLens application.

### 4.2   HoloLens Application

The HoloLens application is developed to be as user-friendly and intuitive as possible. As such it was designed with the newest design language and guiding principles that are offered through MRTK3[1] based on Microsoft research. An additional benefit is that the app itself and its interaction experience feels familiar as first party Microsoft offerings such as Dynamics 365 Guides. The app has been specifically designed and developed to support several maps at once, for instance for multiple shop floors. The correct map file is then loaded seamlessly upon detecting the first tracking marker without any required user input. It is important to stress though that the tracking markers do not need to be constantly in the field of view. They merely serve as an initial help during the interlinking of the map of the HoloLens and the mobile robot.

The main user interface is divided into two interaction modes. The first means of interaction is the so called hand menu. Once the worker looks at the palm of one of his or her hands, a holographic menu appears next to it and offers different sets of actions. A depiction of the hand menu and the mobile robot can be seen in Fig. 3. Each action in the hand menu is highlighted and named in the tooltip. Each of those actions will be described and explained in the following section.

---

[1] Third version of the Mixed Reality Toolkit, which is an open-source software development kit (SDK) for mixed reality developments in Unity.

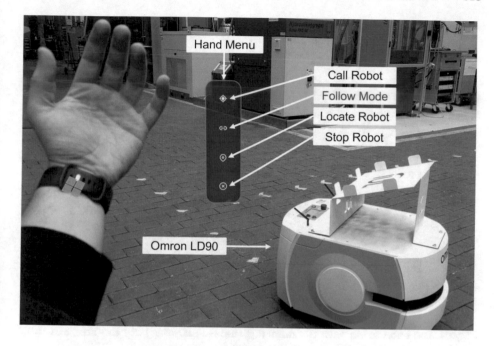

**Fig. 3.** A depiction of the hand menu

## Call Robot

The mobile robot will traverse to the current position of the user. The initial target position is determined by the orientation of the user as such that the robot will present itself in front of the user with easy access to its storage compartment. The final position is visually highlighted and demonstrated using a life size digital twin of the robot. Shall the user desire a different position and / or orientation he or she can simply grab the position indicator and move it to the desired spot using the hand ray feature of the HoloLens. An example of this can be seen in Fig. 5b.

## Follow Mode

The mobile robot will follow the user even with other humans around (Fig. 5a). To initiate the following process the user must step into the displayed area in front of the robot. Once the mobile robot has identified the user and is ready to follow, the user is notified. This process usually takes just a few seconds once the user is roughly in the correct area.

## Locate Robot

The app also provides multiple means of locating the robot shall it ever be desired to walk to the robot and not the other way around. After activating the "where is my robot" mode, a minimap, as seen in Fig. 4, shows up in the hand menu offering two sets of features. On the one hand, the current position of the user as well as the robot is marked on the map. On the other hand predetermined targets, e.g. often used areas, are visible as flags.

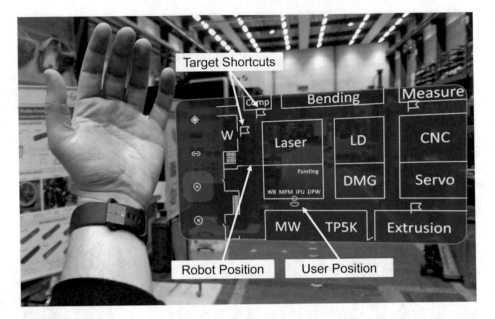

**Fig. 4.** The hand menu with the minimap enabled showing both the user and robot location aswell as shortcuts to predefined targets

Once the user clicks on any of those targets, the mobile robot will start moving towards said target. Upon sending the mobile robot onto a mission to a remote location, a flag indicating its target location will be shown in the real world as a hologram. An example of this with activated robot location features is presented in Fig. 5c. The current status can also be followed over the minimap as it updates in real time. Shall the user decide to walk towards the robot and not be too familiar with the surrounding area, as such that a quick glimpse at the minimap is enough to find the path to the robot, the user is accompanied by a type of indoor navigation system. The mobile robot position is highlighted through walls, see Fig. 5c, and a directional arrow in front of the user guides him towards the mobile robot until the robot is within the field of sight, as shown in Fig. 5d.

**Stop Robot**

This button will halt any action that the robot is currently executing. For instance if the mobile robot should stop midway on its way to a target or stop following the user without doing anything else it can be clicked.

The buttons can either be clicked directly using the other hand, or be used leveraging the HoloLens eye tracking capability. If the user looks directly at a button and focused on it for atleast a specific time, e.g. 2 s, the button action is triggered. This process is accompanied by visual and acoustic feedback to make this action transparent and comprehensible. Additionally, all major functions can also be triggered directly using voice recognition.

(a) Follow Mode

(b) Call Robot

(c) Locate Robot & Target Visualization

(d) Navigate to Robot

**Fig. 5.** Visual representation of multiple different modes of the HoloLens app

Once an action has be activated its execution is confirmed by the HoloLens via audio feedback and if applicable by visual animations just like any robot action that is beneficial for the user. Multiple examples of the in-app visualization can be seen in Fig. 5. The voice feedback includes, but is not limited to:

- "Robot is on its way to your location"
- "Robot is on its way to [target name]"
- "Robot is ready to follow you! Please stand in the highlighted area."
- "Robot is now following you"

A preliminary user study has been carried out based on a demonstration application, which was developed to showcase the devised concept. The user group consisted of six engineers and technicians. Initial tests have already been met with positive feedback. The required familiarization phase in these tests took on average a couple minutes (1–4 min) for the participants to understand the system, its possibilities and take control of the mobile robot. For the majority (5) of the participants, the menu layout and style as well as the voice commands felt intuitive and did not require much explanation. The concept in general was received with great interest.

## 5     Potential Use Case Demonstration

To demonstrate the capabilities of the developed system, this section will portray a demonstration of a potential use case. In this example, an extrusion press of a medium sized enterprise recently broke down and needs urgent repairs. The assigned worker, who is currently working on the machine, previously already send the mobile robot to get some additionally required heavy tools for the repair. Upon discovering that a special replacement part is needed, which can be 3D-printed in-house, the worker tasks the mobile robot with retrieving this part once ready and delivers it to the post-processing area for further adjustments. In the meantime, the worker uses the HoloLens remote assist capabilities to consult an external expert on a different part of the repair process. Once the part is finished the robot, delivers it to the worker at the press. After completing the repair, the worker loads his heavy tools and broken parts onto the mobile robot and walks towards the machine shop, with the robot following him, to put back all the tools and dispose of the broken parts.

## 6     Conclusion and Outlook

A concept for augmented reality based control of autonomous mobile robots was devised and developed. This proposed human-robot-companionship aims to show new human-machine-interaction possibilities of the future, in particular for smaller and medium-sized enterprises. Even given currently available hardware, a possible increase in efficiency and savings (both monetary and time) seems possible. The robot-companion is able to help workers in the field to focus more on their main task and do less courier tasks or heavy lifting over longer periods of time. The developed system can be setup and configured in less than a day and requires no technical knowledge. The results of the first user study indicate an overall positive response regarding the concept and its implementation.

Despite the implementation in this paper being specifically designed for the Microsoft HoloLens 2 and the Omron LD-series, the main concept can be used in multiple different scenarios involving other cobots, such as robot arms. With an increase in AI research and large language models, further advancements in terms of natural interaction between human and machine will be possible. This can help provide the cobot with the ability to directly understand any type of user-request, navigate any surrounding area and executing any desired task without pre-planning.

**Acknowledgements.** This research was funded through the research and development project KORESIL (German abbreviation for: Concepts for the resource-efficient and safe production of lightweight structures) which is funded by the German Federal Ministry of Education and Research (BMBF) (project number 02P20Z004) and supervised by the Project Management Agency Karlsruhe (PTKA).

# References

1. Badia, S.B.I., et al.: Virtual reality for safe testing and development in collaborative robotics: challenges and perspectives. Electronics **11**(11), 1726 (2022)
2. Cogurcu, Y.E., Douthwaite, J.A., Maddock, S.: A comparative study of safety zone visualisations for virtual and physical robot arms using augmented reality. Computers 12(4) (2023)
3. Costa, G.D.M., Petry, M.R., Moreira, A.P.: Augmented reality for human-robot collaboration and cooperation in industrial applications: a systematic literature review. Sensors 22(7) (2022)
4. Cruz Ulloa, C., del Cerro, J., Barrientos, A.: Mixed-reality for quadruped-robotic guidance in SAR tasks. J. Comput. Des. Eng. **10**(4), 1479–1489 (2023)
5. Durrant-Whyte, H., Bailey, T.: Simultaneous localization and mapping: part i. IEEE Robot. Autom. Mag. **13**(2), 99–110 (2006)
6. Grodotzki, J., Müller, B.T., Tekkaya, A.E.: Enhancing manufacturing education based on controller-free augmented reality learning. Manuf. Lett. **35**, 1246–1254 (2023)
7. Inkwood Research: Size of the global market for autonomous mobile robots (amr) from 2016 to 2021, with a forecast through 2028 (2022). https://www.statista.com/statistics/1285835/worldwide-autonomous-robots-market-size/. Accessed 25 Nov 2023
8. Leber, P., Dalm, K.: Using MR to interact with a mobile robot based on ROS. In: Proceedings of the 12th Conference on Learning Factories (CLF 2022) (2022)
9. Microsoft: HoloLens 2 editions — Microsoft Learn. https://learn.microsoft.com/en-us/hololens/hololens2-options. Accessed 25 Nov 2023
10. Microsoft: HoloLens 2 hardware — Microsoft Learn. https://learn.microsoft.com/en-us/hololens/hololens2-hardware Accessed 25 Nov 2023
11. Microsoft: Microsoft HoloLens — Mixed Reality Technology for Business. https://www.microsoft.com/en-us/hololens. Accessed 25 Nov 2023
12. OMRON Corporation: LD-series Fully Autonomous Mobile Robots. https://industrial.omron.eu/en/products/ld-series. Accessed 25 Nov 2023
13. Sievers, T.S., Schmitt, B., Rückert, P., Petersen, M., Tracht, K.: Concept of a mixed-reality learning environment for collaborative robotics. Procedia Manuf. **45**, 19–24 (2020)
14. Statista: Share of traditional and collaborative robot unit sales worldwide from 2018 to 2022. https://www.statista.com/statistics/1018935/traditional-and-collaborative-robotics-share-worldwide/. Accessed 25 Nov 2023

# Scratch-Based Exergame-Educational Approach in Teaching the Course "Physical Education" for IT-Specialties

Oleksandr Blazhko[1]([⊠]) [ID], Viktoriia Podhorna[2] [ID], Anastasiia Kokotieieva[2] [ID], and Nataliia Bashavets[3] [ID]

[1] Department of Information Systems, Odessa Polytechnic National University, Odesa, Ukraine
blazhko@op.edu.ua
[2] Department of Physical Education and Sports, Odessa Polytechnic National University, Odesa, Ukraine
{podgorna.v.v,kokotieieva.a.s}@op.edu.ua
[3] South Ukrainian National Pedagogical University Named After K. D. Ushynsky, Odesa, Ukraine
bashavetsnata@ukr.net

**Abstract.** I's proposed the approach of computer game control of human movements in sport computer games in order to increase the level of physical activity of stu-dents in online classes, which takes into account the initial knowledge of the spe-cialized disciplines of the curriculum. The approach is based on an open-access repository of computer programs created in the Scratch-based block language (47 games for 28 summer sports, 16 games for 10 winter sports), Scratch programs for human motion recognition based on the PoseNet neural network, which allow changing the principle of controlling game characters through the keyboard and mouse. The approach uses static gestures – unchanging position of the human body with fixed relationships of its various parts to each other, dynamic gesture – sequential transition in time between two static gestures and Scratch-block descriptions of recognition of gestures.

**Keywords:** ExerGame · motion recognition · Scratch

## 1 Introduction

Nowadays, there are means of interaction between a person and a computer through human movements' recognition, which allows controlling events in entertaining or edu-cational computer games using ExerGame technologies ("exercise" + "game") [1]. Non-contact motion control sensors, such as the MS Xbox 360/One game console, which contains the MS Kinect sensor with an infrared camera [2] to support Kinect Sport 3D computer sports games have been used for many years to recognize human movements: Tennis, Table Tennis, Golf, Skiing, Baseball, Soccer, Basketball, Boxing, Track and Field (Sprint, Javelin, Discus, Long Jump and Hurdles). Due to COVID restrictions in 2020–2021 and military actions from 2022, students could not come to the sports

© The Author(s), under exclusive license to Springer Nature Switzerland AG 2024
M. E. Auer et al. (Eds.): STE 2024, LNNS 1027, pp. 324–335, 2024.
https://doi.org/10.1007/978-3-031-61891-8_32

facilities of Odesa National Polytechnic University for "Physical Education" classes. Therefore, the "Physical Education and Sports" department together with the "Information Systems" department decided to adapt the discipline program through the ExerGame technologies' introduction. However, the use of these technologies in the first-term students' training program has limitations during online classes: (1) lack of special infrared cameras; (2) lack of access to commercial computer games; (3) lack of knowledge to independently create 3D computer games. The first limitation can be partially removed by using programs for processing human images from a Web-camera based on various machine learning models, for example, PoseNet [3]. The second limitation can be partially removed by using the Scratch programming environment [4] – a tool for imparting computational thinking skills from an early age in schools. Although Scratch allows the creation of computer programs with simple 2D graphics, which cannot be compared to commercial 3D games, Scratch provides access to working computer games created by pupils from around the world and hosted in open-file game repositories [5]. Such a repository allows choosing a game with an interesting scenario, exploring the open source code, and, if desired, making changes to it, and creating a new version of the game. Unfortunately, there are currently no examples of motion control sports games in the Scratch repository. However, the work [6] represents a special program for displaying human movement from the MS Kinect v.1 sensor in Scratch v.1.4 sensor blocks, which allowed the author to quickly create several games with different scenarios [7]: "Alien attack" controls body movements left/right and raising the arms above the head, "Hungry shark" controls the head up/down movements and joining the palms together, "Hungry ant" controls the right hand's movements. During the work [8], new games were created for the stroke patients' rehabilitation. These games can be easily adapted for the webcam instead of the MS Kinect infrared camera if you use the free neural network communication software libraries with the Scratch environment. The PoseBlocks web-program [9] was one of the first examples of free programs, which provided the Scratch environment with new sensory blocks based on the AImodel neural network for tracking body, and based on own images/poses/audio -models from the Google Teachable Machine (GTM) [10]. The PictoBlox [11] desktop program, also supports GTM using the Scratch language. The software library [12] allows representing the human body in the form of reference points based on the PoseNet neural network, as is done by the MS Kinect sensor, and transferring the coordinates of these points to the Scratch environment sensor blocks.

The analysis of works determined the need to modify the "Physical Education" curriculum in the first term for students studying IT specialties in the form of a Scratch-Based Exergame-Educational Approach (exergaming approach) using computer games based on the Olympic sports themes, which should take into account simple technical requirements for students during online classes and low requirements for knowledge of programming languages:

- the presence of a regular web-camera connected to the computer;
- the knowledge of the basics of programming in the Scratch environment;
- the access to the Scratch library for connection with the PoseNet neural network;
- the access to the open Scratch repository of games on the Olympic sports themes.

## 2  Stages of the Scratch-Based Exergame-Approach

The classes conduct by the ExerGame-approach, which includes three stages (see Fig. 1), were proposed.

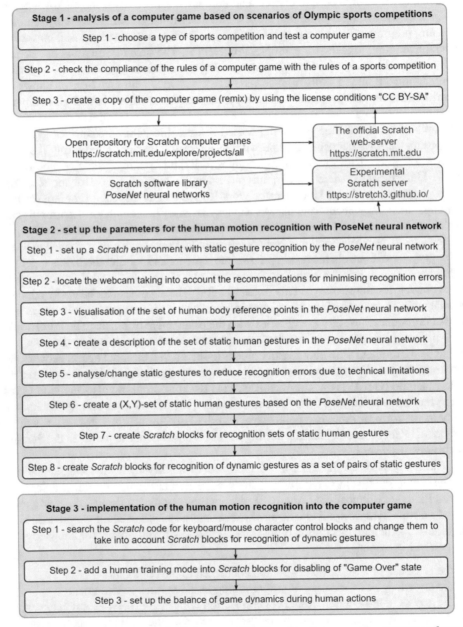

**Fig. 1.** The Exergaming approach of classes conduct based on 3 stages and a sequence of steps.

## 2.1  Stage 1 - Analysis of a Computer Game Based on Scenarios of Olympic Sports Competitions

In order to determine the list of computer games that students will explore and change during online classes, the authors of the ExerGame-approach in the Scratch-repository searched and selected computer games according to the following steps:

1) determination of the English-language names of official summer and winter sports;
2) search in the repository of the list of computer games by the English name of sports;
3) selection of found games by type of single player and by decreasing popularity of their use by the community;
4) selection of computer games by the kind of sports in different planes of the player's view projection: frontal (f-plane) – front/back view, profile (p-plane) – side view, horizontal (h-plane) – top view.

As a result of search the following was found:

- 47 computer games for 28 summer Olympic sports
- 16 computer games for 10 winter Olympic sports

For each game, the following was defined: the projection planes of the player's view, the project code in the Scratch repository, the computer keyboard/computer mouse keys controlling the movements of the computer athlete in the game, and the game screen example. Part of the description of the games found is presented in Table 1.

**Table 1.**  Part of the description of the games found.

| Type of sport, projection plane, project code | Computer keyboard/computer mouse control keys | Game screen example |
|---|---|---|
| Judo, 566395367, p-plane | A, D, W, S, W+Space, S+Space – move forward/backward, raise/lower defense, upper/lower attack | |
| Tennis, 520716879, f-plane | Left, right, forward, backward arrow keys – moving in directions, Space key – racket hits | |
| Swimming, 178358511, h-plane | Right and left arrow keys – movement with the right or left hands | |

## 2.2  Stage 2 – Set up the Parameters for the Human Motion Recognition with PoseNet Neural Network

The Scratch environment can be hosted on any computer due to the open source code, and Scratch v.3 can be hosted on any web-server and new software libraries can be added. One of the web-servers hosts the PoseNet2Scratch software library for recognizing human movements based on the PoseNet [12] neural network.

When using the PoseNet2Scratch software library, students must consider the conditions for working with the webcam:

- the webcam is located at the level of human eyes, but the webcam can be placed lower;
- students stand in front of the even color wall, which contrasts with the person's clothes so that the PoseNet neural network can correctly distinguish the outline (silhouette) of the person's image against the wall image background;
- there are no extra objects in the area monitored by the webcam;
- the webcam is located at a distance sufficient to display the body parts of the students used in the sports' movements (2–5 m);
- the room is illuminated but without the light glare on the wall in front of which the students are standing.

The PoseNet neural network allows representing the human body in the form of 17 reference points (see Fig. 2).

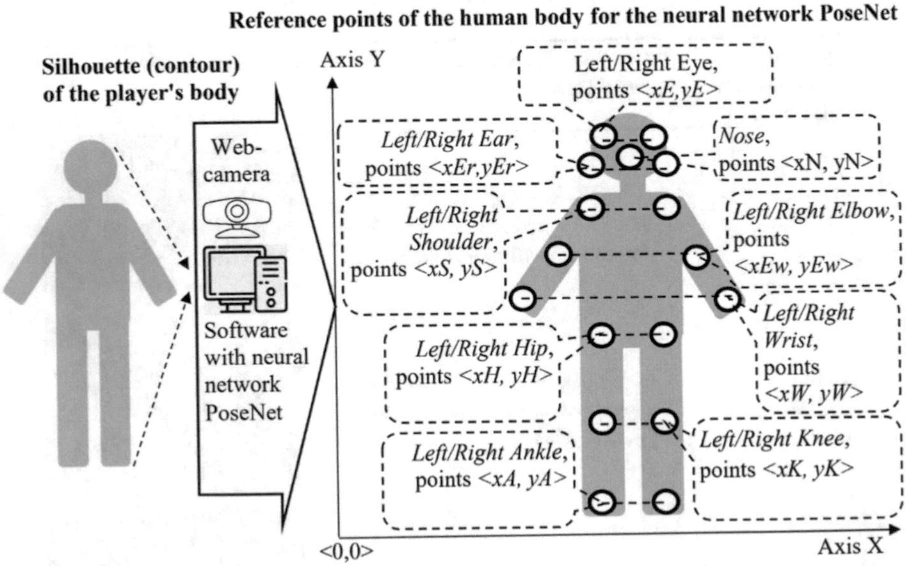

**Fig. 2.** The human body anchor points created in the PoseNet neural network

The PoseNet2Scratch software library adds a section of green blocks to Scratch to control anchor points. Figure 3 shows fragments of the sprite placement rendering software blocks associated with the anchor points of the nose, left shoulder, and right wrist.

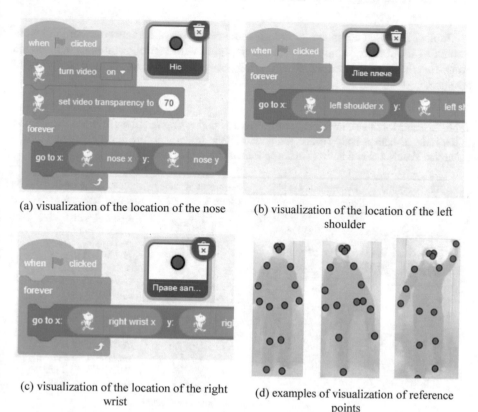

(a) visualization of the location of the nose

(b) visualization of the location of the left shoulder

(c) visualization of the location of the right wrist

(d) examples of visualization of reference points

**Fig. 3.** Fragments of software Scratch-blocks for visualization of the location of sprites.

Each description of the athlete's movement can be considered as a sequence of poses—positions taken by the human body, head, and limbs in relation to each other.

If we talk about the process of human poses recognition by a computer, then a pose can be considered as a gesture – an action or movement of the human body or its part, which has a certain meaning or sense, that is, it is a sign or symbol.

The following terms will be used in the future:

- static gestures (sg) – unchanging position of the human body with fixed relation of its various parts to each other;
- dynamic gesture (dg) – a sequential transition in time between two static gestures.

For static gestures, their coordinate description is proposed, which takes into account the following:

- for each pair of static gestures, anchor points that determine a significant difference between static gestures are searched, and a logical expression comparing the values of the coordinates of these anchor points is described;

- one should strive for a minimum number of comparisons between reference points, but guarantee a significant difference in their description between different static gestures.

Table 2 for three computer games presents examples of screenshots from the web-camera in the form of human body anchor points with static gestures in the Scratch program with a connected PoseNet neural network and a coordinate description of gestures.

**Table 2.** Table for three computer games presents examples of screenshots from the web-camera in the form of human body anchor points with static gestures in the Scratch program with a connected PoseNet neural network and a coordinate description of gestures.

| Kind of sports | Screenshot of static gesture 1 | Screenshot of static gesture 2 | Static gesture coordinate description |
|---|---|---|---|
| Judo | sg1 – defense | sg2 – attack | sg1: $(xK < xH)$ and $(xW < xA)$ and $((ySL - yHL) = (xHR - xHL))$ sg2: $(yWL = yN)$ and $(yWR < yEwR)$ and $(xN = xH)$ |
| Tennis | sg1 – racket ready to hit | sg2 – hit with a racket | sg1: $(yWL > yEwR)$ and $(xWL > xEwL)$ sg2: $(yWL > yEwR)$ and $(xEwL = xEwL)$ |
| Swimming | sg1 – right hand swing | sg2 – left hand swing | sg1: $(yWR > yN)$ and $(yWL < yEwR)$ sg2: $(yWL > yN)$ and $(yWR < yEwL)$ |

To program the recognition conditions of the specified gestures, it is recommended to create variables, for example, *nose_y, left_wrist_y, right_wrist_y*, the values of which are set in the software modules for the anchor points visualization (see Fig. 4).

Figure 5 shows a fragment of the Scratch program where two static gestures "Both hands below nose" and "Both hands above nose" are first recognized using the following steps:

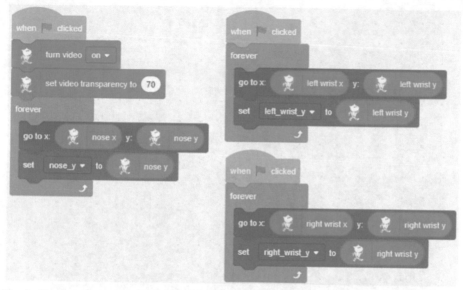

**Fig. 4.** Fragments of the Scratch program that set the values of the *nose_y, left_wrist_y, right_wrist_y* variables for programming static gesture recognition conditions.

1. checking the condition of the first static gesture recognition, which is defined by the logical expression "*sg1: (yN > yWL) and (yN > yWR)*", using the variables *nose_y, leftwrist_y, right_wrist_y*;
2. if the result of the logical expression check of static gesture returns true then set variable *new_static_gesture = 1* as *sg1*;
3. checking the condition of the second static gesture recognition, which is defined by the logical expression "*sg2: (yWL > yN) and (yWR > yN)*", using Scratch blocks *leftwrist_y, right_wrist_y, nose_y*;
4. if the result of the logical expression check of static gesture returns true then set variable *new_static_gesture = 2* as *sg2*;

Scratch blocks of dynamic gesture recognition take into account the transition between a previously recognized static gesture and a new recognized static gesture. Figure 5 also shows a fragment of the Scratch program, which recognizes two dynamic gestures: *dg1* – "*hands raised above the nose*", *dg2* – "*hands lowered below the nose*", through previously programmed recognition processes of two static gestures *sg1, sg2*.

The *dynamic_gesture* variable is used to establish the fact of dynamic gesture recognition, which can later be used in the program code of the game character controlling conditions analysis to replace the classic control version in the game through a computer keyboard or a computer mouse.

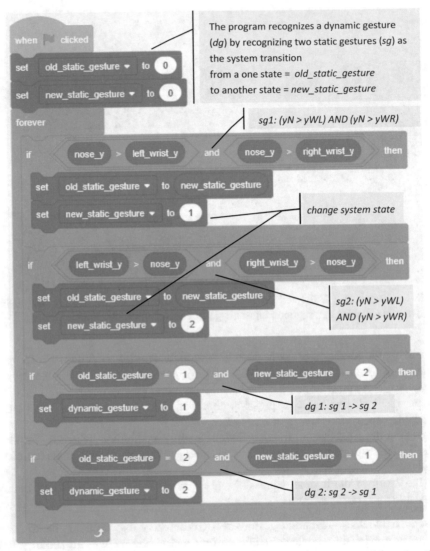

**Fig. 5.** A fragment of the Scratch-program that recognizes two static gestures and, based on them, subsequently recognizes two dynamic gestures.

## 2.3 Stage 3 – Implementation of the Human Motion Recognition into the Computer Game

The existing Scratch program of a computer game uses two classic means to control the game character's movements:

- pressing the keyboard keys, which are programmed by the blue Scratch-block with the name "key pressed";
- pressing the keys of the "mouse" type manipulator, which are programmed by the blue Scratch block called "mouse down";

- horizontal movement of the "mouse" type manipulator, which is programmed by blue Scratch blocks with the names "mouse x", and "mouse y".

Therefore, at the beginning of the method's third stage, it is necessary: (1) to find Scratch blocks with classic computer game controls in the computer game; (2) to replace classic blocks with blocks that take dynamic gesture into account. Figure 6 shows the example of fragments of a Scratch program with the means of a computer game control through the Space key of the keyboard, as well as its replacement with a variable that stores the number of a previously recognized dynamic gesture. After successfully checking that the dynamic gesture matches the specified number, the value of the variable must be cleared to start a new iteration of new dynamic gesture recognition.

(a) a computer game control means through the Space key of the keyboard

(b) replacing the control with a variable that stores the number of a previously recognized dynamic gesture

**Fig. 6.** Fragments of the Scratch program with the computer game control

One of the examples of the psychological and communication theory of games is the Mechanics-Dynamics-Aesthetics (MDA) structure as a tool for analyzing existing games and synthesizing new games based on three components [13]: Mechanics, Dynamics, and Aesthetics. "Mechanics" includes [14]: Game rules – the rules that determine the purpose of the game, for example, "create", "destroy", "avoid", and "match"; Play rules – the rules of manipulation that determine the basic actions that the player can apply in the game, for example, "move", "select", "record", "control", "get lucky", and "shoot". "Dynamics" is a game result of using "Mechanics", caused by:

- the players' actions, which depend on their psychical and physiological abilities, for example, the brain's reaction speed, the fingers' motility speed when pressing the keys of the keyboard/mouse wheel;
- results of different mechanics' interaction with each other.

After switching to game control from using dynamic gestures, the game may become unbalanced because the player does not have time to perform the appropriate movements for the dynamic gesture conditions to work. Therefore, students need to find Scratch blocks that affect the dynamics of the game, for example, determine the character's movement speed, and the events' change speed in the game.

# 3    Analysis of the Exergame-Approach Implementation

In October 2023, 92 first-year students majoring in Computer Science took part in the "Physical Education" online classes based on the Exergaming approach for five weeks, among them 68 students (74%) completed stages 1–3. In order to receive feedback, a questionnaire was conducted among students, the results of which revealed the following:

- 26% of students did not study programming languages at school;
- programming languages studied at school: Scratch – 50%, Python – 25%, Pascal – 15%, C/C++ – 15%, C# – 7%, Java – 6%, Basic – 4%.
- 30% of students did not experience difficulty with completing tasks;
- 20% of students indicated problems with the Internet speed;
- 17% of students had a "weak" computer;
- 15% of students did not have a webcam, but among them only a third could not complete the task, because others did it by temporarily renting webcams;
- 37% of students indicated that the Exergaming approach can be used as an aid for the motor activity increase;
- 26% of students believe that the Exergaming approach can be effective only in the online education period;
- 24% of students do not see the benefit in using the Exergaming approach;
- 57% of students felt motivated to continue practicing Exergaming approach and the desire to achieve a decent result;
- 28% of students felt oppressed while completing tasks, but this may be related to technical difficulties, which were noted in the corresponding item of the survey;

The questionnaire results also showed that the Exergaming approach developed the participants' Soft Skills as the students identified the following priorities for the Soft Skills types:

- 20% – adaptability – the ability to change ways of thinking according to the tasks and the conditions for solving them;
- 15% – creativity – the ability to create and find new original ideas, and solve tasks in a non-standard way;
- 13% – critical thinking – the ability to reason and effectively analyze information.

# 4    Conclusion

The analysis of the results of the first experimental implementation of exergaming approach showed that 1st-year students, starting to study "Physical Education" in the 1st term and not having sufficient skills in using professional programming languages, can already create computer games with 2D graphics where they will test their motor skills. Computer gamification of physical exercise support based on human movement control hardware and software can be introduced as an optional part of "Physical Education" with connections to other disciplines:

- the results of students' current studying in the disciplines they study in parallel, for example, in "Algorithmizing and Programming";

- the review of students' future learning outcomes in such disciplines as "Intelligent Data Analysis"/"Artificial Intelligence Methods and Systems", "Programming of IoT systems" and "Computer Game Systems Development".

For many years, the Scratch programming environment has been the basis for STEM in school education, therefore the proposed exergaming approach can be used in pupils' non-formal education.

# References

1. Krause, J.M., Jenny, S.E.: Physical educators' exergaming integration experiences, attitudes, and self-efficacy beliefs. Phys. Educ. **80**, 91–107 (2023)
2. Kourakli, M., Altanis, I., Retalis, S., Boloudakis, M., Zbainos, D., Antonopoulou, K.: Towards the improvement of the cognitive, motoric and academic skills of students with special educational needs using Kinect learning games. Int. J. Child-Comput. Interact. **11**, 28–39 (2017)
3. Chung, J.-L., Ong, L.-Y., Leow, M.-C.: Comparative analysis of skeleton-based human pose estimation. Futur. Internet **14**, 380 (2022)
4. Grover, S.: Teaching and assessing for transfer from block-to-text programming in middle school computer science. In: Hohensee, C., Lobato, J. (eds.) Transfer of Learning. Research in Mathematics Education, pp. 251–276. Springer, Cham (2021). https://doi.org/10.1007/978-3-030-65632-4_11
5. Scratch project Explore (2021). https://scratch.mit.edu/explore/projects/games/
6. Howell, S.: Install Scratch and Kinect2Scratch. In: St. Jean, J. (ed.), Kinect Hacks: Tips and Tools for Motion and Pattern Detection, pp. 184–206. O'Reilly, Sebastopol, CA (2012)
7. Howell, S.: Kinect2Scratch (Version 1.5) [Computer Software] (2012). https://stephenhowell.github.io/kinect2scratch/
8. Hung, J.-W., et al.: Comparison of kinect2scratch game-based training and therapist-based training for the improvement of upper extremity functions of patients with chronic stroke: a randomized controlled single-blinded trial. Eur. J. Phys. Rehabil. Med. (2019)
9. Jordan, B., Devasia, N., Hong, J., Williams, R., Breazeal, C.: PoseBlocks: a toolkit for creating (and Dancing) with AI. In: AAAI, vol. 35, no. 17, pp. 15551–15559 (2021)
10. Carney, M., et al.: Teachable machine: approachable web-based tool for exploring machine learning classification. In: Extended Abstracts of the 2020 CHI Conference on Human Factors in Computing Systems, CHI EA '20, pp. 1–8. Association for Computing Machinery, New York, NY, USA (2020)
11. Cruz, S., Bento, M., Lencastre, J.A.: Computational thinking training using PICTOBLOX. In: Kommers, P., et al. (ed.), Proceedings of International Conferences Internet Technologies & Society 2021, pp. 53–60 (2021)
12. Junya Ishihara: PoseNet2Scratch (2021). https://github.com/champierre/posenet2scratch
13. Duarte, L.C.S., Battaiola, A.L.: Distinctive features and game design. Entertain. Comput. **21**, 83–93 (2017)
14. Ben-Sadoun, G., Alvarez, J.: Gameplay bricks model, a theoretical framework to match game mechanics and cognitive functions. Games Cult. **18**(1), 79–101 (2023)

# A Virtual Lab of Matrix-Assisted Laser Desorption Ionization MALDI-TOF Protocols for Microbiology Students

Soma Datta[(✉)] [iD] and Ibrahim Ibaad Syed[iD]

University of Houston-Clear Lake, Houston, TX 77058, USA
datta@uhcl.edu

**Abstract.** This study is a practical way of teaching software engineering and microbiology in a virtual setting, which is beneficial during natural calamities, including the recent pandemic. This project details how the software engineering students developed a virtual lab for the microbiology students that started as a class project. The benefit of creating the project for the software engineering students was that it helped to develop their soft skills and the details of collecting requirements. Similarly, it allowed the microbiology students to collaborate with software engineering students to educate them on the protocol and eventually practice the MALDI-TOF experiment before they do it in their labs. Input from microbiology students seems crucial in ensuring that the virtual lab accurately reflects the real-world lab environment and meets the needs of those conducting experiments. This study explains how students learned to communicate in developing real-world software as a class project. The study exemplifies how technology can bridge the gap between theoretical knowledge and practical skills.

**Keywords:** Virtual Lab · Software Engineering · Microbiology · Virtual Reality

## 1 Introduction

This study investigates a university setting where almost half of the undergraduates are non-traditional students. These students may have work and family obligations that leave them little free time outside regularly scheduled classrooms. The virtual lab develops flexible learning strategies to make the research opportunities productive for diverse, non-traditional students. Virtual laboratory simulations (VLS) can supplement traditional teaching laboratories and allow students to learn at a distance at their own pace. These systems can also help students continue their learning of non-cognitive science skills during campus disruption, like the recent pandemic.

The evolution of traditional microbiology education into a cutting-edge Virtual Lab for Microbiology involved transforming a pre-existing PC-based application into an immersive VR experience. This section delves into the development process, challenges encountered, and the iterative improvements made based on valuable feedback.

This project aims to develop a virtual lab to teach students microbiology laboratory techniques remotely. VL was essential during this pandemic, but these simulations will supplement traditional laboratory training exercises and facilitate sharing protocols between institutions. This project will be an innovative technique of interdisciplinary teaching that will mimic a real-world situation for microbiology and software engineering. This will encourage and supplement microbiology students to learn and practice the lab work by doing virtual lab work in an informal setting.

## 2 Background

### 2.1 Literature Review

VLS range from immersive to observational models. Immersive models involve 3D simulations; however, this immersive approach doesn't necessarily improve student learning [1, 2], and broad implementation would require students to acquire specialized equipment or use on-campus facilities. In the observational model, students watch videos showing step-by-step protocols. This video approach effectively improves cognitive skills [3]; however, there is little evidence that the observational models improve manual (non-cognitive) laboratory skills. PC-based laboratory simulators teach manual laboratory skills like a flight simulator trains pilots and appear effective in preparing microbiology students for laboratory work [4]. Labster (Denmark) and other vendors have developed laboratory simulations; however, the company controls the content, limiting the curriculum available to instructors. Further, many students and universities need help to afford to pay for these commercial products.

Collaborate with software engineering (SE) and microbiology (MB) students to develop VLS of MALDI-TOF protocols. PC-based VLSs teach manual laboratory skills the way a flight simulator trains pilots. These simulators appear effective in preparing microbiology students for laboratory work [5].

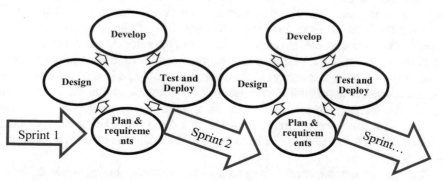

**Fig. 1.** The Iterative Process how each component was developed, starting from the core element (Agile process)

### 2.2 The Goal

The goal was to help SE students understand software development, teamwork, and communication with other domain experts and users. A collaboration between SE and

MB students to develop a virtual lab of MALDI-TOF protocols helps with the goal. These simulators appear effective in preparing microbiology students for laboratory work.

This project is an innovative interdisciplinary teaching technique mimicking a real-world situation for both SE and MD students. It will advance the SE field by integrating two methods, i.e., agile (Fig. 1, which had seven sprints) and design thinking frameworks, in developing software [5]. The VR-lab will encourage and supplement microbiology students to learn and practice the lab work by doing virtual lab work in an informal setting.

## 3   Laptop Version of the MALDI-TOF Virtual Lab

This initial application had limitations: relying on keyboard inputs, lacking hands-on experiments, and resembling a quiz rather than an interactive learning experience. The virtual lab(VL) concept started during COVID-19 in April 2020 when microbiology students could not come to the lab to do lab work.

The idea of building it in-house was because buying a commercially made software application was expensive both for the student and the university. Although, that would have been perfect software. On the other hand, not every experiment has the same protocol. Every professor or instructor teaches an experiment in various ways [6]. Hence, it is reasonable to develop the software on campus.

This VL was first built using Unity. The VL then had a mouse and keyboard. Students were able to install it on their laptops. It was not three-dimensional as in virtual reality, but it was a version that students could still practice the experiment using their laptops and understand the protocol. The VR-Lab was assigned as a class project to a team of four.

Since this started during the Pandemic in September 2020, requirements were collected by either joining the Zoom session or reviewing the video recording from the instructor. Students interested in doing something using Unity were teamed into one group, and they started working on it. A prototype of the software was developed by the end of the semester. In the next semester, another team of students continued to work on the project. During this semester, the team was asked to add a new feature to the product. A research assistant started converting the prototype. During the semester, students developed new features that were later integrated into the product. Some features added in this version were a virtual lab instructor, a demo option by the lab instructor, a whiteboard with instructions, and a knowledge test. Features that were not added were the lab book.

Another student researcher working on this one was a microbiology student. The MB students could be guided with their domain knowledge about what was right because all the others developing were computer science/software engineering students. Thus, it went through several iterations, leading to the beta version. Microbiology students used the beta version before their actual lab experiment.

Microbiology students pointed out some significant changes that were required in beta versions are mentioned here: The robot lab instructor had to have a lab coat, the shoes of the lab instructor were not appropriate, and the lab instructor had to wear protective glasses during the demonstration. SE students learned the importance of collecting detailed requirements. A screenshot of the beta version of this software is shown in Fig. 2.

**Fig. 2.** Screenshot of virtual lab simulation developed to prepare students to implement the ethanol/formic acid extraction protocol. This VLS was encoded in Unity (2020.1.5) and C#.

## 4 Virtual Reality Version of the MALDI-TOF Virtual Lab

### 4.1 Introduction

In response to the favorable reception of the initial version, our aim was to elevate the project's immersive qualities. To achieve this, we sought to transform the project from a desktop application into a Virtual Reality (VR) application compatible with Oculus Quest devices. This strategic shift was prompted by valuable feedback from students and professors who had experienced the first version of the project. Rather than starting from scratch, our objective was to convert the existing application into a VR format, harnessing assets from the original version.

During the development process, it became evident that a restructuring of the application code was imperative to seamlessly transition it into a Virtual Reality application. This decision was driven by a desire to enhance the overall user experience and cater to the specific needs and preferences expressed by our user base.

### 4.2 Development Journey

The objective of this version was to transition from a PC-based application to a VR environment. However, the process faced initial setbacks due to resource limitations, including the absence of essential equipment such as a gaming PC, Unity software, and an Oculus Quest. Logistic challenges, such as issues with hand tracking and hardware malfunctions, further prolonged the development phase.

The first version of the VR application had to use controllers due to these initial technical constraints. We integrated the lab environment from the previous application to expedite development, maintaining user familiarity. The product initially occurred in Unity version 2021.3.16f1, later transitioning to Unity 2022.3.2f1.

**Fig. 3.** The Water Station (Station 1) marks the inception of experiments. It features a beaker containing water and essential apparatus such as a pipette, pipette tips, and test tubes. The pipette, a crucial tool in the experimental process, allows users to extract a predetermined quantity of water and transfer it into designated test tubes. The pipette tips are designed for single use, emphasizing hygienic practices and necessitating ejection after each application

**Fig. 4.** At the Bacteria Station (Station 2), a disk container harboring a diverse array of bacterial samples is provided. This station also incorporates a loop mechanism, enabling users to introduce bacteria into the experimental solution with accuracy and control.

**Fig. 5.** Moving on to the Ethanol Station (Station 3), students encounter a dedicated beaker containing ethanol. The station is equipped with a pipette, enabling users to extract a precise amount of ethanol from the beaker into smaller test tubes, crucial for experiment protocols.

**Fig. 6.** Vortex Machine Station (Station 5) introduces a vortex machine, a vital apparatus for specific experimental processes. A user-friendly switch mechanism allows students to activate and deactivate the vortex as needed, enhancing the overall experiential learning process.

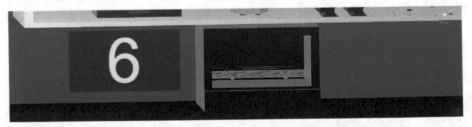

**Fig. 7.** The Refrigerator Station (Station 6) houses a virtual refrigerator, serving as a valuable resource for students when the need arises to cool a particular solution as part of experimental procedures.

The Virtual Reality (VR) Lab, meticulously crafted for seamless use on any Oculus Quest device and developed utilizing Unity 3D, represents an immersive replication of the microbiology laboratory. Comprising six intricately designed stations, each is pivotal in facilitating a comprehensive and hands-on learning experience. Designed to utilize Quest controllers, each control is meticulously mapped using Unity Input Manager, ensuring precise and intuitive interaction within the virtual environment. This thoughtful integration enhances the user experience, allowing students to seamlessly navigate and engage with the VR Lab, fostering a dynamic and effective learning environment for microbiology experiments.

Figures 3, 4, 5, 6 and 7 shows various stations present in the VR application. Each station is meticulously crafted to replicate the physical attributes of a microbiology lab and incorporates interactive elements that amplify the educational experience. This detailed and immersive design ensures that students can seamlessly navigate and engage with the VR Lab, fostering a dynamic and effective learning environment for microbiology experiments.

Additionally, a Virtual Lab Assistant has been integrated into the VR Lab environment to enhance the overall learning experience. This digital guide is designed to provide step-by-step instructions, offering insightful guidance to students navigating through the experiment procedures. The Virtual Lab Assistant plays a pivotal role in ensuring the correctness of each student's action, promptly identifying and correcting any deviations from the prescribed protocol. Through interactive prompts and real-time feedback, the assistant not only assists students in understanding the intricacies of the experiment but also serves as a valuable tool for reinforcing proper laboratory techniques. This innovative feature adds an element of personalized guidance, creating a supportive and immersive learning environment within the virtual microbiology laboratory.

For the controls in the VR Lab, user interaction is seamlessly facilitated through intuitive controls mapped to the Oculus Quest controllers. The left analog stick allows users to navigate the virtual environment, providing fluid and responsive movement. Meanwhile, the right analog stick enables users to look around and explore the intricacies of the laboratory freely. Essential for manipulating experiment parameters, the 'A' button is designated to increase values, while the 'B' button serves to decrease them, ensuring precise adjustments during the hands-on learning experience. The functionality of holding and interacting with objects, such as the refrigerator handle or vortex switch, is seamlessly executed using the right and left grip buttons. The right trigger is employed to utilize the pipette, offering precise control and responsiveness. On the other hand, the left trigger is designated for the critical function of ejecting the pipette tip, enhancing the overall realism and functionality of the virtual microbiology laboratory experience. These carefully mapped controls mirror real-world actions and contribute to an immersive and educational journey within the VR Lab.

## 4.3 Experiment Protocol

In adherence to the official procedure, the first step involves meticulously labeling the frosted side of each Eppendorf microfuge tube with assigned positions, denoted as B1, B2, and so forth, alongside the researcher's initials. Following this, 300 µL of water is methodically added to each tube at the designated Water Station. Moving to the next phase, a large, singular colony of microorganisms is carefully transferred to the tube, with potential variations based on the microorganism's size, emphasizing the selection of isolated colonies. This step involves labeling the tube's top with its position (e.g., B01) and the researcher's initials for clear identification. Subsequently, a thorough vortex is executed to ensure homogeneity within the solution. The procedure advances to the Ethanol Station, where 900 µL of ethanol is meticulously added to the tube, followed by another round of thorough vertexing. Placing the tube in its designated position within a 96-well microtube rack (e.g., B1 in B1), researchers are presented with the option to either store the prepared sample at 4 °C for a period of up to two weeks or proceed with the subsequent phases of the experiment. This comprehensive and sequential approach ensures precision and adherence to the established microbiology experiment protocol.

The primary objective of our VR simulation is to replicate the Formic Acid/Ethanol Tube Extraction (TE) Method, a crucial laboratory technique employed in microbiology. The underlying principle of this experiment revolves around addressing the challenge posed by microorganisms with naturally thick cell walls, such as yeast, and older isolates like Staphylococcus sp. And Corynebacteria sp., which may develop thicker cell walls over time. The necessity arises to effectively break down these robust cell walls and separate ribosomal proteins before proceeding to spectra analysis. In cases where the conventional ethanol-formic acid (eDT) procedure fails to yield the desired scores, the TE procedure becomes a vital last resort. The VR simulation is an innovative educational tool that allows students to practice this specialized laboratory method hands-on. It not only offers a safe and accessible learning environment but also provides a dynamic platform for students to familiarize themselves with the intricate steps of the TE method, promoting a comprehensive understanding of microbiological techniques.

During the development phase, one of the primary challenges revolved around devising controls for each device incorporated into the virtual laboratory setting. Devices with relatively simple functionalities, such as the vortex with only one switch, posed minimal developmental hurdles. However, more complex instruments like the pipette, which entails many actions a student is expected to perform, presented a considerably more significant challenge. The intricacies of the pipette's functions made developing its controller notably more complicated in the current application version. Moreover, devices that demanded heightened precision, such as the pipette, required a substantial investment of time in the development process. Navigating these challenges was essential to ensuring each laboratory instrument's realistic and practical integration within the virtual environment, contributing to an authentic and educational user experience.

## 5   Conclusions, Feedback, and Future Work

Feedback from students and other stakeholders, given anonymously, played a crucial role in enhancing our system. Users pointed out a learning curve, stressed the importance of precise instructions, and suggested additional hygiene considerations. The feedback also highlighted challenges such as speech clarity, navigation issues, and understanding instructions. Specific concerns included difficulty recognizing pipette tasks, the need for pop-ups, and suggestions for better lighting. The administration recommended starting with a familiar VR experience like Beat Saber to ease students into the virtual environment. Additionally, as MB students collaborate with SE students to communicate their requirements, they foster team building and gain insights into real-world situations.

Two main challenges surfaced: SE students faced time constraints in developing the VR Lab for the experiment, and testing the lab involved both SE and MB students. With only two sets of equipment available in the library, students had to sign up for their turn to practice or test the system.

In response to the received feedback, future iterations of the VR application will prioritize transitioning from controllers to a more intuitive hand interaction system. These improvements aim to enhance instruction clarity, address hygiene concerns, and refine the overall learning experience. This iterative approach ensures the evolution of the Virtual Lab for Microbiology into a practical and user-friendly educational tool. Another upcoming project involves developing a VR lab for a Deoxyribonucleic acid (DNA) prep protocol for Polymerase chain reaction (PCR) tests, with the requirements collection for this project currently underway.

# References

1. Rosenberg, J., Lawson, M.: An investigation of students' use of a computational science simulation in an online high school physics class. Educ. Sci. **9**, 49 (2019). https://doi.org/10.3390/educsci9010049
2. Wimpenny, K., et al.: Unpacking frames of reference to inform the design of virtual world learning in higher education. AJET 28 (2012). https://doi.org/10.14742/ajet.848
3. Paxinou, E., et al.: An IRT-based approach to assess the learning gain of a virtual reality lab students' experience. Intell. Decis. Technol. **15**, 487–496 (2021). https://doi.org/10.3233/IDT-200216
4. Holly, M., et al.: STEM learning experiences in VR – challenges and recommendations from the learners' and teachers' perspective. J. Educ. Technol. Soc. **24** (2021)
5. Damasceno, E.F., Fernandes, L., Silva, A.P.D., Días, J.B.: A Systematic mapping study on low-cost immersive virtual reality for microbiology. In: IEEE Revista Iberoamericana de Tecnologias del Aprendizaje, vol. 18, no. 1, pp. 62–69 (2023). https://doi.org/10.1109/RITA.2023.3250551
6. Clark, A.E., Kaleta, E.J., Arora, A., Wolk, D.M.: Matrix-assisted laser desorption ionization-time of flight mass spectrometry: a fundamental shift in the routine practice of clinical microbiology. Clin. Microbiol. Rev. **26**, 547–603 (2013). https://doi.org/10.1128/CMR.00072-12

# How Can Pneumatics Be Trained with Augmented Reality in the Context of Training for Industrial-Technical Professions?

Jan-Niklas Terschüren[(✉)] [iD], Valérie Varney [iD], and Larissa Müller

Process Engineering, Energy and Mechanical Systems, Cologne University of Applied Science, Cologne, Germany
{jan-niklas.terschueren,valerie.varney,
larissa.mueller1}@th-koeln.de

**Abstract.** Augmented reality (AR) holds promise as a technology for creating hands-on learning environments. This paper presents the development of an AR-based training program for pneumatics teaching, aiming to investigate the effects of different elements of AR visualization on learning and knowledge acquisition. Two concepts for AR visualization elements were developed: a realistic replica of the pneumatic system and a concept that incorporates additional AR elements, such as airflow tracking and an X-ray view. A mixed-methods approach study was conducted with fifteen first-year industrial mechanics apprentices. The participants were divided into three groups (traditional learning method, concept 1, concept 2) and performed two practical exercises on pneumatics. The results indicated that AR-based training can be an effective method for imparting knowledge in the field of pneumatics. However, the type of visualization used significantly impacts the learner's knowledge absorption. Visualizations conveying additional information, such as airflow tracking and an X-ray view, proved to help learners better understand and internalize complex relationships and functionalities. This study suggests that AR-based training is an effective way to teach pneumatics, but the choice of visualization is crucial. Visualizations providing additional information, such as airflow tracking and an X-ray view, enhance learners' understanding and internalization of complex relationships and functionalities. More research is needed to validate these findings and explore additional ways to use AR to improve the learning experience in pneumatics and other technical fields.

**Keywords:** augmented reality · AR-training · education · pneumatics · didactics

## 1 Context

The training of industrial-technical professions is an important component for the successful integration of young skilled workers into the industry. Due to the ongoing digital transformation and its impact on workplaces, the acquisition of specialist knowledge through innovative teaching and learning methods is becoming increasingly important. One promising way to make learning more practice-oriented and effective is the use of

M. E. Auer et al. (Eds.): STE 2024, LNNS 1027, pp. 344–355, 2024.
https://doi.org/10.1007/978-3-031-61891-8_34

augmented reality (AR) technologies. These make it possible to combine the real and virtual worlds by superimposing virtual content on reality. This allows complex issues to be conveyed more clearly and intuitively. This can have a positive effect on the learning process, as learners get a better idea of the content and can understand connections more easily [1].

The teaching of pneumatics is part of the training for industrial-technical professions, such as industrial mechanic or mechatronics technician, and is relevant for examinations in some of these professions. Pneumatics refers to the science of fluid technology, which focuses on gaseous media, in particular compressed air. In contrast to hydraulics, which uses liquid media, pneumatics uses compressed air as a fluid for energy transmission. This technology is primarily used in automation technology to control, regulate, drive, and move machines and systems [2].

A pneumatic system consists of several components for power supply, signal input, signal processing, signal output and command execution. The energy supply, in the form of compressed air, is generated by compressors and stored in a compressed air reservoir or fed directly into the compressed air network. Additional elements are used to maintain and process the compressed air. Buttons and switches are used for signal input. The signals are then processed by various valves, including directional control valves and pressure valves, which serve as control elements. Various actuators, especially directional control valves, are used for signal output. Finally, the actuated drive and working elements, including motors and cylinders, are driven in the command execution [2].

## 2  Purpose and Goal

During the training of industrial-technical professions, the acquisition of knowledge and skills in the field of pneumatics is offered in practice-oriented exercises, e.g. by setting up pneumatic systems. The framework curriculum for each training program, provided by the Federal Institute for Vocational Education and Training (BIBB), a recognized competence center for research and further development of vocational education and training in Germany, specifies the learning objectives [3]. When examining various framework curricula, it is obvious that in most cases, trainees must learn to identify pneumatic components and functional sequences [4]. Furthermore, trainees must be able to plan, install and commission pneumatic systems using circuit diagrams [4].

Current practice-oriented teaching methods in the field are based on textbooks, simulation software, training laboratories and learning systems, which is often mostly theoretical. The simulation software is predominantly two-dimensional (2D) and does not represent reality, since pneumatic components are usually only indicated or represented by circuit symbols. Training laboratories and learning systems are practical options for teaching. However, they are cost-intensive, which is why not all training companies or vocational schools are equipped with such systems. For this reason, courses are offered in special pneumatics training workshops, but these are time-consuming and expensive. In addition, such systems take up a lot of space and are location-dependent due to the compressed air supply [5, 6].

The subject area of pneumatics is therefore a suitable use case for the integration of AR in teaching. Augmented reality-based training offers the chance of a location-independent and cost-effective alternative for practice-oriented teaching methods.

In the field of training and teaching, AR technologies generally offer the potential to improve the learning process by conveying content in a more practical and interactive way. This is particularly relevant for the teaching of technical content, as this can often be abstract and difficult to understand [7, 8].

The use of AR technologies in education or teaching has already been widely researched and shows that the usage of technology produces many positive effects, such as increasing motivation and engagement [8]. However, the ways in which AR elements can be used for visualization and the effects that different types of visualization can have on the learning process of learners in this area have not yet been investigated. The current state of art in this field of research therefore provides an opportunity for further scientific investigation [9].

For this reason, an AR-based training course in the field of pneumatics was developed and examined for trainees in industrial-technical professions. The aim was to gain new insights into how visualizations in this area influence learners' knowledge absorption in AR-based training courses. To this end, two concepts for visualizations in AR were examined:

- **Concept 1:** A realistic simulation of the pneumatic system
- **Concept 2:** A realistic simulation of the pneumatic system with additional AR elements for the visualization of pneumatic processes

The results of this work can contribute to the establishment of AR technologies in the training of industrial-technical professions.

## 3   Related Work

For the development and investigation of the AR app and the visualizations, already developed augmented reality learning applications in the field of pneumatics are re-searched and analyzed. The aim is to gain an overview of existing AR applications and to record the current state of research in this area. This allows conclusions to be drawn about successful implementations and concepts. The following relevant applications are listed alphabetically and briefly described.

The "Festo Didactic AR" app was developed to facilitate the commissioning of mechatronic learning systems. Users can use markers to access information, data sheets, and videos. The app offers intuitive visualizations for setting up the controls and commissioning real systems, but requires the use of a paid learning system from Festo Didactic. There are no user tests or test reports on the application. Therefore, no statement can be made about the effectiveness of the application [10].

The "Augmented Reality Pneumatic System" app, developed at the University of Technology Malaysia (2020), uses Unity for a brand-based AR application. The application allows the user to display three-dimensional models of pneumatic components in AR with the help of brands and a smartphone camera and to show a description of these on the display. In an evaluation, the teachers rated the app positively and the application showed an increased effect on the students' understanding and cognitive abilities [11, 12].

In summary, the applications researched illustrate the frequent use of the Unity game engine for mobile AR applications. The visualization of 3D models using brand-based AR also proved to be effective. However, in some cases there is a lack of evaluations regarding the exact acceptance and effectiveness of the applications in teaching. In addition, the focus of the applications is on the visualization of 3D models and their explanation. There is no interactive use of the models or animations of the pneumatic processes.

The current research shows that there are currently no interactive AR-based training courses for the field of pneumatics as part of training for a technical trade. In addition, the question arises as to how different AR elements for visualization in such training affect the learners. For a more efficient use of AR elements in the field of pneumatics, it would therefore be necessary to investigate how the visualizations in AR-based training affect the learners.

## 4   AR-Based Training App

AR elements and serious game elements are used for the development of the AR-based training, which combined can enable practice-oriented training in the field of pneumatics. The development is based on the ADDIE model, a widely used and systematic procedure for developing digital learning applications and serious games. Development is divided into five steps: Analyze, Design, Develop, Implement and Evaluate [13]. To this end, a concept is first designed on the basis of the research and this concept is implemented as an AR application for mobile devices using the Unity programming environment. The concept of the app and the visualizations is developed independently due to the specific application in the field of pneumatics and the little-researched field of research. The learning objectives to be achieved, which are defined in the BiBB training program for the field of pneumatics, serve as an aid in the conceptual design [3].

The realized application with the name A.R.P.S., an acronym for Augmented Reality Pneumatic Simulation, represents a first solution for augmented reality-based training in the field of pneumatics for mobile devices. In the app, the user must set up a pneumatic control system using a circuit diagram and a functional description with the help of 3D models and virtual hoses. Two exercises on the subject of pneumatics are integrated into the app for this purpose. The exercises are taken from the textbook "Steuerungstechnik Metall" which is used in vocational schools [14]. This ensures that the content is adapted to the framework curricula and meets the requirements of industrial and technical professions in industry and skilled trades.

The app offers the user the option of choosing between a tutorial level and level 1. The tutorial level serves to familiarize the user with pneumatic controls, while level 1 involves a more complex task with a larger number of pneumatic components. The task, a description and the circuit diagram are displayed before each level. During the level, the camera of the mobile device is activated and the real-time recording is projected onto the screen, accompanied by a user interface with buttons, in-game menu, and sliders. The immersion is enhanced by 3D-printed components that are based on real components in a pneumatic system and are overlaid with the AR visualizations in the form of 3D models of the pneumatic components. The user selects and places the components according to the task. Tracking is carried out by capturing markers in the form of QR codes with the camera of the mobile device, whereby printed components are overlaid with 3D models. The components can then be connected to each other via virtual hoses in order to simulate the pneumatic processes and test the circuit.

As can be seen in Fig. 1, two different concepts for AR visualizations were implemented in the app. The first concept, the realistic simulation of the pneumatic system, is the visualization of the 3D models of the pneumatic components via the QR codes. The second concept, in which additional AR elements are used to visualize pneumatic processes, consists of two visualization aids. These can be switched on and off via the user interface. The first visualization aid allows the user to track the air flows between the components within the hoses. The second visualization aid enables an X-ray view, which allows the user to look inside the components and track the internal air flows. These are designed to help the user understand complex circuits and learn the functions and structure of pneumatic components.

**Fig. 1.** Capturing a screenshot of the application, depicting the augmented reality (AR) visualization concepts in action. Left: Concept 1. Right: Concept 2.

After connecting the components, the user can test the circuit and feedback in the form of a rating is given based on the use of the visualization aids. The aim of the feedback is self-reflection and replay value to deepen knowledge. After completing the exercise, the user can return to the menu and select another level. To examine the concepts, the AR application is provided on a mobile device for iOS.

## 5   Study

To investigate the effectiveness of AR elements in relation to knowledge acquisition, a mixed-methods study with a sequential design was conducted. This approach combined quantitative and qualitative research methods, whereby a quantitative phase is conducted first, followed by a qualitative phase. Quantitative research consists of two knowledge tests before and after AR app exercises to measure knowledge uptake. The qualitative phase includes a survey to gather additional information on the user experience and to better understand the quantitative results. The study participants are deliberately selected based on characteristics of trainees in industrial-technical professions. The test subjects are divided into three groups depending on the comparison. Figure 2 shows the different groups performing an exercise in the study.

- **Group 1 (control group):** Learning with the traditional learning method (simulation of the pneumatic system from 3D printed components)
- **Group 2:** Learning with AR-based concept 1 (realistic simulation of the pneumatic system)
- **Group 3:** Learning with AR-based concept 2 (additional AR elements for visualizing pneumatic processes)

**Fig. 2.** Illustration of how to perform the exercises with and without the app. Left: Group 1, without app. Center: Group 2 with app and concept 1. Right: Group 3 with app and concept 2.

In this case, the control group (group 1) with the traditional learning method is represented by a replica of the pneumatic system consisting of 3D printed components.

The study begins with the collection of demographic data and knowledge levels through an online survey and a knowledge test before and after the exercises. The knowledge test is based on the fundamentals of pneumatics listed in the book "Steuerungstechnik Metall" and is developed with the help of a teacher of pneumatics at a vocational college [10]. The learning objectives, which are defined in the framework curricula and are to be taught by the app, are queried. The participants then carry out two practical exercises on the subject of pneumatics. In the exercises, a pneumatic system has to be set up and tested. The participants set up the controls in the exercises with the 3D-printed components and connect them physically or virtually with the hoses provided, depending on the group. The time taken to perform the exercises with and without the app is measured in order to analyze the effects of the time required. After the exercises have been carried out, the same knowledge test is repeated in order to compare the knowledge acquired. Quantitative research is followed by a qualitative survey using semistructured interviews to gather detailed experiences and opinions from the participants, taking into account group membership. The data collected are then analyzed and interpreted to answer the research question.

For the study, a random sample is drawn from trainees in the industrial mechanic occupation at a private vocational college of [company, location].The test subjects were selected from two different classes in the first year of training. A total of n = 15 trainees took part in the study, divided into 3 groups of 5 participants each. Participants were anonymized and given numbers. The study included the use of hardware (MacBook, iPad, ballpoint pen), software (AR app), paper exercises, 3D printed components, and a survey document with QR codes. The process consisted of an information phase, an initial questionnaire and knowledge test about the tablet, instructions for the exercises, completion of the exercises (10 min per task or level), repetition of the knowledge test and a qualitative survey with audio recording.

# 6  Results

The results are derived from the quantitative and qualitative data. For this purpose, quantitative data are evaluated using descriptive statistics, and qualitative survey are analyzed using a content analysis based on Mayring's [15] content analysis approach. Subsequently, correlations and patterns were identified in the analyzed data to gain insights into the subject of the research.

The study results include the participation of a total of 15 trainees, 14 of whom were male and 1 female, with an average age of approximately 20 years. Participants are in their first year of training in the industrial-technical occupation of a industrial mechanic. The majority of participants have a high school diploma, followed by a secondary school diploma and a primary/middle school diploma. Overall, the participants showed a high level of interest in AR technologies and had an affinity for technical systems. Most of the participants regularly use mobile devices, such as cell phones, for learning. However, very few have experience with learning applications in the classroom. The participants have had little or no experience with AR applications, particularly in relation to teaching, but are open to the use of such applications for teaching purposes.

As the participants are in their first year of training, no prior knowledge of pneumatics is required, although some participants stated that they already had some prior knowledge. As can be seen in Fig. 3, it is striking that the practical skills are rated better than the theoretical skills. Overall, the demographic data, as well as the knowledge and skills are well distributed between the various groups. Individual outliers can be recognized, which can be attributed to the sample size. The distribution of the data for theoretical knowledge in group 3 is striking, with two participants rating their knowledge as above average. Using a correlation matrix, there seems to be a trend that the assessed knowledge has no direct influence on the results of the knowledge tests. Using a correlation matrix, there seems to be a trend that the assessed knowledge has no direct influence on the results of the knowledge tests. A low correlation between the information on practical skills and the results of the knowledge tests is suspected, but its influence is estimated to be low due to the even distribution in this study.

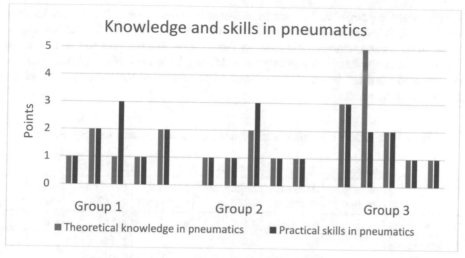

**Fig. 3.** Mean values of knowledge and skills in pneumatics

In the knowledge tests, 10 tasks with one point per task were performed before and after the exercises. The results, shown in Fig. 4, show that 5 participants performed better before the exercises than after the exercises. In group 3, which used the app with visualization aids, 4 out of 5 participants improved after the exercises.

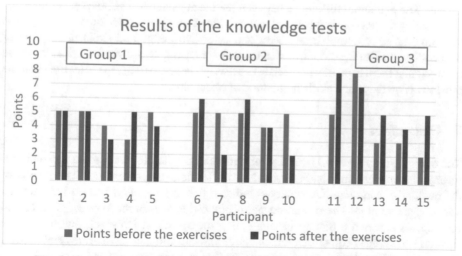

**Fig. 4.** Results of the participants' knowledge tests before and after the exercises

Table 1 shows the mean values, differences in the mean values and the standard deviations for each group. It can be seen that the mean value of group 1 did not change before and after the exercises were carried out. In group 2, it is noticeable that the mean score before the exercises is 4.8 points and is lower after the exercises at 4 points. In group

3, an increase in points can be seen from the mean values. The mean values increase from 4.2 points before the exercises to 5.8 points after the exercises. The differences in the mean values and standard deviations in Table 1 show that Group 1 shows no change, Group 2 performs worse (−0.8 difference) and Group 3 performs better (1.6 difference). The rounded standard deviations in relation to the mean values are a measure of the dispersion of the results.

**Table 1.** Results of the participants' knowledge tests before and after the exercises

|  | Group 1 | | Group 2 | | Group 3 | |
|---|---|---|---|---|---|---|
|  | Before the exercises | After the exercises | Before the exercises | After the exercises | Before the exercises | After the exercises |
| Mean values | 4,4 | 4,4 | 4,8 | 4 | 4,2 | 5,8 |
| Differences of the mean values | 0 | | −0,8 | | 1,6 | |
| Standard deviation | 0,89 | 0,89 | 0,45 | 2 | 2,39 | 1,64 |

The mean values for the total time for the exercises show that group 3 took the least time overall at 17.4 min and group 1 the longest at 16.6 min. Group 2 is in between with 17 min. The maximum difference between the mean values is less than one minute.

In order to determine whether there is a correlation between the test results, the total time taken to complete the tasks and prior knowledge and skills in the field of pneumatics, a scatter plot matrix was created for each group. Based on this, it can be assumed that there is no clear correlation between the test results, the total time and the previous knowledge and skills in the field of pneumatics.

## 7 Discussion

The control group (group 1) consists of participants who did not use the app and represents the conventional learning method in the field of pneumatics. The results of the qualitative and quantitative survey indicate that the exercises had no influence on knowledge retention. There were no significant differences in the knowledge tests, and participants reported difficulty in completing the exercises. However, one participant showed knowledge acquisition, which could indicate variance in learning preferences. The traditional learning method without teacher support does not seem to be optimal for practical tasks in the field of pneumatics. For this reason, teaching support is usually provided by the instructor.

Group 2, who used the app without visualization aids, also showed no significant improvement in knowledge retention. A negative mean difference in the knowledge tests indicated that the app alone was not sufficiently effective. Although some participants claimed to have had a learning effect, the results were not consistent and the standard deviation increased after the exercises. The participants in group 2 had less difficulty in completing the tasks than the control group, indicating a positive effect of the app design and implementation. Nevertheless, there was no clear pattern indicating an influence of the concept of visualization on knowledge acquisition.

Group 3, who used the app with visualization aids, showed positive signs of knowledge acquisition. The mean difference in the results was positive and the participants had fewer difficulties with the tasks. The participants' self-assessments and their positive reactions to the app and visualization aids supported the assumption of knowledge acquisition. The high standard deviation of the knowledge test results can be attributed to the different prior knowledge of the participants.

Overall, the comparison of all groups shows that the app was supportive in solving the exercises. However, the second visualization concept could be more effective as it led to a significant difference in the knowledge tests. The positive experiences and evaluations of the participants are in line with the literature on AR and serious games in education, with motivation and engagement being promoted by AR elements.

Based on the findings from the study, the following two hypotheses are derived about the influence of knowledge absorption by the concepts.

- **Hypothesis 1:** The first concept, the realistic simulation of a pneumatic system in AR-based training, has no significant influence on the acquisition of knowledge acquisition in the field of pneumatics.
- **Hypothesis 2:** The second concept, the use of additional AR elements to visualize the pneumatic system, such as airflow tracking and an X-ray view, in an AR-based training, has a significant influence on the trainees' knowledge absorption in the field of pneumatics.

With the hypotheses help of the derived and the findings from the study as well as the literature research, a generalized theory about the research object is established.

The use of AR-based training can be an effective method to impart knowledge to learners in the field of pneumatics. However, this depends on the type of visualization of the pneumatic system that has an impact on the learner's knowledge absorption. Visualizations that can convey additional information, such as air flow tracing and an X-ray view, can help to better understand and internalize complex relationships and modes of operation.

## 7.1 Future Work

The study conducted in this thesis provides hypotheses and a general theory of how different visualization concepts in AR-based training in the field of pneumatics can affect learners' knowledge retention. It therefore remains to be seen whether the hypotheses and theory prove to be true. To test these, the next step is to verify or refute them using quantitative research methods such as significance testing.

Limitations, particularly with regard to sample size and the limited implementation time, require further research. The control group used 3D printed components for the exercises, which could affect the validity of the results. Despite the subjective realism of the exercises, future studies should avoid this limitation by using a real pneumatic system. Furthermore, a long-term study is needed to investigate the long-term effects of the concepts and the impact of the learning environment on learners.

Since the AR-based training was developed to explore the concepts, there is a possibility that the AR app has an impact on the results. Therefore, future studies could investigate the extent to which the explored concepts impact learners in other AR-based trainings. It is worth noting that the present study focuses solely on knowledge retention, therefore future research could explore the concepts in relation to other relevant variables such as learner engagement.

Finally, it should be noted that only the concepts for visualization in AR developed in this thesis were examined. However, these represent only a part of the possible AR elements, which is why future research should investigate alternative concepts.

## 8   Conclusion

This thesis presents the development and evaluation of a serious game in the form of an augmented reality app, designed for trainees in industrial-technical professions, especially in the field of pneumatics. The main objective of this research was to create an AR-based training that uses various AR elements for visualization in order to investigate their influence on the acquisition of knowledge by learners.

The results of the study show that both concepts and the app were positively received by the participants overall. However, the second concept was significantly better, especially in terms of knowledge retention. Compared to the traditional learning method and the first concept, the additional AR elements, in particular the airflow tracking and the X-ray view, were perceived as supportive and helpful.

The hypotheses derived and the findings from the study provide initial indications of how visualizations in AR can influence learners' knowledge transfer in the field of pneumatics. The type of visualization of the pneumatic system has a decisive influence on learners' knowledge acquisition. Despite the insights gained, limitations, in particular the small number of participants, must be taken into account. Nevertheless, the study provides important findings for the field of research. Overall, this work lays the foundation.

Finally, it should be mentioned that the use of AR and digital learning applications in teaching is becoming increasingly relevant. It can be assumed that this research is a reason to support other teaching areas in education with AR and digital learning applications. The concept and realization of the AR application set up in this work, as well as the investigation carried out, provide a point of reference for further investigations into similar teaching areas, such as hydraulics or electro pneumatics.

**Acknowledgement.** Thanks to Dario Luipers and Anja Richert for supervising my Master's thesis, which is an important part of this paper.

## References

1. Thomas, O.: Digitalisierung in der Aus- und Weiterbildung: Virtual und Augmented Reality für Industrie 4.0. Springer, Berlin (2018). https://doi.org/10.1007/978-3-662-56551-3

2. Grollius, H.-W.: Grundlagen der Pneumatik, 5th edn. Hanser, München (2020)
3. Bundesinstitut für Berufsbildung Homepage. https://www.bibb.de/datenreport/de/2019/101 256.php. Accessed 01 Feb 2023
4. Bekanntmachung der Neufassung der Verordnung über die Berufsausbildung in den industriellen Metallberufen (2018)
5. Busse, F., et al.: Industrielle Metallberufe; Ausbildung Gestalten. Bundesinstitut für Berufsbildung, Bonn (2018)
6. Festo Vertrieb GmbH & Co. KG: FESTO Didactic. Impulsgeber der technischen Bildung. https://www.festo.com/media/cms/central/media/de/editorial/downloads/did_1/FESTO_Didactic_Digital_DE.pdf. Accessed 05 Jan 2023
7. Popescu, E., et al.: Innovations in Smart Learning. Springer, Singapore (2017). https://doi.org/10.1007/978-981-10-2419-1
8. Wang, M., Callaghan, V., Bernhardt, J., et al.: Augmented reality in education and training: pedagogical approaches and illustrative case studies. J. Ambient Intell. Hum. Comput. **9**, 1391–1402 (2018)
9. Akçayır, M., Akçayır, G.: Advantages and challenges associated with augmented reality for education: a systematic review of the literature. Educ. Res. Rev. **20**, 1–11 (2017)
10. Festo Vertrieb GmbH & Co. KG, Homepage. https://www.festo.com/de/e/technische-bildung/digitales-lernen/mixedreality-id_31287/. Accessed 11 Jan 2023
11. Jabar, A.R., Nohseth, N.H., Jambari, H., Pairan, M.R., Ahyan, N.A.M.N., Lokman, H.: Exploring the potential of augmented reality teaching aid for vocational subject. In: IC4E 2020: 2020 the 11th International Conference on E-Education, E-Business, E-Management, and E-Learning, Japan, pp. 54–58 (2020)
12. Bajči, B., et al.: Augmented reality as an advanced learning tool for pneumatic control. In: 2019 5th Experiment Conference, Funchal (Madeira Island), Portugal, pp. 415–418 (2019)
13. Dörner, R., Effelsberg, W., Göbel, S., Wiemeyer, J.: Serious Games: Foundations, Concepts and Practice, 1st edn. Springer, Cham (2016). https://doi.org/10.1007/978-3-319-40612-1
14. Von der Heide, V., Hölken, F.-J.: Steuerungstechnik Metall: Lernsituationen: Arbeitsbuch, 17th edn. Bildungsverlag EINS, Köln (2019)
15. Hedderich, J.: Angewandte Statistik: Methodensammlung mit R, 15th edn. Springer, Heidelberg (2016). https://doi.org/10.1007/978-3-662-45691-0

# Author Index

## A

Adolfsson, Kristoffer Kuvaja  106
Alptekin, Mesut  297
Alves, Gustavo R.  122
Al-Zoubi, Abdallah  122
Amer, Muhieddin  114
Arras, Peter  169
Aubel, Ines  206

## B

Bashavets, Nataliia  324
Biström, Dennis  106
Blazhko, Oleksandr  324
Block, Brit-Maren  86, 146
Boettcher, Konrad  44, 206
Bustillo, Andres  267

## C

Chardonnet, Jean-Rémy  267
Chevalier, Amélie  178
Ciolacu, Monica I.  122
Cuperman, D.  226

## D

Dani, Raghuveer Rajesh  198
Datta, Soma  336
Dethmann, Jannis  146

## E

Erceg, Igor  186
Espinosa-Leal, Leonardo  32, 245

## F

Franke, Daniel  275
Franke, Susanne  275

## G

Gebrehiwot, Silas  245
Geiger, Benjamin  198
Gherman-Dolhăscu, Elena-Cătălina  98

## G

Greenholts, M.  217
Grodotzki, Joshua  305, 313
Groš, Stjepan  186
Guerne, Marie Gillian  86
Günther, Norman  155

## H

Haase, Jan  206
Havelka Mestrovic, Ana  114
Haus, Benedikt  146
Helbing, Pierre  206
Hellbach, Sven  275
Herrmann, Franziska  206
Hills, Catherine  135
Hu, Wenshan  67

## J

Joshi, Abhishek S.  234

## K

Kaiser, Doreen  206
Karnalim, Oscar  56
Katulić, Filip  186
Kaufhold, Nils  44, 206
Kist, Alexander A.  122, 135
Kobras, Louis  206
Kokotieieva, Anastasiia  324

## L

Lei, Zhongcheng  67
Liu, Guo-Ping  67

## M

Maiti, Ananda  135
Mayer, Anjela  267
Meussen, Bernhard  206
Miguel-Alonso, Ines  267
Modran, Horia  98
Mounsef, Jinane  114

M. E. Auer et al. (Eds.): STE 2024, LNNS 1027, pp. 357–358, 2024.
https://doi.org/10.1007/978-3-031-61891-8

Müller, Benedikt Tobias    305, 313
Müller, Larissa    344

**N**
Nathasya, Rossevine Artha    56
Nau, Johannes    206

**O**
Ohashi, Yutaro    255
Ortelt, Tobias R.    44, 206
Ovtcharova, Jivka    267

**P**
Parkhomenko, Anzhelika    169
Parkhomenko, Illia    169
Patabendige, Thumula Madduma    245
Pautzke, Friedbert    198
Podhorna, Viktoriia    324
Polishuk, A.    217
Potsepaiev, Valerii    285
Pozzo, María Isabel    122
Prabhu, Prajwal    3
Prell, Bastian    155

**R**
Ram, B. Kalyan    234
Rana, Vighneshkumar    15
Reiff-Stephan, Jörg    155
Riedel, Ralph    275
Roos, Kim    32
Rosen, U.    226

**S**
Samardzija, Jasminka    114
Samoilă, Cornel    98
Schade, Marcel    44, 206
Sharma, Nitin    234
Shcherbakov, Andrey    77

Singh, Vishal    3, 15
Stocchetti, Matteo    106
Streitferdt, Detlef    206
Stupak, Hlib    285
Stupak, Maryna    285
Su, Jiuzheng    67
Sujadi, Sendy Ferdian    56
Sumina, Damir    186
Syed, Ibrahim Ibaad    336

**T**
Tabunshchyk, Galyna    198
Teich, Tobias    275
Tekkaya, A. Erman    305, 313
Telychko, Hanna    285
Temmen, Katrin    297
Terkowsky, Claudius    44, 206
Terschüren, Jan-Niklas    344
Trommer, Martin    275

**U**
Ullah, Rizwan    245
Ursuțiu, Doru    98

**V**
Varney, Valérie    344
Vermani, Advik    234
Verner, I.    217, 226

**W**
Waller, Matias    32
Westerlund, Magnus    77
Wilbers, Simon    155
Wolf, Carsten    198

**Z**
Zagar, Martin    114
Zhou, Xingwei    67

Printed in the United States
by Baker & Taylor Publisher Services